College Physics

College Physics

Elora Melton

Willford Press
www.willfordpress.com

Published by Willford Press,
118-35 Queens Blvd., Suite 400,
Forest Hills, NY 11375, USA

ISBN: 978-1-68285-749-6

Cataloging-in-Publication Data

College physics / Elora Melton.
p. cm.
Includes bibliographical references and index.
ISBN 978-1-68285-749-6
1. Physics--Textbooks. I. Melton, Elora.
QC21.3 .C65 2019
530--dc23

For information on all Willford Press publications
visit our website at www.willfordpress.com

Willford Press

Contents

Preface

In the natural science discipline of physics, matter is studied in the fabric of space and time. The core theories of classical mechanics, thermodynamics, quantum mechanics, electromagnetism, special relativity and statistical mechanics are integral to the understanding of physics. The behavior of matter and energy under different conditions are addressed from the twin approaches of classical and modern physics. Classical physics is ideal for the study of phenomena at normal scales of observation. Modern physics studies matter at extreme conditions, such as very small or very large scales. Nuclear physics, optical physics, particle physics, atomic physics and astrophysics are some of the rapidly expanding branches of physics. The principles of physics are of immense importance in the varied engineering fields. This book presents the complex subject of physics in the most comprehensible and easy to understand language. While understanding the long-term perspectives of the topics, the book makes an effort in highlighting their impact as a modern tool for the growth of the discipline. It will serve as a valuable source of reference for those interested in this field.

Given below is the chapter wise description of the book:

Chapter 1- Physics is a fundamental science that deals with the study of the universe and how it behaves. It focuses on the study of matter, and its motion through space and time. The various approaches to the study of physics fall within the chief categories of theoretical physics, applied physics, computational physics and mathematical physics, which have been discussed in detail in this chapter.

Chapter 2- Physics studies matter and energy. The chief branches of physics are classical mechanics, thermodynamics, magnetism, astrophysics, condensed matter physics and particle physics, among others. This chapter closely examines these branches along with prominent principles of thermodynamics, magnetism, optics, etc. to develop a holistic understanding of the subject.

Chapter 3- Thermodynamics is a branch of physics that studies the nature of heat and temperature and their relation to work and energy. Its applications are in the fields of chemical engineering, physical chemistry, mechanical engineering, etc. This chapter includes detailed information on the core topics of heat transfer, temperature gradient, thermodynamic temperature, work and entropy.

Chapter 4- A wave is a disturbance in space or matter, which involves the transfer of energy without any mass transport. It is generated by oscillations or vibrations in a field or physical medium. The different types of waves have been discussed in this chapter such as, mechanical waves, seismic waves, harmonic waves and standing waves. Other concepts discussed within the chapter include the phenomena of wave interference, Doppler effect and resonance.

Chapter 5- Any vibration, which propagates as an audible pressure wave through a solid, liquid or gas, is known as sound. The human ear can only perceive sound between the frequencies of 20 Hz and 20 kHz. This chapter elaborates the different characteristics of sound such as speed, timbre, pitch and loudness.

Chapter 6- Light is an electromagnetic wave propagating in space with the speed of 3×10^8 m/s. The understanding of the different phenomena of light is approached from the wave-like or particle-like nature of light. The concepts of reflection and refraction of light are fundamental to the study of light, which have been discussed in detail in this book.

Indeed, my job was extremely crucial and challenging as I had to ensure that every chapter is informative and structured in a student-friendly manner. I am thankful for the support provided by my family and colleagues during the completion of this book.

Elora Melton

1

Introduction to Physics

Physics is a fundamental science that deals with the study of the universe and how it behaves. It focuses on the study of matter, and its motion through space and time. The various approaches to the study of physics fall within the chief categories of theoretical physics, applied physics, computational physics and mathematical physics, which have been discussed in detail in this chapter.

Physics is the scientific study of matter and energy and how they interact with each other. This energy can take the form of motion, light, electricity, radiation, gravity — just about anything, honestly. Physics deals with matter on scales ranging from sub-atomic particles (i.e. the particles that make up the atom and the particles that make up those particles) to stars and even entire galaxies.

Energy may manifest itself in many forms, such as light, motion, gravity, radiation, electricity, and others. Matter, on the other hand, covers any and every physical manifestation, from the smallest particles such as atoms and sub-atomic particles, further into larger physical groups such as stars, universes, and galaxies.

Physics can also be described as the science dealing with physical quantities. In this regard, physics is widely considered to be the most fundamental (and important) of all the natural sciences. After all, physics pertains to the quantification of almost all matter that exists in this world. It is any aspect of nature that can be expressed, measured or calculated in specific terms. An appreciation of the world we live in and all its elements that can be observed by the five physical senses could be considered as physics in its most general sense.

In this regard, physics and mathematics are closely related to one another. It could be said that mathematics is the language of physics. That is another way of better understanding what is physics. Numerical values, units of measurement, and similar concepts are all mathematical in nature, and are used to describe physical quantities in the most accurate and precise manner.

Thus, these physical quantities form the basis of study of other branches of science, such as chemistry, biology, engineering, and others. The information, facts, and concepts covered in these sciences can all be appreciated and explained only through physical quantities and physical laws. Biology as the study of living things would also necessitate an understanding of the physical laws that govern nature, humans, animals, their body parts, organs, systems-after all, they are all physical manifestations of their existence here on earth. Chemistry, on the other hand, is the study of processes, reactions, and interaction among physical elements on the subatomic level-which is again a realm of physics.

The importance of physics as a fundamental science can also be appreciated by its basic processes, which are either experimental or theoretical. Experimental physics makes use of the scientific method of creating a hypothesis, testing the hypothesis, observing the results, and coming to a conclusion regarding the natural world. The end-goal of these experiments is to establish scientific laws, which seek to understand and predict phenomena and occurrences in the natural world.

Major Concepts in Physics

Because physics covers so much area, it is divided into several specific fields of study, such as electronics, quantum physics, astronomy, and biophysics.

Physics includes the study of astronomy, and in many ways astronomy was humanity's first organized field of science. Ancient peoples looked to the stars and recognized patterns there, then began using mathematical precision to make predictions about what would happen in the heavens based on those patterns. Whatever flaws there were in these specific predictions, the method of trying to understand the unknown was a worthy one.

Trying to understand the unknown is still a central problem in human life. Despite all of our advancements in science and technology, being a human being means that you are able to understand some things and also that there are things you do not understand. Science teaches you a methodology for approaching the unknown and asking questions that get to the heart of what is unknown and how to make it known.

Physics, in particular, focuses on some of the most fundamental of questions about our physical universe. Pretty much the only more fundamental questions that could be asked fall in the philosophical realm of "metaphysics" (named for being literally "beyond physics"), but the problem is that these questions are so fundamental that many of the questions in the metaphysical realm remain unresolved even after centuries or millennia of inquiry by most of history's greatest minds. Physics, on the other hand, has resolved many fundamental issues, even though those resolutions tend to open up whole new types of questions.

Core Theories

Though physics deals with a wide variety of systems, certain theories are used by all physicists. Each of these theories were experimentally tested numerous times and found to be an adequate approximation of nature. For instance, the theory of classical mechanics accurately describes the motion of objects, provided they are much larger than atoms and moving at much less than the speed of light. These theories continue to be areas of active research today. Chaos theory, a remarkable aspect of classical mechanics was discovered in the 20th century, three centuries after the original formulation of classical mechanics by Isaac Newton (1642–1727).

These central theories are important tools for research into more specialised topics, and any physicist, regardless of their specialisation, is expected to be literate in them. These include classical mechanics, quantum mechanics, thermodynamics and statistical mechanics, electromagnetism, and special relativity.

Classical Physics

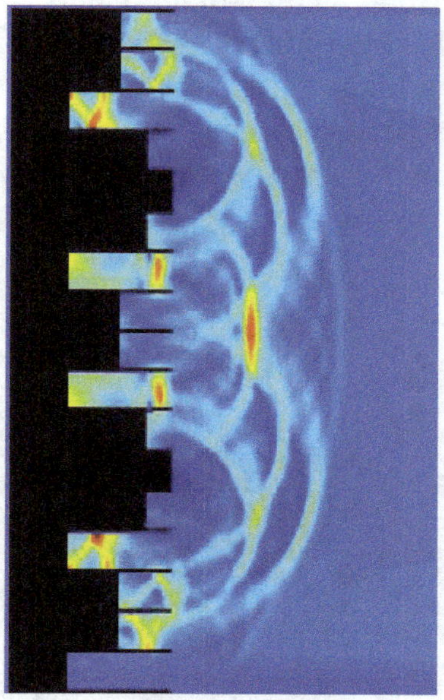

Classical physics implemented in an acoustic engineering model of sound reflecting from an acoustic diffuser

Classical physics includes the traditional branches and topics that were recognised and well-developed before the beginning of the 20th century—classical mechanics, acoustics, optics, thermodynamics, and electromagnetism. Classical mechanics is concerned with bodies acted on by forces and bodies in motionand may be divided into statics (study of the forces on a body or bodies not subject to an acceleration), kinematics (study of motion without regard to its causes), and dynamics (study of motion and the forces that affect it); mechanics may also be divided into solid mechanics and fluid mechanics (known together as continuum mechanics), the latter include such branches as hydrostatics, hydrodynamics, aerodynamics, and pneumatics. Acoustics is the study of how sound is produced, controlled, transmitted and received. Important modern branches of acoustics include ultrasonics, the study of sound waves of very high frequency beyond the range of human hearing; bioacoustics, the physics of animal calls and hearing, and electroacoustics, the manipulation of audible sound waves using electronics.

Optics, the study of light, is concerned not only with visible light but also with infrared and ultraviolet radiation, which exhibit all of the phenomena of visible light except visibility, e.g., reflection, refraction, interference, diffraction, dispersion, and polarization of light. Heat is a form of energy, the internal energy possessed by the particles of which a substance is composed; thermodynamics deals with the relationships between heat and other forms of energy. Electricity and magnetism have been studied as a single branch of physics since the intimate connection between them was discovered in the early 19th century; an electric current gives rise to a magnetic field, and a changing magnetic field induces an electric current. Electrostatics deals with electric charges at rest, electrodynamics with moving charges, and magnetostatics with magnetic poles at rest.

Modern Physics

Classical physics is generally concerned with matter and energy on the normal scale of observation, while much of modern physics is concerned with the behavior of matter and energy under extreme conditions or on a very large or very small scale. For example, atomic and nuclear physicsstudies matter on the smallest scale at which chemical elements can be identified. The physics of elementary particles is on an even smaller scale since it is concerned with the most basic units of matter; this branch of physics is also known as high-energy physics because of the extremely high energies necessary to produce many types of particles in particle accelerators. On this scale, ordinary, commonsense notions of space, time, matter, and energy are no longer valid.

Solvay Conference, with prominent physicists such as Albert Einstein, Werner Heisenberg, Max Planck, Hendrik Lorentz, Niels Bohr, Marie Curie, Erwin Schrödinger and Paul Dirac

The two chief theories of modern physics present a different picture of the concepts of space, time, and matter from that presented by classical physics. Classical mechanics approximates nature as continuous, while quantum theory is concerned with the discrete nature of many phenomena at the atomic and subatomic level and with the complementary aspects of particles and waves in the description of such phenomena. The theory of relativity is concerned with the description of phenomena that take place in a frame of reference that is in motion with respect to an observer; the special theory of relativity is concerned with relative uniform motion in a straight line and the general theory of relativity with accelerated motion and its connection with gravitation. Both quantum theory and the theory of relativity find applications in all areas of modern physics.

Difference between Classical and Modern Physics

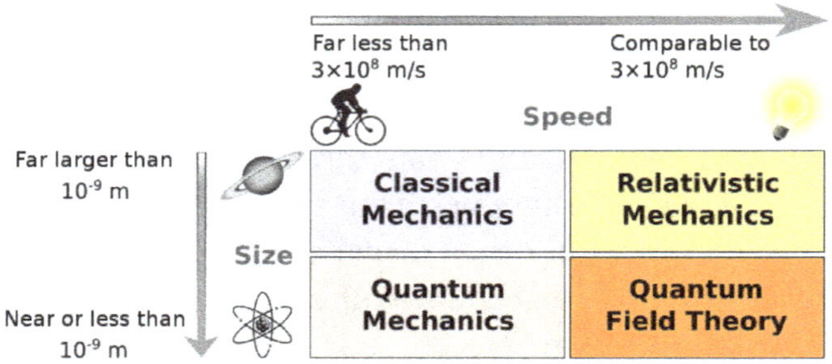

The basic domains of physics

While physics aims to discover universal laws, its theories lie in explicit domains of applicability. Loosely speaking, the laws of classical physics accurately describe systems whose important length scales are greater than the atomic scale and whose motions are much slower than the speed of light. Outside of this domain, observations do not match predictions provided by classical mechanics. Albert Einstein contributed the framework of special relativity, which replaced notions of absolute time and space with spacetime and allowed an accurate description of systems whose components have speeds approaching the speed of light. Max Planck, Erwin Schrödinger, and others introduced quantum mechanics, a probabilistic notion of particles and interactions that allowed an accurate description of atomic and subatomic scales. Later, quantum field theory unified quantum mechanics and special relativity. General relativityallowed for a dynamical, curved spacetime, with which highly massive systems and the large-scale structure of the universe can be well-described. General relativity has not yet been unified with the other fundamental descriptions; several candidate theories of quantum gravityare being developed.

Research

Scientific Method

Physicists use the scientific method to test the validity of a physical theory. By using a methodical approach to compare the implications of a theory with the conclusions drawn from its related experiments and observations, physicists are better able to test the validity of a theory in a logical, unbiased, and repeatable way. To that end, experiments are performed and observations are made in order to determine the validity or invalidity of the theory.

A scientific law is a concise verbal or mathematical statement of a relation which expresses a fundamental principle of some theory, such as Newton's law of universal gravitation.

Theory and Experiment

Theorists seek to develop mathematical models that both agree with existing experiments and successfully predict future experimental results, while experimentalists devise and perform experiments to test theoretical predictions and explore new phenomena. Although theory and e xperiment are developed separately, they are strongly dependent upon each other. Progress in physics frequently comes about when experimentalists make a discovery that existing theories cannot explain, or when new theories generate experimentally testable predictions, which inspire new experiments.

The astronaut and Earth are both in free-fall

Lightning is an electric current

Physicists who work at the interplay of theory and experiment are called phenomenologists, who study complex phenomena observed in experiment and work to relate them to a fundamental theory.

Theoretical physics has historically taken inspiration from philosophy; electromagnetism was unified this way. Beyond the known universe, the field of theoretical physics also deals with hypothetical issues, such as parallel universes, a multiverse, and higher dimensions. Theorists invoke these ideas in hopes of solving particular problems with existing theories. They then explore the consequences of these ideas and work toward making testable predictions.

Experimental physics expands, and is expanded by, engineering and technology. Experimental physicists involved in basic research design and perform experiments with equipment such as particle accelerators and lasers, whereas those involved in applied research often work in industry developing technologies such as magnetic resonance imaging (MRI) and transistors. Feynman has noted that experimentalists may seek areas which are not well-explored by theorists.

Scope and Aims

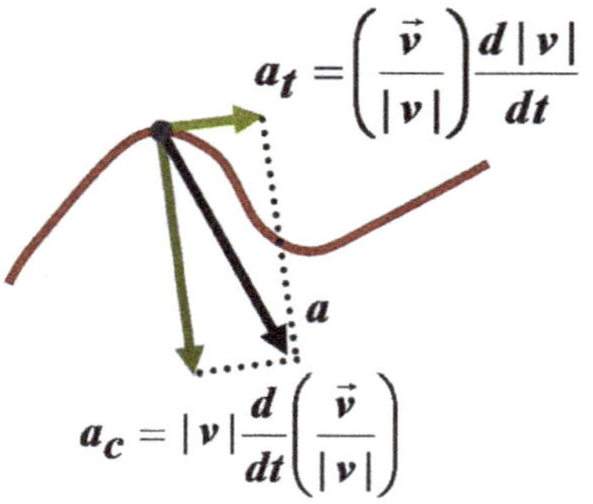

Physics involves modeling the natural world with theory, usually quantitative. Here, the path of a particle is modeled with the mathematics of calculus to explain its behavior: the purview of the branch of physics known as mechanics

Physics covers a wide range of phenomena, from elementary particles (such as quarks, neutrinos, and electrons) to the largest superclusters of galaxies. Included in these phenomena are the most basic objects composing all other things. Therefore, physics is sometimes called the "fundamental science". Physics aims to describe the various phenomena that occur in nature in terms of simpler phenomena. Thus, physics aims to both connect the things observable to humans to root causes, and then connect these causes together.

For example, the ancient Chinese observed that certain rocks (lodestone and magnetite) were attracted to one another by an invisible force. This effect was later called magnetism, which was first rigorously studied in the 17th century. But even before the Chinese discovered magnetism, the ancient Greeks knew of other objects such as amber, that when rubbed with fur would cause a similar invisible attraction between the two. This was also first studied rigorously in the 17th century and came to be called electricity. Thus, physics had come to understand two observations of nature in terms of some root cause (electricity and magnetism). However, further work in the 19th century revealed that these two forces were just two different aspects of one force—electromagnetism. This process of "unifying" forces continues today, and electromagnetism and the weak nuclear force are now considered to be two aspects of the electroweak interaction. Physics hopes to find an ultimate reason (Theory of Everything) for why nature is as it is.

Research Fields

Contemporary research in physics can be broadly divided into nuclear and particle physics; condensed matter physics; atomic, molecular, and optical physics; astrophysics; and applied physics. Some physics departments also support physics education research and physics outreach.

Since the 20th century, the individual fields of physics have become increasingly specialised, and today most physicists work in a single field for their entire careers. "Universalists" such as Albert Einstein (1879–1955) and Lev Landau (1908–1968), who worked in multiple fields of physics, are now very rare.

The major fields of physics, along with their subfields and the theories and concepts they employ, are shown in the following table.

Field	Subfields	Major theories	Concepts
Nuclear and particle physics	Nuclear physics, Nuclear astrophysics, Particle physics, Particle astrophysics, Particle physics phenomenology	Standard Model, Quantum field theory, Quantum electrodynamics, Quantum chromodynamics, Electroweak theory, Effective field theory, Lattice field theory, Lattice gauge theory, Gauge theory, Supersymmetry, Grand unification theory, Superstring theory, M-theory	Fundamental force(gravitational, electromagnetic, weak, strong), Elementary particle, Spin, Antimatter, Spontaneous symmetry breaking, Neutrino oscillation, Seesaw mechanism, Brane, String, Quantum gravity, Theory of everything, Vacuum energy
Atomic, molecular, and optical physics	Atomic physics, Molecular physics, Atomic and Molecular astrophysics, Chemical physics, Optics, Photonics	Quantum optics, Quantum chemistry, Quantum information science	Photon, Atom, Molecule, Diffraction, Electromagnetic radiation, Laser, Polarization (waves), Spectral line, Casimir effect

Condensed matter physics	Solid state physics, High pressure physics, Low-temperature physics, Surface Physics, Nanoscale and Mesoscopic physics, Polymer physics	BCS theory, Bloch wave, Density functional theory, Fermi gas, Fermi liquid, Many-body theory, Statistical Mechanics	Phases (gas, liquid, solid), Bose-Einstein condensate, Electrical conduction, Phonon, Magnetism, Self-organization, Semiconductor, superconductor, superfluid, Spin,
Astrophysics	Astronomy, Astrometry, Cosmology, Gravitation physics, High-energy astrophysics, Planetary astrophysics, Plasma physics, Solar physics, Space physics, Stellar astrophysics	Big Bang, Cosmic inflation, General relativity, Newton's law of universal gravitation, Lambda-CDM model, Magnetohydrodynamics	Black hole, Cosmic background radiation, Cosmic string, Cosmos, Dark energy, Dark matter, Galaxy, Gravity, Gravitational radiation, Gravitational singularity, Planet, Solar System, Star, Supernova, Universe
Applied Physics	Accelerator physics, Acoustics, Agrophysics, Biophysics, Chemical Physics, Communication Physics, Econophysics, Engineering physics, Fluid dynamics, Geophysics, Laser Physics, Materials physics, Medical physics, Nanotechnology, Optics, Optoelectronics, Photonics, Photovoltaics, Physical chemistry, Physics of computation, Plasma physics, Solid-state devices, Quantum chemistry, Quantum electronics, Quantum information science, Vehicle dynamics		

Nuclear and Particle Physics

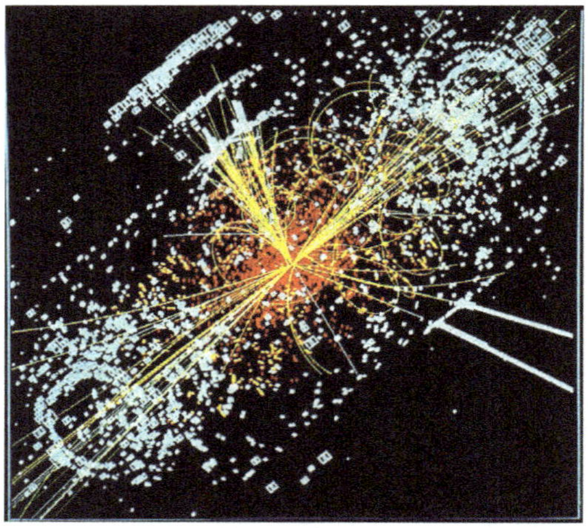

A simulated event in the CMS detector of the Large Hadron Collider, featuring a possible appearance of the Higgs boson

Particle physics is the study of the elementaryconstituents of matter and energy and the interactions between them. In addition, particle physicists design and develop the high energy accelerators, detectors, and computer programs necessary for this research. The field is also called "high-energy physics" because many elementary particles do not occur naturally but are created only during high-energy collisions of other particles.

Currently, the interactions of elementary particles and fields are described by the Standard Model. The model accounts for the 12 known particles of matter (quarks and leptons) that interact via the strong, weak, and electromagnetic fundamental forces. Dynamics are described in terms of matter particles exchanging gauge bosons (gluons, W and Z bosons, and photons, respectively). The

Standard Model also predicts a particle known as the Higgs boson. In July 2012 CERN, the European laboratory for particle physics, announced the detection of a particle consistent with the Higgs boson, an integral part of a Higgs mechanism.

Nuclear physics is the field of physics that studies the constituents and interactions of atomic nuclei. The most commonly known applications of nuclear physics are nuclear powergeneration and nuclear weapons technology, but the research has provided application in many fields, including those in nuclear medicine and magnetic resonance imaging, ion implantation in materials engineering, and radiocarbon dating in geology and archaeology.

Atomic, Molecular and Optical Physics

Atomic, molecular, and optical physics (AMO) is the study of matter–matter and light–matter interactions on the scale of single atoms and molecules. The three areas are grouped together because of their interrelationships, the similarity of methods used, and the commonality of their relevant energy scales. All three areas include both classical, semi-classical and quantum treatments; they can treat their subject from a microscopic view (in contrast to a macroscopic view).

Atomic physics studies the electron shells of atoms. Current research focuses on activities in quantum control, cooling and trapping of atoms and ions, low-temperature collision dynamics and the effects of electron correlation on structure and dynamics. Atomic physics is influenced by the nucleus (e.g., hyperfine splitting), but intra-nuclear phenomena such as fission and fusion are considered part of nuclear physics.

Molecular physics focuses on multi-atomic structures and their internal and external interactions with matter and light. Optical physics is distinct from optics in that it tends to focus not on the control of classical light fields by macroscopic objects but on the fundamental properties of optical fields and their interactions with matter in the microscopic realm.

Condensed Matter Physics

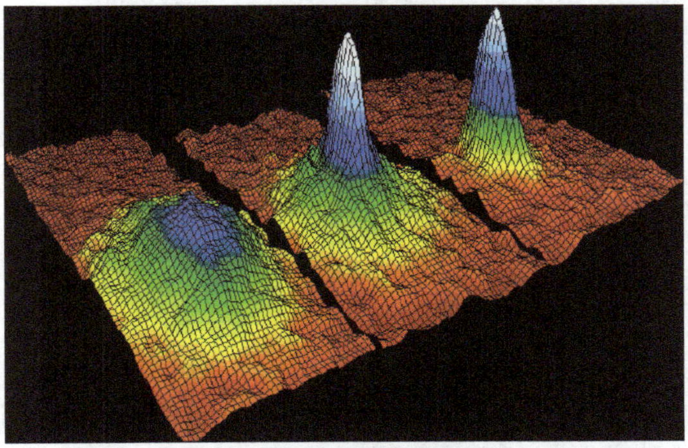

Velocity-distribution data of a gas of rubidium atoms, confirming the discovery of a new phase of matter, the Bose–Einstein condensate

Condensed matter physics is the field of physics that deals with the macroscopic physical properties of matter. In particular, it is concerned with the "condensed" phases that appear whenever the number of particles in a system is extremely large and the interactions between them are strong.

The most familiar examples of condensed phases are solidsand liquids, which arise from the bonding by way of the electromagnetic force between atoms. More exotic condensed phases include the superfluid and the Bose–Einstein condensate found in certain atomic systems at very low temperature, the superconducting phase exhibited by conduction electrons in certain materials, and the ferromagnetic and antiferromagnetic phases of spins on atomic lattices.

Condensed matter physics is the largest field of contemporary physics. Historically, condensed matter physics grew out of solid-state physics, which is now considered one of its main subfields. The term *condensed matter physics* was apparently coined by Philip Anderson when he renamed his research group—previously *solid-state theory*—in 1967.In 1978, the Division of Solid State Physics of the American Physical Society was renamed as the Division of Condensed Matter Physics. Condensed matter physics has a large overlap with chemistry, materials science, nanotechnology and engineering.

Astrophysics

The deepest visible-light image of the universe, the Hubble Ultra Deep Field

Astrophysics and astronomy are the application of the theories and methods of physics to the study of stellar structure, stellar evolution, the origin of the Solar System, and related problems of cosmology. Because astrophysics is a broad subject, astrophysicists typically apply many disciplines of physics, including mechanics, electromagnetism, statistical mechanics, thermodynamics, quantum mechanics, relativity, nuclear and particle physics, and atomic and molecular physics.

The discovery by Karl Jansky in 1931 that radio signals were emitted by celestial bodies initiated the science of radio astronomy. Most recently, the frontiers of astronomy have been expanded by space exploration. Perturbations and interference from the earth's atmosphere make space-based observations necessary for infrared, ultraviolet, gamma-ray, and X-ray astronomy.

Physical cosmology is the study of the formation and evolution of the universe on its largest scales. Albert Einstein's theory of relativity plays a central role in all modern cosmological theories. In the early 20th century, Hubble's discovery that the universe is expanding, as shown by the Hubble diagram, prompted rival explanations known as the steady stateuniverse and the Big Bang.

The Big Bang was confirmed by the success of Big Bang nucleosynthesis and the discovery of the cosmic microwave background in 1964. The Big Bang model rests on two theoretical pillars: Albert Einstein's general relativity and the cosmological principle. Cosmologists have recently established the ΛCDM model of the evolution of the universe, which includes cosmic inflation, dark energy, and dark matter.

Numerous possibilities and discoveries are anticipated to emerge from new data from the Fermi Gamma-ray Space Telescope over the upcoming decade and vastly revise or clarify existing models of the universe. In particular, the potential for a tremendous discovery surrounding dark matter is possible over the next several years. Fermi will search for evidence that dark matter is composed of weakly interacting massive particles, complementing similar experiments with the Large Hadron Collider and other underground detectors.

IBEX is already yielding new astrophysical discoveries: "No one knows what is creating the ENA (energetic neutral atoms) ribbon" along the termination shock of the solar wind, "but everyone agrees that it means the textbook picture of the heliosphere—in which the Solar System's enveloping pocket filled with the solar wind's charged particles is plowing through the onrushing 'galactic wind' of the interstellar medium in the shape of a comet—is wrong."

Relation to other Fields

Prerequisites

Mathematics provides a compact and exact language used to describe of the order in nature. This was noted and advocated by Pythagoras,Plato, Galileo, and Newton.

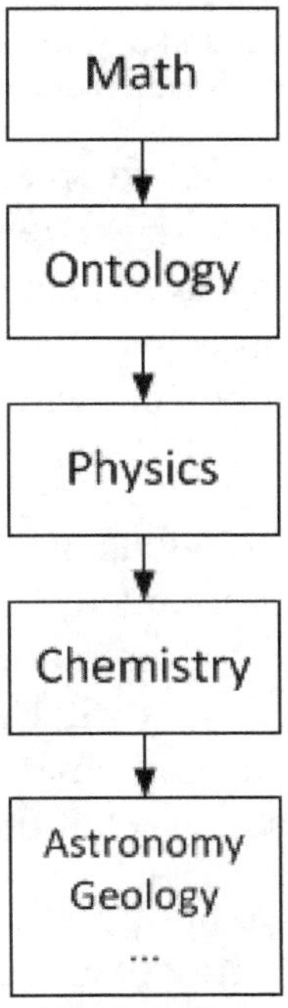

Mathematics and ontology are used in physics. Physics is used in chemistry and cosmology

This parabola-shaped lava flowillustrates the application of mathematics in physics—in this case, Galileo's law of falling bodies

Physics uses mathematicsto organise and formulate experimental results. From those results, precise or estimated solutions, quantitative results from which new predictions can be made and experimentally confirmed or negated. The results from physics experiments are numerical measurements. Technologies based on mathematics, like computationhave made computational physics an active area of research.

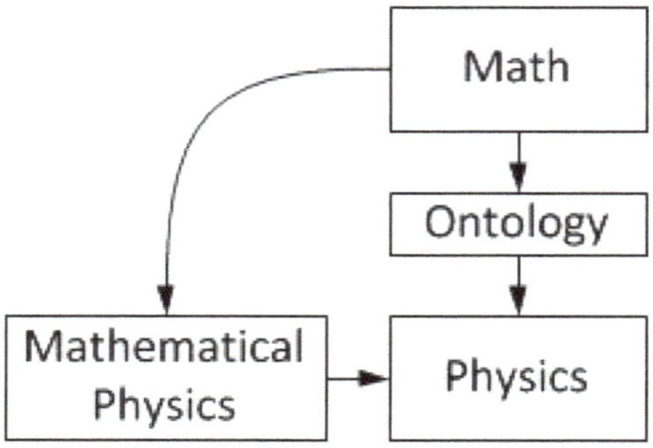

The distinction between mathematics and physics is clear-cut, but not always obvious, especially in mathematical physics.

Ontology is a prerequisite for physics, but not for mathematics. It means physics is ultimately concerned with descriptions of the real world, while mathematics is concerned with abstract patterns, even beyond the real world. Thus physics statements are synthetic, while mathematical statements are analytic. Mathematics contains hypotheses, while physics contains theories. Mathematics statements have to be only logically true, while predictions of physics statements must match observed and experimental data.

The distinction is clear-cut, but not always obvious. For example, mathematical physics is the application of mathematics in physics. Its methods are mathematical, but its subject is physical. The problems in this field start with a "mathematical model of a physical situation" (system) and a "mathematical description of a physical law" that will be applied to that system. Every mathematical statement used for solving has a hard-to-find physical meaning. The final mathematical solution has an easier-to-find meaning, because it is what the solver is looking for.

Physics is a branch of fundamental science, not practical science. Physics is also called "the fundamental science" because the subject of study of all branches of natural science like chemistry, astronomy, geology, and biology are constrained by laws of physics, similar to how chemistry is often called the central science because of its role in linking the physical sciences. For example, chemistry studies properties, structures, and reactions of matter (chemistry's focus on the atomic scale distinguishes it from physics). Structures are formed because particles exert electrical forces on each other, properties include physical characteristics of given substances, and reactions are bound by laws of physics, like conservation of energy, mass, and charge.

Physics is applied in industries like engineering and medicine.

Application and Influence

Applied physics is a general term for physics research which is intended for a particular use. An applied physics curriculumusually contains a few classes in an applied discipline, like geology or electrical engineering. It usually differs from engineering in that an applied physicist may not be designing something in particular, but rather is using physics or conducting physics research with the aim of developing new technologies or solving a problem.

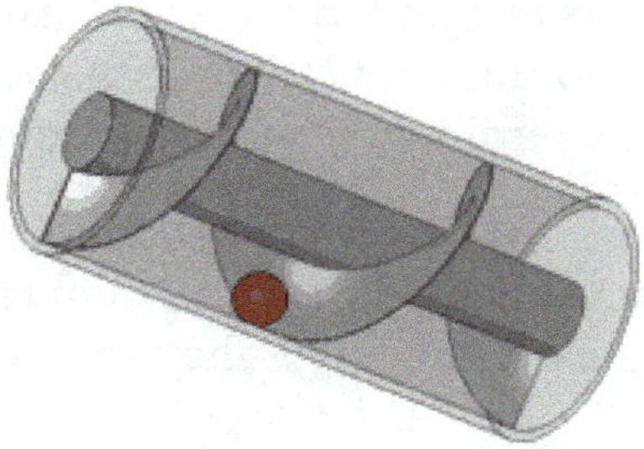

Archimedes' screw, a simple machine for lifting

The application of physical laws in lifting liquids

The approach is similar to that of applied mathematics. Applied physicists use physics in scientific research. For instance, people working on accelerator physics might seek to build better particle detectors for research in theoretical physics.

Physics is used heavily in engineering. For example, statics, a subfield of mechanics, is used in the building of bridges and other static structures. The understanding and use of acousticsresults in sound control and better concert halls; similarly, the use of optics creates better optical devices. An understanding of physics makes for more realistic flight simulators, video games, and movies, and is often critical in forensic investigations.

With the standard consensus that the laws of physics are universal and do not change with time, physics can be used to study things that would ordinarily be mired in uncertainty. For example, in the study of the origin of the earth, one can reasonably model earth's mass, temperature, and rate of rotation, as a function of time allowing one to extrapolate forward or backward in time and so predict future or prior events. It also allows for simulations in engineering which drastically speed up the development of a new technology.

But there is also considerable interdisciplinarity in the physicist's methods, so many other important fields are influenced by physics (e.g., the fields of econophysics and sociophysics).

Theoretical Physics

Theoretical physics is the development of mathematical formalisms and computational protocols for describing all aspects of objects found in the world around us and their interaction. This can involve both providing models for understanding empirical results or constructing self-logical theories for explain phenomena beyond current experiments.

It is perhaps the highest-impact, lowest-cost area of basic research. The field advances our fundamental understanding of the universe, and seeds the technologies of tomorrow. Solar cells, computers, wireless technologies, diagnostic imaging – they are all rooted in breakthroughs made by theoretical physicists.

Time and again, breakthroughs in our fundamental understanding of how the universe works have led to transformative technologies, with innumerable benefits to society. *The reason is simple: since any technology relies on the laws of nature, the better we understand those laws, the more powerful the technologies we can create.*

For example, electricity and magnetism were once thought to be unrelated forces. Building largely on the experimental work of Michael Faraday, theorist James Clerk Maxwell discovered electromagnetism – a single, deeper theory that unified these two forces. Faraday and Maxwell's work eventually led to the development of electric motors, generators, and a safe electric power distribution system ("the grid"), the cleanest and most versatile form of energy the world has ever seen.

As a further example, when Guglielmo Marconi applied Maxwell's electromagnetism to send the first wireless message between continents, the Communications Age was born, changing our lives with radio, television, smartphones and wireless internet. By pondering the nature of space and time, Albert Einstein unexpectedly discovered that matter is a concentrated form of energy, which soon led to an understanding of how stars like our Sun work, by a process called fusion.

These ideas are currently being applied or contemplated in some of the largest-scale science and engineering projects in human history to harness the power of fusion. Over the 21st century, these projects may play a role in helping to solve the world's energy problems, and as a by-product, reducing human impact on the environment and global warming. Even further, Einstein's deeper understanding of the nature of gravity provided the know-how required to build the Global Positioning System (GPS), a profoundly useful technology across a wide spectrum of human endeavours.

To unravel the mysteries of the atom, theoretical physicists developed quantum theory, which has had innumerable spinoff technologies ranging from the transistor (at the heart of computers and virtually every electronic device on the planet), the laser (used everywhere from home DVD players to fibre optic communication networks spanning the globe), Magnetic Resonance Imaging (MRI) and other life-saving medical technologies, and many more.

This is how fundamental physics drives innovation. Theoretical physics is always at the root of it. Perhaps no other field of science can have as deep and as broad of an impact on society and how we understand the universe. Beyond the technologies that may emerge from physics is the intellectual journey of discovery.

Scientific research is a global effort, and the pursuit of new knowledge – from theory to experiment – provides a framework in which past lessons and new ideas are transferred throughout a great research chain. This includes the deep understanding that is passed along to new generations of critical thinkers who emerge throughout academia, industry, and all parts of society.

Physical theory is a model of physical events. It is judged by the extent to which its predictions agree with empirical observations. The quality of a physical theory is also judged on its ability to make new predictions which can be verified by new observations. A physical theory differs from a mathematical theorem in that while both are based on some form of axioms, judgment of mathematical applicability is not based on agreement with any experimental results. A physical theory similarly differs from a mathematical theory, in the sense that the word "theory" has a different meaning in mathematical terms.

> "$Ric = kg$
>
> *The equations for an Einstein manifold, used in general relativity to describe the curvature of spacetime*"

A physical theory involves one or more relationships between various measurable quantities. Archimedes realized that a ship floats by displacing its mass of water, Pythagoras understood the relation between the length of a vibrating string and the musical tone it produces. Other examples include entropy as a measure of the uncertainty regarding the positions and motions of unseen particles and the quantum mechanical idea that (action and) energy are not continuously variable.

Theoretical physics consists of several different approaches. In this regard, theoretical particle physics forms a good example. For instance: "phenomenologists" might employ (semi-) empirical formulas to agree with experimental results, often without deep physical understanding. "Modelers" (also called "model-builders") often appear much like phenomenologists, but try to model speculative theories that have certain desirable features (rather than on experimental data), or apply the techniques of mathematical modeling to physics problems. Some attempt to create approximate theories, called *effective theories*, because fully developed theories may be regarded as unsolvable or too complicated. Other theorists may try to unify, formalise, reinterpret or generalise extant theories, or create completely new ones altogether. Sometimes the vision provided by pure mathematical systems can provide clues to how a physical system might be modeled; e.g., the notion, due to Riemann and others, that space itself might be curved. Theoretical problems that need computational investigation are often the concern of computational physics.

Theoretical advances may consist in setting aside old, incorrect paradigms (e.g., aether theory of light propagation, caloric theory of heat, burning consisting of evolving phlogiston, or astronomical bodies revolving around the Earth) or may be an alternative model that provides answers that are more accurate or that can be more widely applied. In the latter case, a correspondence principle will be required to recover the previously known result. Sometimes though, advances may proceed along different paths. For example, an essentially correct theory may need some conceptual or factual revisions; atomic theory, first postulated millennia ago (by several thinkers in Greece and India) and the two-fluid theory of electricity are two cases in this point. However, an exception to all the above is the wave–particle duality, a theory combining aspects of different, opposing models via the Bohr complementarity principle.

Physical theories become accepted if they are able to make correct predictions and no (or few) incorrect ones. The theory should have, at least as a secondary objective, a certain economy and elegance (compare to mathematical beauty), a notion sometimes called "Occam's razor". English philosopher William of Occam (or Ockham), in which the simpler of two theories that describe the same matter just as adequately is preferred (but conceptual simplicity may mean mathematical complexity). They are also more likely to be accepted if they connect a wide range of phenomena. Testing the consequences of a theory is part of the scientific method.

Physical theories can be grouped into three categories: *mainstream theories*, *proposed theories* and *fringe theories*.

Mainstream Theories

Mainstream theories (sometimes referred to as *central theories*) are the body of knowledge of both factual and scientific views and possess a usual scientific quality of the tests of repeatability, consistency with existing well-established science and experimentation. There do exist mainstream theories that are generally accepted theories based solely upon their effects explaining a wide variety of data, although the detection, explanation, and possible composition are subjects of debate.

Examples

- Classical mechanics
- Continuum mechanics

- Solid mechanics
- Mathematical physics
- Statistical physics
- Statistical mechanics
- Condensed matter physics (including solid state physics and the electronic structure of materials)
- Computational physics
- Fundamental interaction
- Weak interaction
- Electroweak interaction
- Analog models of gravity
- Conservation law
- Conservation of energy
- Conservation of mass
- Conservation of momentum
- Conservation of angular momentum
- Thermodynamics
- Laws of thermodynamics
- Newton's laws of motion
- Dynamics
- Dynamo theory
- Kinetic theory of gases
- Electromagnetism
- Field theory
- Classical field theory
- Scattering theory
- Quantum field theory
- Quantum field theory in curved spacetime
- Conformal field theory
- Fluid mechanics
- Fluid dynamics
- Information theory

- Algorithmic information theory
- Quantum information theory
- Theoretical computer science
- Computational complexity theory
- Quantum complexity theory
- Theory of relativity
- General relativity
- Special relativity
- Physical cosmology
- Quantum mechanics
- Fractional quantum mechanics
- Relativistic quantum mechanics
- Quantum dynamics
- Quantum chromodynamics
- Quantum computing
- Quantum electrochemistry
- Quantum electrodynamics
- Quantum geometry
- Quantum chaos
- Chaos theory
- Perturbation theory
- Perturbation theory (quantum mechanics)
- Dark matter
- Cosmological Constant
- Gravitational constant
- Standard Model
- Spontaneous symmetry breaking

Proposed Theories

The proposed theories of physics are usually relatively new theories which deal with the study of physics which include scientific approaches, means for determining the validity of models and new types of reasoning used to arrive at the theory. However, some proposed theories include theories that have been around for decades and have eluded methods of discovery and testing. Proposed theories can include fringe theories in the process of becoming established (and, sometimes, gaining wider acceptance).

Examples

- Black hole thermodynamics
- Causal Sets
- Dark fluid
- Doubly special relativity
- Einstein–Rosen Bridge
- Emergence
- Hamiltonian mechanics
- Supervenience
- Theory of everything
- Mathematical universe hypothesis
- Grand unification theory
- Loop quantum gravity
- Loop quantum cosmology
- Quantum gravity
- Canonical quantum gravity
- Scale relativity
- Supersymmetry
- Supersymmetry breaking
- Supergravity
- String theory
- Superstring theory
- M-theory
- Brane cosmology
- Eigenstate thermalization hypothesis
- Experimental physics
- Unparticle physics
- Digital physics
- Hidden variable theory
- Pilot wave theory

- CPT symmetry
- Event symmetry

Fringe Theories

Fringe theories include any new area of scientific endeavor in the process of becoming established and some proposed theories. It can include speculative sciences. This includes physics fields and physical theories presented in accordance with known evidence, and a body of associated predictions have been made according to that theory.

Some fringe theories go on to become a widely accepted part of physics. Other fringe theories end up being disproven. Some fringe theories are a form of protoscience and others are a form of pseudoscience. The falsification of the original theory sometimes leads to reformulation of the theory.

Examples

- Electrogravitics
- Tesla's dynamic theory of gravity
- Aether (classical element)
- Luminiferous aether
- Orgone
- Prana
- Qi
- Odic force

Thought Experiments vs Real Experiments

"Thought" experiments are situations created in one's mind, asking a question akin to "suppose you are in this situation, assuming such is true, what would follow?". They are usually created to investigate phenomena that are not readily experienced in every-day situations. Famous examples of such thought experiments are Schrödinger's cat, the EPR thought experiment, simple illustrations of time dilation, and so on. These usually lead to real experiments designed to verify that the conclusion (and therefore the assumptions) of the thought experiments are correct. The EPR thought experiment led to the Bell inequalities, which were then tested to various degrees of rigor, leading to the acceptance of the current formulation of quantum mechanics and probabilism as a working hypothesis.

Experimental Physics

At the core of the understanding of the natural processes in the universe is the spirit of scientific inquiry. Knowledge is acquired through a systematic method of scientific research in the natural sciences. This process extends across multiple stages. The first stage is observation, which can be qualitative or quantitative, involving analysis of a phenomenon with the senses alone or measured or recorded using tools and instruments. The next stage is the formulation of a hypothesis that strives to offer an explanation for the observation. This follows a stage of rigorous scientific experimentation that either validates or disproves the original hypothesis. At this point, the hypothesis is either accepted or modified. The accepted theory is even then subject to scrutiny. If the theory is supported but further experimental evidence proves it to the contrary, it is considered invalid or subject to modification.

In the field of physics, each scientific theory and law is guided by empirical evidence and scientific research. Experimentation is the centerpiece of the scientific method in physics. This is evident in the emergence of the myriad disciplines within physics that have benefited from experimentation during the Scientific Revolution of 17th and 18th century Europe, by great physicists like Galileo, Christiaan Huygens, Kepler, Pascal and Issac Newton, besides many others. Experimentation continues to this day and this is reflected in the great strides in not just the understanding of the universe at large, but in the development of modern technologies as well.

The development of classical mechanics is founded on the scientific verification of Galileo's law of inertia, Newton's laws of motion and the law of universal gravitation. The drive for the creation of vacuum and the creation of the air pump allowed the formulation of Boyle's law, which is a fundamental principle of thermodynamics. A major initiative in experimental physics was focused on the investigation of the nature of electricity and magnetism. The field of electromagnetism grew as a result of the work of researchers and experimenters like Stephen Gray, Robert Boyle and Benjamin Franklin, whose work later proved to be the foundation for later work by Hans Christian Ørsted, Ampère, Coulomb, Michael Faraday and Maxwell.

Many famous experiments have dotted the history of experimental physics. From Galileo demonstrating that objects when dropped from a height fall with the same speed, to the modern discovery of the Higgs boson at the CERN's Large Hadron Collider, and from Rutherford's gold foil experiment for understanding the structure of an atom to the discovery of gravitational waves by LIGO as an experimental proof of general relativity, experimental physics has occurred in tandem with theoretical innovations in physics. Many of the theoretical predictions of physics would have remained mere postulates and conjectures had the experimental equivalents of these theories not been supported with evidence. Significant examples of these that have an incredible standing in modern physics are the idea of quarks, the neutrino, gravitational lensing, besides many others. As such, the study of experimental physics is crucial to developing a holistic understanding of the history of physics development.

Applied Physics

Applied physics is the study of physics for a practical purpose, as opposed to physics motived solely for an improved fundamental understanding. This includes technological advances such as the development of electronics, photonics and device physics, or improving practical investigations, such as experimental nuclear physics and experimental particle physics.

Experimentation is the gateway to discovering more about our world. Applied Physics focuses on using scientific principles in practical devices and systems, and on the application of physics in other fields. A great deal of the work done by physicists is not immediately visible. But without recent breakthroughs in physics, the mobile phone and the laptop would have been power-guzzling machines that weigh a ton and lab-on-a-chip technology would still be science fiction.

It is usually considered as a bridge or a connection between physics and engineering. "Applied" is distinguished from "pure" by a subtle combination of factors such as the motivation and attitude of researchers and the nature of the relationship to the technology or science that may be affected by the work.

Applied physics may not be directly involved with developing new technology, however, it studies physics concepts with the aim of understanding them and applying them in such a way as to be able to build or help in building new technology. Applied physics is closely related with engineering, as they both are concerned with applying physics that is understood and useful to new technologies.

In other words, applied physics is rooted in the fundamental truths and basic concepts of the physical sciences but is concerned with the utilization of these scientific principles in practical devices and systems.

Applied physicists can also be interested in the use of physics for scientific research. For instance, the field of accelerator physics can contribute to research in theoretical physics by working with engineers enabling design and construction of high-energy colliders.

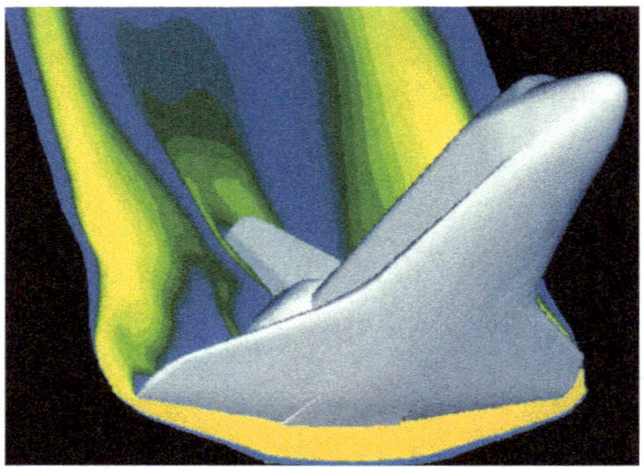

Computer modeling of the space shuttle during re-entry

Examples of Research and Development Areas

- The transistor, which was first invented by physicists John Bardeen, Walter Brattain and William Shockley in 1947
- Lasers, such as Vertical-cavity surface-emitting lasers
- Photonic crystals and quantum optics
- Magnetic resonance imaging
- Microscopy
- Semiconductors
- Accelerator physics

- Quantum information science
- Quantum technology
- Astrodynamics
- Electromagnetic propulsion
- Stealth technology
- Nuclear engineering
- Engineering Physics
- Electronics
- Sonar
- Radar
- Lidar
- Biophysics
- Chemical Physics
- Geophysics

Computational Physics

Computational physics can be stated as the study and the implementation of the numerical analysis for solving problems in physics for which there is the existence of a quantitative theory. Some time ago, computational physics was used as a application of modern computers in the field of science but now, it has become a subset of computational science.

It is basically an intermediate branch between the experimental physics and the theoretical physics which helps in supplementing both theory and experiment. In simple words, it is using computers for solving the problems of physics, for explaining the theories of physics along with its stimulations. It makes the life of computational physicists easier.

In the field of physics, various theories based on the mathematical models seem to predict the behavior of the systems. Unfortunately, many a times, solving the mathematical model for predicting a useful outcome is not always possible.

This occurs when the solution is complicated or does not have closed-form expression. In these cases, a numerical representation is needed. Computational physics deals with such numerical approximations. A computer is used for writing the solution in simple mathematical algorithm along with computing a solution with respective error.

There is more often than not a debate about the status of computers in physics. Many scientists regard the simulation of computers as mere computer experiments while some treat it as theoretical physics. The main purpose of using computers in physics is because the calculations get easier and the stored data is secure.

Computational physics helps in broadening and deepening of the understanding of physics in the human mind and also increases the range of calculations that can be performed mathematically.

Some of the software used in this field is MATHEMATICA, MAPLE, MATLAB etc. NAG, 4 LINPACK, 5 ODEPACK, 6 are some of the libraries that are used for performing all the real numerical work.

Computational physics is used in a wide area of physics like astrophysics, lattice field theory, fluid mechanics, plasma physics, weather prediction, soft condensed matter, hyper velocity impact physics etc. For example, computational solid state physics used the density functional theory for calculating the properties of solids in a method similar to the one used by the chemists for studying the molecules.

Problems of physics are not very easy to solve owing to various mathematical reasons but computational physics make it easy to solve. The problems that cannot be solve using computers are the ones involving chaos which renders any solution meaningless.

Challenges in Computational Physics

Physics problems are in general very difficult to solve exactly. This is due to several (mathematical) reasons: lack of algebraic and/or analytic solubility, complexity, and chaos.

For example, - even apparently simple problems, such as calculating the wavefunction of an electron orbiting an atom in a strong electric field (Stark effect), may require great effort to formulate a practical algorithm (if one can be found); other cruder or brute-force techniques, such as graphical methods or root finding, may be required. On the more advanced side, mathematical perturbation theory is also sometimes used (a working is shown for this particular example here).

In addition, the computational cost and computational complexity for many-body problems (and their classical counterparts) tend to grow quickly. A macroscopic system typically has a size of the order of constituent particles, so it is somewhat of a problem. Solving quantum mechanical problems is generally of exponential order in the size of the system and for classical N-body it is of order N-squared.

Finally, many physical systems are inherently nonlinear at best, and at worst chaotic: this means it can be difficult to ensure any numerical errors do not grow to the point of rendering the 'solution' useless.

Methods and Algorithms

Because computational physics uses a broad class of problems, it is generally divided amongst the different mathematical problems it numerically solves, or the methods it applies. Between them, one can consider:

- root finding (using e.g. Newton-Raphson method)
- system of linear equations (using e.g. LU decomposition)
- ordinary differential equations (using e.g. Runge–Kutta methods)
- integration (using e.g. Romberg method and Monte Carlo integration)
- partial differential equations (using e.g. finite difference method and relaxation method)
- matrix eigenvalue problem (using e.g. Jacobi eigenvalue algorithm and power iteration)

All these methods (and several others) are used to calculate physical properties of the modeled systems.

Computational physics also borrows a number of ideas from computational chemistry - for example, the density functional theory used by computational solid state physicists to calculate properties of solids is basically the same as that used by chemists to calculate the properties of molecules.

Furthermore, computational physics encompasses the tuning of the software/hardware structure to solve the problems (as the problems usually can be very large, in processing power need or in memory requests).

Divisions

It is possible to find a corresponding computational branch for every major field in physics, for example computational mechanics and computational electrodynamics. Computational mechanics consists of computational fluid dynamics (CFD), computational solid mechanics and computational contact mechanics. One subfield at the confluence between CFD and electromagnetic modelling is computational magnetohydrodynamics. The quantum many-body problem leads naturally to the large and rapidly growing field of computational chemistry.

Computational solid state physics is a very important division of computational physics dealing directly with material science.

A field related to computational condensed matter is computational statistical mechanics, which deals with the simulation of models and theories (such as percolation and spin models) that are difficult to solve otherwise. Computational statistical physics makes heavy use of Monte Carlo-like methods. More broadly, (particularly through the use of agent based modeling and cellular automata) it also concerns itself with (and finds application in, through the use of its techniques) in the social sciences, network theory, and mathematical models for the propagation of disease (most notably, the SIR Model) and the spread of forest fires.

On the more esoteric side, numerical relativity is a (relatively) new field interested in finding numeric solutions to the field equations of general (and special) relativity, and computational particle physics deals with problems motivated by particle physics.

Computational astrophysics is the application of these techniques and methods to astrophysical problems and phenomena.

Computational biophysics is a branch of biophysics and computational biology itself, applying methods of computer science and physics to large complex biological problems.

Applications

Due to the broad class of problems computational physics deals, it is an essential component of modern research in different areas of physics, namely: accelerator physics, astrophysics, fluid mechanics (computational fluid dynamics), lattice field theory/lattice gauge theory (especially lattice quantum chromodynamics), plasma physics, simulating physical systems (using e.g. molecular dynamics), nuclear engineering computer codes, protein structure prediction, weather prediction, solid state physics, soft condensed matter physics, hypervelocity impact physics etc.

Computational solid state physics, for example, uses density functional theory to calculate properties of solids, a method similar to that used by chemists to study molecules. Other quantities of interest in solid state physics, such as the electronic band structure, magnetic properties and charge densities can be calculated by this and several methods, including the Luttinger-Kohn/k.p method and ab-initio methods.

Mathematical Physics

An area of science concerned with the application of mathematical concepts to the physical sciences and the development of mathematical ideas in response to the needs of physics. Historically, the concept of mathematical physics was synonymous with that of theoretical physics. In present-day terminology, however, a distinction is made between the two. Whereas most of theoretical physics uses a large amount of mathematics as a tool and as a language, mathematical physics places greater emphasis on mathematical rigor, and devotes attention to the development of areas of mathematics that are, or show promise to be, useful to physics. The results obtained by pure mathematicians, with no thought to applications, are almost always found to be both useful and effective in formulating physical theories.

Mathematical physics forms the bridge between physics as the description of nature and its structure on the one hand, and mathematics as a construction of pure logical thought on the other. This bridge between the two disciplines benefits and strengthens both fields enormously.

The methods employed in mathematical physics range over most of mathematics, the areas of analysis and algebra being the most commonly used. Partial differential equations and differential geometry, with heavy use of vector and tensor methods, are of particular importance in the formulation of field theories, and functional analysis as well as operator theory in quantum mechanics.

Group theory has become an especially valuable tool in the construction of quantum field theories and in elementary-particle physics. There has also been an increase in the use of general geometrical approaches and of topology. For solution methods and the calculation of quantities that are amenable to experimental tests, of particular prominence are Fourier analysis, complex analysis, variational methods, the theory of integral equations, and perturbation theory.

Mathematical Physics

The theory of mathematical models of physical phenomena, which occupies a special place in both mathematics and physics.

Mathematical physics is closely connected with physics inasmuch as it deals with the construction of mathematical models; at the same time it is a branch of mathematics inasmuch as the methods used to investigate the models are mathematical. The concept of mathematical physics also includes those mathematical methods that are used to set up and study mathematical models that describe large classes of physical phenomena.

The methods of mathematical physics as the theories of mathematical models in physics were first developed intensively in I. Newton's works dealing with the formulation of the foundations of classical mechanics, universal gravitation, and the theory of light. The further development of the methods of mathematical physics and their successful application to a wide range of physical phenomena are associated with J. Lagrange, L. Euler, P. Laplace, J. Fourier, K. Gauss, B. Riemann, and M. V. Ostrogradskii, among others. A. M. Liapunov and V. A. Steklov made major contributions to the development of the methods of mathematical physics.

Beginning in the second half of the 19th century, the methods of mathematical physics were used successfully in studying mathematical models of physical phenomena related to various physical fields and wave functions in electrodynamics, acoustics, the theory of elasticity, hydrodynamics, aerodynamics, and other fields related to the study of physical phenomena in continuous media. The mathematical models of this class of phenomena are described most often by means of partial differential equations called the equations of mathematical physics. In addition, integral and integrodifferential equations, variational and probability theory methods, potential theory, the theory of functions of a complex variable, and many other branches of mathematics are also used in describing the mathematical models of physics. With the rapid development of computer mathematics direct numerical methods based on the use of computers and especially finite-difference methods of solving boundary value problems have acquired particular importance in the investigation of mathematical models in physics. Theoretical investigations in quantum electrodynamics, the axiomatic theory of fields, and many other branches of modern physics have led to the creation of a new class (including the theory of generalized functions and the theory of continuous-spectra operators) of mathematical models constituting an important branch of mathematical physics.

The formulation of the problems of mathematical physics consists in the setting up of mathematical models describing the basic regularities shown by the class of physical phenomena being studied. Such a formulation involves the derivation of equations (differential, integral, integrodifferential, or algebraic) that are satisfied by the quantities characterizing the particular physical process. In so doing we proceed from the fundamental physical laws that take account of only the most significant features of the phenomenon, disregarding its secondary characteristics. These laws are usually the conservation laws (the laws of the conservation of momentum, energy, and the number of particles). Thus, the same mathematical model can be used to describe different physical processes having common features. For example, mathematical questions associated with the simplest hyperbolic equation

$$\frac{\partial^2 u}{\partial t^2} = a^2 \frac{\partial^2 u}{\partial x^2}$$

which was initially obtained by J. D'Alembert in 1747 to describe the free vibrations of a uniform string, are also pertinent in describing a broad range of wave processes in acoustics, hydrodynamics, and electrodynamics. Similarly, the equation

$$\frac{\partial^2 u}{\partial x^2} + \frac{\partial^2 u}{\partial y^2} + \frac{\partial^2 u}{\partial z^2} = 0$$

the boundary value problems for which were originally studied by Laplace (at the end of the 18th century) in connection with the development of the theory of gravitation, subsequently found application in the solution of many problems of electrostatics, the theories of elasticity, and the steady motion of an ideal fluid. An entire class of physical processes corresponds to every mathematical model used in physics.

Another feature of mathematical physics is that many general methods used to solve the problems of mathematical physics were developed from particular methods of solving specific physical problems and in their original form did not have a rigorous mathematical basis or lacked completeness. This is true of such well-known methods as the Ritz and Galerkin methods, the theory of perturbations, Fourier transformations, and the separation of variables. The effective application of all these methods to the solution of specific problems is one reason for their rigorous mathematical substantiation and generalization, which has often led to the emergence of new mathematical disciplines.

The impact of mathematical physics on various branches of mathematics is also evident from the fact that the development of mathematical physics, which reflects the requirements of the natural sciences and technology, entails reorientation of research in some branches of mathematics that have already taken shape. The formulation of the problems of mathematical physics, which was connected with the elaboration of mathematical models of real physical phenomena, has led to a change in the basic approach to the theory of partial differential equations. The theory of boundary value problems emerged, which subsequently made it possible to relate partial differential equations to integral equations and variational methods.

The study of the mathematical models of physics by mathematical methods not only makes it possible to obtain the quantitative characteristics of physical phenomena and to compute with a given degree of accuracy the course of real processes, but also provides the possibility of gaining insight into the very nature of physical phenomena, revealing hidden laws and predicting new effects. The striving for a more detailed study of physical phenomena leads to the formulation of increasingly complex mathematical models. This in turn makes it impossible to use analytic methods to investigate the models. This impossibility arises since the mathematical models of real physical processes generally are nonlinear, that is, are described by nonlinear equations of mathematical physics.

However, direct numerical methods involving the use of computers are used successfully in the detailed investigation of these models. The use of numerical models for typical problems of mathematical physics reduces to replacing the equations of mathematical physics for functions of a continuous argument with algebraic equations for finite-difference functions that are defined in a discrete set of points (in a network). In other words, a discrete analog is introduced in place of a continuous model of the medium. In many cases, the use of numerical methods makes it possible to replace a complex, laborious, and costly physical experiment with a less costly mathematical (numerical) experiment. A mathematical numerical experiment, if it is complete enough, can serve as the basis in the selection of optimal conditions for a real physical experiment, the selection of the parameters of complex physical installations and the determination of the conditions under which new physical effects would appear. Numerical methods thus greatly expand the range of effective use of mathematical models of physical phenomena.

A mathematical model of a physical phenomenon, like any model, cannot convey all the characteristics of the phenomenon. The adequacy of the adopted model for the phenomenon being studied can be established only through practice and a comparison of the results of theoretical investigations of the model adopted with experimental data.

In many cases, the adequacy of the model adopted can be assessed on the basis of the solution of inverse problems of mathematical physics, when the properties of natural phenomena that are inaccessible to direct observation are ascertained from indirect physical manifestations.

A characteristic feature of mathematical physics is the setting up of mathematical models that not only provide a description and explanation of physical principles already established for the range of phenomena under study but also make it possible to predict as yet undiscovered principles. A classic example of such a model is Newton's theory of universal gravitation, which made it possible not only to explain the motion of bodies in the solar system that were known at the time the theory was developed but also to predict the existence of new planets. On the other hand, new experimental data that appear cannot always be explained within the framework of the model adopted and therefore the development of a more complex model is required.

References

- Leggett, A. J. (2006). "What DO we know about high T_c?" (PDF). Nature Physics. 2 (3): 134–136. Bibcode:2006NatPh...2..134L. doi:10.1038/nphys254. Archived from the original(PDF) on 10 June 2010
- "physics". Online Etymology Dictionary. Archived from the original on 24 December 2016. Retrieved 1 November 2016
- Noll notes that some universities still use this title —Noll, Walter (23 June 2006). "On the Past and Future of Natural Philosophy" (PDF). Journal of Elasticity. 84 (1): 1–11. doi:10.1007/s10659-006-9068-y. Archived (PDF) from the original on 18 April 2016
- Wolf, S. A.; Chtchelkanova, A. Y.; Treger, D. M. (2006). "Spintronics—A retrospective and perspective". IBM Journal of Research and Development. 50: 101. doi:10.1147/rd.501.0101

- Weinberg, S. (1993). Dreams of a Final Theory: The Search for the Fundamental Laws of Nature. Hutchinson Radius. ISBN 0-09-177395-4
- Gill, N.S. "Atomism - Pre-Socratic Philosophy of Atomism". About Education. Archivedfrom the original on 10 July 2014. Retrieved 1 April 2014
- "Journal of Mathematical Physics". ResearchGate. Archived from the original on 18 August 2014. Retrieved 31 March 2014
- Singiresu S. Rao (2007). Vibration of Continuous Systems (illustrated ed.). John Wiley & Sons. 5,12. ISBN 04 71771716. ISBN 9780471771715
- Gibney, E. (2015). "LHC 2.0: A new view of the Universe". Nature. 519 (7542): 142–143. Bibcode:2015Natur.519..142G. doi:10.1038/519142a. PMID 25762263. Archived from the original on 21 December 2016
- Eli Maor (2007). The Pythagorean Theorem: A 4,000-year History (illustrated ed.). Princeton University Press. pp. 18–20. ISBN 0691125260. ISBN 9780691125268

2

Prominent Principles and Branches of Physics

Physics studies matter and energy. The chief branches of physics are classical mechanics, thermodynamics, magnetism, astrophysics, condensed matter physics and particle physics, among others. This chapter closely examines these branches along with prominent principles of thermodynamics, magnetism, optics, etc. to develop a holistic understanding of the subject.

Mechanics

Mechanics is the science concerned with the motion of bodies under the action of forces, including the special case in which a body remains at rest. Of first concern in the problem of motion are the forces that bodies exert on one another. This leads to the study of such topics as gravitation, electricity, and magnetism, according to the nature of the forces involved. Given the forces, one can seek the manner in which bodies move under the action of forces; this is the subject matter of mechanics proper.

Historically, mechanics was among the first of the exact sciences to be developed. Its internal beauty as a mathematical discipline and its early remarkable success in accounting in quantitative detail for the motions of the Moon, the Earth, and other planetary bodies had enormous influence on philosophical thought and provided impetus for the systematic development of science into the 20th century.

Mechanics may be divided into three branches: statics, which deals with forces acting on and in a body at rest; kinematics, which describes the possible motions of a body or system of bodies; and kinetics, which attempts to explain or predict the motion that will occur in a given situation. Alternatively, mechanics may be divided according to the kind of system studied. The simplest mechanical system is the particle, defined as a body so small that its shape and internal structure are of no consequence in the given problem. More complicated is the motion of a system of two or more particles that exert forces on one another and possibly undergo forces exerted by bodies outside of the system.

Classical versus Quantum

Historically, classical mechanics came first, and quantum mechanics is a comparatively recent invention. Classical mechanics originated with Isaac Newton's laws of motion in Philosophiæ Naturalis Principia Mathematica; Quantum Mechanics was discovered in the early 20th century. Both are commonly held to constitute the most certain knowledge that exists about physical nature.

Classical mechanics has especially often been viewed as a model for other so-called exact sciences. Essential in this respect is the extensive use of mathematics in theories, as well as the decisive role played by experiment in generating and testing them.

Quantum mechanics is of a bigger scope, as it encompasses classical mechanics as a sub-discipline which applies under certain restricted circumstances. According to the correspondence principle, there is no contradiction or conflict between the two subjects, each simply pertains to specific situations. The correspondence principle states that the behavior of systems described by quantum theories reproduces classical physics in the limit of large quantum numbers. Quantum mechanics has superseded classical mechanics at the foundation level and is indispensable for the explanation and prediction of processes at the molecular, atomic, and sub-atomic level. However, for macroscopic processes classical mechanics is able to solve problems which are unmanageably difficult in quantum mechanics and hence remains useful and well used. Modern descriptions of such behavior begin with a careful definition of such quantities as displacement (distance moved), time, velocity, acceleration, mass, and force. Until about 400 years ago, however, motion was explained from a very different point of view. For example, following the ideas of Greek philosopher and scientist Aristotle, scientists reasoned that a cannonball falls down because its natural position is in the Earth; the sun, the moon, and the stars travel in circles around the earth because it is the nature of heavenly objects to travel in perfect circles.

Often cited as father to modern science, Galileo brought together the ideas of other great thinkers of his time and began to calculate motion in terms of distance traveled from some starting position and the time that it took. He showed that the speed of falling objects increases steadily during the time of their fall. This acceleration is the same for heavy objects as for light ones, provided air friction (air resistance) is discounted. The English mathematician and physicist Isaac Newton-improved this analysis by defining force and mass and relating these to acceleration. For objects traveling at speeds close to the speed of light, Newton's laws were superseded by Albert Einstein's theory of relativity. [A sentence illustrating the computational complication of Einstein's theory of relativity.] For atomic and subatomic particles, Newton's laws were superseded by quantum theory. For everyday phenomena, however, Newton's three laws of motion remain the cornerstone of dynamics, which is the study of what causes motion.

Relativistic versus Newtonian

In analogy to the distinction between quantum and classical mechanics, Einstein's general and specialtheories of relativity have expanded the scope of Newton and Galileo's formulation of mechanics. The differences between relativistic and Newtonian mechanics become significant and even dominant as the velocity of a massive body approaches the speed of light. For instance, in Newtonian mechanics, Newton's laws of motion specify that $\mathbf{F} = m\mathbf{a}$, whereas in Relativistic mechanics and Lorentz transformations, which were first discovered by Hendrik Lorentz, $\mathbf{F} = \gamma m\mathbf{a}$ (where γ is the Lorentz factor, which is almost equal to 1 for low speeds).

General Relativistic versus Quantum

Relativistic corrections are also needed for quantum mechanics, although general relativity has not been integrated. The two theories remain incompatible, a hurdle which must be overcome in developing a theory of everything.

Types of Mechanical Bodies

The often-used term body needs to stand for a wide assortment of objects, including particles, projectiles, spacecraft, stars, parts of machinery, parts of solids, parts of fluids (gases and liquids), etc.

Other distinctions between the various sub-disciplines of mechanics, concern the nature of the bodies being described. Particles are bodies with little (known) internal structure, treated as mathematical points in classical mechanics. Rigid bodies have size and shape, but retain a simplicity close to that of the particle, adding just a few so-called degrees of freedom, such as orientation in space.

Otherwise, bodies may be semi-rigid, i.e. elastic, or non-rigid, i.e. fluid. These subjects have both classical and quantum divisions of study.

For instance, the motion of a spacecraft, regarding its orbit and attitude (rotation), is described by the relativistic theory of classical mechanics, while the analogous movements of an atomic nucleus are described by quantum mechanics.

Fundamental Concepts

Units and Dimensions

Quantities have both dimensions, which are an expression of their fundamental nature, and units, which are chosen by convention to express magnitude or size. For example, a series of events have a certain duration in time. Time is the dimension of the duration. The duration might be expressed as 30 minutes or as half an hour. Minutes and hours are among the units in which time may be expressed. One can compare quantities of the same dimensions, even if they are expressed in different units (an hour is longer than a minute). Quantities of different dimensions cannot be compared with one another.

The fundamental dimensions used in mechanics are time, mass, and length. Symbolically, these are written as t, m, and l, respectively. The study of electromagnetism adds an additional fundamental dimension, electric charge, or q. Other quantities have dimensions compounded of these. For example, speed has the dimensions distance divided by time, which can be written as l/t, and volume has the dimensions distance cubed, or l^3. Some quantities, such as temperature, have units but are not compounded of fundamental dimensions.

There are also important dimensionless numbers in nature, such as the number $\pi = 3.14159$. Dimensionless numbers may be constructed as ratios of quantities having the same dimension. Thus, the number π is the ratio of the circumference of a circle (a length) to its diameter (another length). Dimensionless numbers have the advantage that they are always the same, regardless of what set of units is being used.

Governments have traditionally been responsible for establishing and enforcing standard units for the sake of orderly commerce, navigation, science, and, of course, taxation. Today all such units are established by international treaty, revised every few years in light of scientific findings. The units used for most scientific measurements are those designated the International System of Units(Système International d'Unités), or SI for short. They are based on the metric system, first adopted officially by France in 1795. Other units, such as those of the British engineering system, are still in use in some places, but these are now defined in terms of the SI units.

The fundamental unit of length is the metre. A metre used to be defined as the distance between two scratch marks on a metal bar kept in Paris, but it is now much more precisely defined as the distance that light travels in a certain time interval (1/299,792,458 of a second). By contrast, in the British system, units of length have a clear human bias: the foot, the inch (the first joint of the thumb), the yard (distance from nose to outstretched fingertip), and the mile (one thousand standard paces of a Roman legion). Each of these is today defined as some fraction or multiple of a metre (one yard is nearly equal to one metre). In the SI or the metric system, lengths are expressed as decimal fractions or multiples of a metre (a millimetre = one-thousandth of a metre; a centimetre = one-hundredth of a metre; a kilometre = one thousand metres).

Times longer than one second are expressed in the units seconds, minutes, hours, days, weeks, and years. Times shorter than one second are expressed as decimal fractions (a millisecond = one-thousandth of a second, a microsecond = one-millionth of a second, and so on). The fundamental unit of time (i.e., the definition of one second) is today based on the intrinsic properties of certain kinds of atoms (an excitation frequency of the isotope cesium-133).

Units of mass are also defined in a way that is technically sound, but in common usage they are the subject of some confusion because they are easily confused with units of weight, which is a different physical quantity. The weight of an object is the consequence of the Earth's gravity operating on its mass. Thus, the mass of a given object is the same everywhere, but its weight varies slightly if it is moved about the surface of the Earth, and it would change a great deal if it were moved to the surface of another planet. Also, weight and mass do not have the same dimensions (weight has the dimensions ml/t^2). The Constitution of the United States, which calls on the government to establish uniform "weights and measures," is oblivious to this distinction, as are merchants the world over, who measure the weight of bread or produce but sell it in units of kilograms, the SI unit of mass. (The kilogram is equal to 1,000 grams; 1 gram is the mass of 1 cubic centimetre of water—under appropriate conditions of temperature and pressure.)

Vectors

The equations of mechanics are typically written in terms of Cartesian coordinates. At a certain time t, the position of a particle may be specified by giving its coordinates $x(t)$, $y(t)$, and $z(t)$ in a particular Cartesian frame of reference. However, a different observer of the same particle might choose a differently oriented set of mutually perpendicular axes, say, x', y', and z'. The motion of the particle is then described by the first observer in terms of the rate of change of $x(t)$, $y(t)$, and $z(t)$, while the second observer would discuss the rates of change of $x'(t)$, $y'(t)$, and $z'(t)$. That is, both observers see the same particle executing the same motion and obeying the same laws, but they describe the situation with different equations. This awkward situation may be avoided by means of a mathematical construction called a vector. Although vectors are mathematically simple and extremely useful in discussing mechanics, they were not developed in their modern form until late in the 19th century, when J. Willard Gibbs and Oliver Heaviside (of the United States and Britain, respectively) each applied vector analysis in order to help express the new laws of electromagnetism proposed by James Clerk Maxwell.

A vector is a quantity that has both magnitude and direction. It is typically represented symbolically by an arrow in the proper direction, whose length is proportional to the magnitude of the vector. Although a vector has magnitude and direction, it does not have position. A vector is not altered if it is displaced parallel to itself as long as its length is not changed.

By contrast to a vector, an ordinary quantity having magnitude but not direction is known as a scalar. In printed works vectors are often represented by boldface letters such as $\boldsymbol{A}$ or $\boldsymbol{X}$, and scalars are represented by lightface letters, A or X. The magnitude of a vector, denoted $|\mathbf{A}|$, is itself a scalar—i.e., $|\mathbf{A}| = A$.

Because vectors are different from ordinary (i.e., scalar) quantities, all mathematical operations involving vectors must be carefully defined. Addition, subtraction, three kinds of multiplication, and differentiation will be discussed here. There is no mathematical operation that corresponds to division by a vector.

If vector $\boldsymbol{A}$ is added to vector $\boldsymbol{B}$, the result is another vector, $\boldsymbol{C}$, written $\boldsymbol{A} + \boldsymbol{B} = \boldsymbol{C}$. The operation is performed by displacing $\boldsymbol{B}$ so that it begins where $\boldsymbol{A}$ ends. $\boldsymbol{C}$ is then the vector that starts where $\boldsymbol{A}$ begins and ends where $\boldsymbol{B}$ ends.

Vector addition is defined to have the (nontrivial) property $\boldsymbol{A} + \boldsymbol{B} = \boldsymbol{B} + \boldsymbol{A}$. There do exist quantities having magnitude and direction that do not obey this requirement. An example is finite rotations in space. Two finite rotations of a body about different axes do not necessarily result in the same orientation if performed in the opposite order.

Vector subtraction is defined by $\boldsymbol{A} - \boldsymbol{B} = \boldsymbol{A} + (-\boldsymbol{B})$, where the vector $-\boldsymbol{B}$ has the same magnitude as $\boldsymbol{B}$ but the opposition direction.

A vector may be multiplied by a scalar. Thus, for example, the vector $2\boldsymbol{A}$ has the same direction as $\boldsymbol{A}$ but is twice as long. If the scalar has dimensions, the resulting vector still has the same direction as the original one, but the two cannot be compared in magnitude. For example, a particle moving with constant velocity $\boldsymbol{v}$ suffers a displacement $\boldsymbol{s}$ in time t given by $\boldsymbol{s} = \boldsymbol{v}t$. The vector $\boldsymbol{v}$ has been multiplied by the scalar t to give a new vector, $\mathbf{s}$, which has the same direction as $\boldsymbol{v}$ but cannot be compared to $\boldsymbol{v}$ in magnitude (a displacement of one metre is neither bigger nor smaller than a velocity of one metre per second). This is a typical example of a phenomenon that might be represented by different equations in differently oriented Cartesian coordinate systems but that has a single vector equation (for all observers not moving with respect to one another).

The dot product (also known as the scalar product, or sometimes the inner product) is an operation that combines two vectors to form a scalar. The operation is written $\boldsymbol{A} \cdot \boldsymbol{B}$. If θ is the (smaller) angle between $\boldsymbol{A}$ and $\boldsymbol{B}$, then the result of the operation is $\boldsymbol{A} \cdot \boldsymbol{B} = AB \cos\theta$. The dot product measures the extent to which two vectors are parallel. It may be thought of as multiplying the magnitude of one vector (either one) by the projection of the other upon it. If the two vectors are perpendicular, the dot product is zero.

The cross product (also known as the vector product) combines two vectors to form another vector, perpendicular to the plane of the original vectors. The operation is written $\boldsymbol{A} \times \boldsymbol{B}$. If θ is the (smaller) angle between $\boldsymbol{A}$ and $\boldsymbol{B}$, then $|A \times B| = AB \sin\theta$. The direction of $\boldsymbol{A} \times \boldsymbol{B}$ is given by the right-hand rule: if the fingers of the right hand are made to rotate from $\boldsymbol{A}$ through θ to $\boldsymbol{B}$, the thumb points in the direction of $\boldsymbol{A} \times \boldsymbol{B}$. The cross product is zero if the two vectors are parallel, and it is maximum in magnitude if they are perpendicular.

The derivative, or rate of change, of a vector is defined in perfect analogy to the derivative of a scalar: if the vector $\boldsymbol{A}$ changes with time t, then

$$\frac{dA}{} \quad \underset{\Delta \rightarrow}{\text{im}} \frac{A(t+\Delta t) \ - \ (\ )}{} .$$

Before going to the limit on the right-hand side of the equation, the operations described are vector subtraction $\left[A(t+\Delta t) - A(t)\right]$ and scalar multiplication $(by\ 1/\Delta t)$. The result, $d\boldsymbol{A}/dt$, is therefore itself a vector. Notice that the difference between two vectors, in this case $A(t+\Delta t) - A(t)$, may be in quite a different direction than either of the vectors from which it is formed, here $A(t+\Delta t)$ and $\boldsymbol{A}(t)$. As a result, $d\boldsymbol{A}/dt$ may be in a different direction than $\boldsymbol{A}(t)$.

$$\frac{dA}{dt} = \lim_{\Delta t \rightarrow 0} \frac{A(t+\Delta t) - A(t)}{\Delta t} .$$

Newton's Laws of Motion and Equilibrium

In his *Principia*, Newton reduced the basic principles of mechanics to three laws:

- Every body continues in its state of rest or of uniform motion in a straight line, unless it is compelled to change that state by forces impressed upon it.
- The change of motion of an object is proportional to the force impressed and is made in the direction of the straight line in which the force is impressed.
- To every action there is always opposed an equal reaction; or, the mutual actions of two bodies upon each other are always equal and directed to contrary parts.

Newton's first law is a restatement of the principle of inertia, proposed earlier by Galileo and perfected by Descartes.

The second law is the most important of the three; it may be understood very nearly to summarize all of classical mechanics. Newton used the word "motion" to mean what is today called momentum—that is, the product of mass and velocity, or $\boldsymbol{p} = m\boldsymbol{v}$, where $\boldsymbol{p}$ is the momentum, m the mass, and $\boldsymbol{v}$ the velocity of a body. The second law may then be written in the form of the equation $\boldsymbol{F} = d\boldsymbol{p}/dt$, where $\boldsymbol{F}$ is the force, the time derivative expresses Newton's "change of motion," and the vector form of the equation assures that the change is in the same direction as the force, as the second law requires.

For a body whose mass does not change,

$$\frac{dA}{dt} = m \frac{dv}{dt} ma,$$

where $\boldsymbol{a}$ is the acceleration. Thus, Newton's second law may be put in the following form:

$$F = ma.$$

It is probably fair to say that equation is the most famous equation in all of physics.

$$F = ma.$$

Newton's third law assures that when two bodies interact, regardless of the nature of the interaction, they do not produce a net force acting on the two-body system as a whole. Instead, there is an action and reaction pair of equal and opposite forces, each acting on a different body (action and reaction forces never act on the same body). The third law applies whether the bodies in question are at rest, in uniform motion, or in accelerated motion.

If a body has a net force acting on it, it undergoes accelerated motion in accordance with the second law. If there is no net force acting on a body, either because there are no forces at all or because all forces are precisely balanced by contrary forces, the body does not accelerate and may be said to be in equilibrium. Conversely, a body that is observed not to be accelerated may be deduced to have no net force acting on it.

Consider, for example, a massive object resting on a table. The object is known to be acted on by the gravitational force of the Earth; if the table were removed, the object would fall. It follows therefore from the fact that the object does not fall that the table exerts an upward force on the object, equal and opposite to the downward force of gravity. This upward force is not a mere physicist's bookkeeping device but rather a real physical force. The table's surface is slightly deformed by the weight of the object, causing the surface to exert a force analogous to that exerted by a coiled spring.

It is useful to recall the following distinction: the massive object exerts a downward force on the table that is equal and opposite to the upward force exerted by the table (owing to its deformation) on the object. These two forces are an action and reaction pair operating on different bodies (one on the table, the other on the object) as required by Newton's third law. On the other hand, the upward force exerted on the object by the table is balanced by a downward force exerted on the object by the Earth's gravity. These two equal and opposite forces, acting on the same body, are not related to or by Newton's third law, but they do produce the equilibrium immobile state of the body.

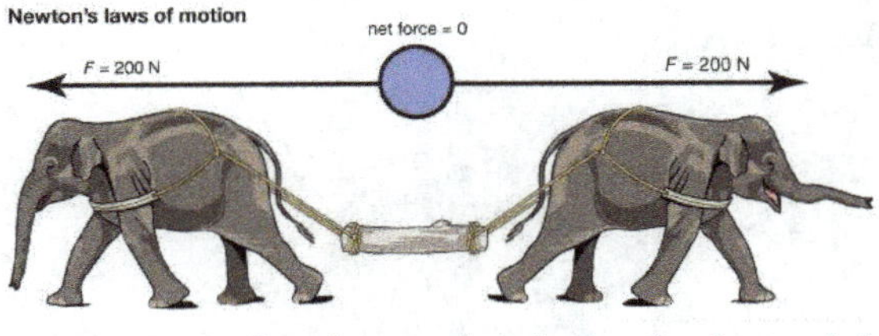

When an object is acted on by equal and opposite forces, it is at rest.

Classical Mechanics

Classical mechanics is the study of the *motion* of bodies (including the special case in which bodies remain at rest) in accordance with the general principles first enunciated by Sir Isaac Newton in his *Philosophiae Naturalis Principia Mathematica* (1687), commonly known as the *Principia*. Classical mechanics was the first branch of Physics to be discovered, and is the foundation upon which all other branches of Physics are built. Moreover, classical mechanics has many important applications in other areas of science, such as Astronomy (*e.g.*, celestial mechanics), Chemistry (*e.g.*, the dynamics of molecular collisions), Geology (*e.g.*, the propagation of seismic waves, generated by earthquakes, through the Earth's crust), and Engineering (*e.g.*, the equilibrium and stability of structures). Classical mechanics is also of great significance outside the realm of science. After all, the sequence of events leading to the discovery of classical mechanics--starting with the ground-breaking work of Copernicus, continuing with the researches of Galileo, Kepler, and Descartes, and culminating in the monumental achievements of Newton--involved the complete overthrow of the Aristotelian picture of the Universe, which had previously prevailed for more than a millennium, and its replacement by a recognizably modern picture in which humankind no longer played a privileged role.

Some mathematical laws and principles at the core of classical mechanics include the following:

- Newton's First Law of Motion: A body at rest will remain at rest, and a body in motion will remain in motion unless it is acted upon by an external force.
- Newton's Second Law of Motion: The net force acting on an object is equal to the mass of that object times its acceleration.
- Newton's Third Law of Motion: For every action, there is an equal and opposite reaction.
- Newton's Law of Universal Gravitation: The pull of gravity between two objects will be proportional to the masses of the objects and inversely proportional to the square of the distance between their centers of mass.
- Law of Conservation of Energy: Energy cannot be created nor destroyed, and instead changes from one form to another; for example, mechanical energy turning into heat energy.
- Law of Conservation of Momentum: In the absence of external forces such as friction, when objects collide, the total momentum before the collision is the same as the total momentum after the collision.
- Bernoulli's Principle: Within a continuous streamline of fluid flow, a fluid's hydrostatic pressure will balance in contrast to its speed and elevation.

Theories of Classical Mechanics

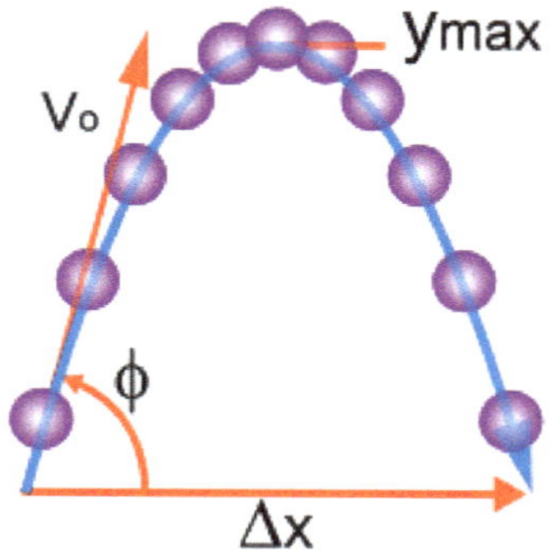

The analysis of projectile motion is a part of classical mechanics.

The following introduces the basic concepts of classical mechanics. For simplicity, it often models real-world objects as point particles (objects with negligible size). The motion of a point particle is characterized by a small number of parameters: its position, mass, and the forces applied to it. Each of these parameters is discussed in turn.

In reality, the kind of objects that classical mechanics can describe always have a non-zero size. (The physics of *very* small particles, such as the electron, is more accurately described by quantum mechanics.) Objects with non-zero size have more complicated behavior than hypothetical point particles, because of the additional degrees of freedom, e.g., a baseball can spin while it is moving. However, the results for point particles can be used to study such objects by treating them as composite objects, made of a large number of collectively acting point particles. The center of mass of a composite object behaves like a point particle.

Classical mechanics uses common-sense notions of how matter and forces exist and interact. It assumes that matter and energy have definite, knowable attributes such as location in space and speed. Non-relativistic mechanics also assumes that forces act instantaneously.

Position and its Derivatives

The SI derived "mechanical" (that is, not electromagnetic or thermal) units with kg, m and s	
position	m
angular position/angle	unitless (radian)
velocity	$m \cdot s^{-1}$
angular velocity	s^{-1}
acceleration	$m \cdot s^{-2}$
angular acceleration	s^{-2}
jerk	$m \cdot s^{-3}$
"angular jerk"	s^{-3}
specific energy	$m^2 \cdot s^{-2}$
absorbed dose rate	$m^2 \cdot s^{-3}$
moment of inertia	$kg \cdot m^2$
momentum	$kg \cdot m \cdot s^{-1}$
angular momentum	$kg \cdot m^2 \cdot s^{-1}$
force	$kg \cdot m \cdot s^{-2}$
torque	$kg \cdot m^2 \cdot s^{-2}$
energy	$kg \cdot m^2 \cdot s^{-2}$
power	$kg \cdot m^2 \cdot s^{-3}$
pressure and energy density	$kg \cdot m^{-1} \cdot s^{-2}$
surface tension	$kg \cdot s^{-2}$
spring constant	$kg \cdot s^{-2}$
irradiance and energy flux	$kg \cdot s^{-3}$
kinematic viscosity	$m^2 \cdot s^{-1}$
dynamic viscosity	$kg \cdot m^{-1} \cdot s^{-1}$
density (mass density)	$kg \cdot m^{-3}$
density (weight density)	$kg \cdot m^{-2} \cdot s^{-2}$
number density	m^{-3}
action	$kg \cdot m^2 \cdot s^{-1}$

The *position* of a point particle is defined in relation to a coordinate system centered on an arbitrary fixed reference point in space called the origin *O*. A simple coordinate system might describe the position of a particle *P* with a vector notated by an arrow labeled **r**that points from the origin *O* to point P. In general, the point particle does not need to be stationary relative to O. In cases where P is moving relative to O, $\mathbf{r}$ is defined as a function of t, time. In pre-Einstein relativity (known as Galilean relativity), time is considered an absolute, i.e., the time interval that is observed to elapse between any given pair of events is the same for all observers. In addition to relying on absolute time, classical mechanics assumes Euclidean geometry for the structure of space.

Velocity and Speed

The *velocity*, or the rate of change of position with time, is defined as the derivative of the position with respect to time:

$$\mathbf{v} = \frac{\mathrm{d}\mathbf{r}}{\mathrm{d}t}.$$

In classical mechanics, velocities are directly additive and subtractive. For example, if one car travels east at 60 km/h and passes another car traveling in the same direction at 50 km/h, the slower car perceives the faster car as traveling east at 60 − 50 = 10 km/h. However, from the perspective of the faster car, the slower car is moving 10 km/h to the west, often denoted as -10 km/h where the sign implies opposite direction. Velocities are directly additive as vector quantities; they must be dealt with using vector analysis.

Mathematically, if the velocity of the first object in the previous discussion is denoted by the vector $\mathbf{u} = u\mathbf{d}$ and the velocity of the second object by the vector $\mathbf{v} = v\mathbf{e}$, where u is the speed of the first object, v is the speed of the second object, and $\mathbf{d}$ and $\mathbf{e}$ are unit vectors in the directions of motion of each object respectively, then the velocity of the first object as seen by the second object is

$$\mathbf{u}' = \mathbf{u} - \mathbf{v}.$$

Similarly, the first object sees the velocity of the second object as

$$\mathbf{v}' = \mathbf{v} - \mathbf{u}.$$

When both objects are moving in the same direction, this equation can be simplified to

$$\mathbf{u}' = (u - v)\mathbf{d}.$$

Or, by ignoring direction, the difference can be given in terms of speed only:

$$u' = u - v.$$

Acceleration

The *acceleration*, or rate of change of velocity, is the derivative of the velocity with respect to time (the second derivative of the position with respect to time):

$$\mathbf{a} = \frac{\mathrm{d}\mathbf{v}}{\mathrm{d}t} = \frac{\mathrm{d}^2\mathbf{r}}{\mathrm{d}t^2}.$$

Acceleration represents the velocity's change over time. Velocity can change in either magnitude or direction, or both. Occasionally, a decrease in the magnitude of velocity "v" is referred to as *deceleration*, but generally any change in the velocity over time, including deceleration, is simply referred to as acceleration.

Frames of Reference

While the position, velocity and acceleration of a particle can be described with respect to any observer in any state of motion, classical mechanics assumes the existence of a special family of reference frames in which the mechanical laws of nature take a comparatively simple form. These special reference frames are called inertial frames.

An inertial frame is a frame of reference within which an object interacting with no forces (an idealized situation) appears either at rest or moving uniformly in a straight line. This is the fundamental definition of an inertial frame. These are characterized by the requirement that all forces entering the observer's physical laws originate from identifiable sources caused by fields, such as electro-static field (caused by static electrical charges), electro-magnetic field (caused by moving charges), gravitational field (caused by mass), and so forth.

A key concept of inertial frames is the method for identifying them. For practical purposes, reference frames that do not accelerate with respect to distant stars (an extremely distant point) are regarded as good approximations to inertial frames. Non-inertial reference frames accelerate in relation to an existing inertial frame. They form the basis for Einstein,s relativity. Due to the relative motion, particles in the non-inertial frame appear to move in ways not explained by forces from existing fields in the reference frame. Hence, it appears that there are other forces that enter the equations of motion solely as a result of the relative acceleration. These forces are referred to as fictitious forces, inertia forces, or pseudo-forces.

Consider two reference frames S and S'. For observers in each of the reference frames an event has space-time coordinates of (x,y,z,t) in frame S and (x',y',z',t') in frame S'. Assuming time is measured the same in all reference frames, and if we require $x = x'$ when $t = 0$, then the relation between the space-time coordinates of the same event observed from the reference frames S' and S, which are moving at a relative velocity of u in the x direction is:

$$x' = x - ut$$

$$y' \quad y$$

$$z' = z$$

$$t' = t.$$

This set of formulas defines a group transformation known as the Galilean transformation (informally, the *Galilean transform*). This group is a limiting case of the Poincaré group used in special relativity. The limiting case applies when the velocity u is very small compared to c, the speed of light.

The transformations have the following consequences:

- $\mathbf{v}' = \mathbf{v} - \mathbf{u}$ (the velocity $\mathbf{v}'$ of a particle from the perspective of S' is slower by $\mathbf{u}$ than its velocity $\mathbf{v}$ from the perspective of S).

- $\mathbf{a}' = \mathbf{a}$ (the acceleration of a particle is the same in any inertial reference frame).
- $\mathbf{F}' = \mathbf{F}$ (the force on a particle is the same in any inertial reference frame).
- The speed of light is not a constant in classical mechanics, nor does the special position given to the speed of light in relativistic mechanics have a counterpart in classical mechanics.

For some problems, it is convenient to use rotating coordinates (reference frames). Thereby one can either keep a mapping to a convenient inertial frame, or introduce additionally a fictitious centrifugal force and Coriolis force.

Forces: Newton's Second Law

Newton was the first to mathematically express the relationship between force and momentum. Some physicists interpret Newton's second law of motion as a definition of force and mass, while others consider it a fundamental postulate, a law of nature. Either interpretation has the same mathematical consequences, historically known as "Newton's Second Law":

$$\mathbf{F} = \frac{\mathrm{d}\mathbf{p}}{\mathrm{d}t} = \frac{\mathrm{d}(m\mathbf{v})}{\mathrm{d}t}.$$

The quantity $m\mathbf{v}$ is called the (canonical) momentum. The net force on a particle is thus equal to the rate of change of the momentum of the particle with time. Since the definition of acceleration is $\mathbf{a} = \mathrm{d}\mathbf{v}/\mathrm{d}t$, the second law can be written in the simplified and more familiar form:

$$\mathbf{F} = m\mathbf{a}.$$

So long as the force acting on a particle is known, Newton's second law is sufficient to describe the motion of a particle. Once independent relations for each force acting on a particle are available, they can be substituted into Newton's second law to obtain an ordinary differential equation, which is called the *equation of motion*.

As an example, assume that friction is the only force acting on the particle, and that it may be modeled as a function of the velocity of the particle, for example:

$$\mathbf{F}_{\mathrm{R}} = -\lambda\mathbf{v},$$

where λ is a positive constant, the negative sign states that the force is opposite the sense of the velocity. Then the equation of motion is

$$-\lambda\mathbf{v} = m\mathbf{a} = m\frac{\mathrm{d}\mathbf{v}}{\mathrm{d}t}.$$

This can be integrated to obtain

$$\mathbf{v} = \mathbf{v}_0 e^{\frac{-\lambda t}{m}}$$

where $\mathbf{v}_0$ is the initial velocity. This means that the velocity of this particle decays exponentially to zero as time progresses. In this case, an equivalent viewpoint is that the kinetic energy of the particle is absorbed by friction (which converts it to heat energy in accordance with the conservation of energy), and the particle is slowing down. This expression can be further integrated to obtain the position **r** of the particle as a function of time.

Important forces include the gravitational force and the Lorentz force for electromagnetism. In addition, Newton's third law can sometimes be used to deduce the forces acting on a particle: if it is known that particle *A* exerts a force **F** on another particle *B*, it follows that *B* must exert an equal and opposite *reaction force*, −**F**, on *A*. The strong form of Newton's third law requires that **F** and −**F** act along the line connecting *A* and *B*, while the weak form does not.

Work and Energy

If a constant force **F** is applied to a particle that makes a displacement $\Delta\mathbf{r}$, the *work done* by the force is defined as the scalar product of the force and displacement vectors:

$$W = \mathbf{F} \cdot \Delta\mathbf{r}.$$

More generally, if the force varies as a function of position as the particle moves from $\mathbf{r}_1$ to $\mathbf{r}_2$ along a path *C*, the work done on the particle is given by the line integral

$$W = \int_C \mathbf{F}(\mathbf{r}) \cdot \mathrm{d}\mathbf{r}.$$

If the work done in moving the particle from $\mathbf{r}_1$ to $\mathbf{r}_2$ is the same no matter what path is taken, the force is said to be conservative. Gravity is a conservative force, as is the force due to an idealized spring, as given by Hooke's law. The force due to friction is non-conservative.

The kinetic energy E_k of a particle of mass *m* travelling at speed *v* is given by

$$E_k = \tfrac{1}{2} m v^2.$$

For extended objects composed of many particles, the kinetic energy of the composite body is the sum of the kinetic energies of the particles.

The work–energy theorem states that for a particle of constant mass *m*, the total work *W* done on the particle as it moves from position $\mathbf{r}_1$ to $\mathbf{r}_2$ is equal to the change in kinetic energy E_k of the particle:

$$W = \Delta E_k = E_{k,2} - E_{k,1} = \tfrac{1}{2} m \left(v_2^2 - v_1^2 \right).$$

Conservative forces can be expressed as the gradient of a scalar function, known as the potential energyand denoted E_p:

$$\mathbf{F} = -\nabla E_p.$$

If all the forces acting on a particle are conservative, and E_p is the total potential energy (which is defined as a work of involved forces to rearrange mutual positions of bodies), obtained by summing the potential energies corresponding to each force:

$$\mathbf{F}\cdot\Delta\mathbf{r} = -\nabla E_{\mathrm{p}}\cdot\Delta\mathbf{r} = -\Delta E_{\mathrm{p}}.$$

The decrease in the potential energy is equal to the increase in the kinetic energy,

$$-\Delta E_{\mathrm{p}} = \Delta E_{\mathrm{k}} \Rightarrow \Delta(E_{\mathrm{k}} + E_{\mathrm{p}}) = 0.$$

This result is known as *conservation of energy* and states that the total energy,

$$\sum E = E_{\mathrm{k}} + E_{\mathrm{p}},$$

is constant in time. It is often useful, because many commonly encountered forces are conservative.

Beyond Newton's Laws

Classical mechanics also describes the more complex motions of extended non-pointlike objects. Euler's laws provide extensions to Newton,s laws in this area. The concepts of angular momentum rely on the same calculus used to describe one-dimensional motion. The rocket equation extends the notion of rate of change of an object,s momentum to include the effects of an object "losing mass".

There are two important alternative formulations of classical mechanics: Lagrangian mechanics and Hamiltonian mechanics. These, and other modern formulations, usually bypass the concept of "force", instead referring to other physical quantities, such as energy, speed and momentum, for describing mechanical systems in generalized coordinates.

The expressions given above for momentum and kinetic energy are only valid when there is no significant electromagnetic contribution. In electromagnetism, Newton's second law for current-carrying wires breaks down unless one includes the electromagnetic field contribution to the momentum of the system as expressed by the Poynting vector divided by c^2, where c is the speed of light in free space.

Limits of Validity

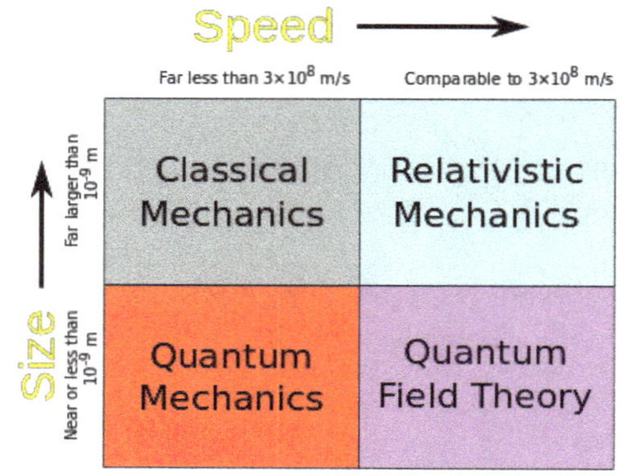

Domain of validity for Classical Mechanics

Many branches of classical mechanics are simplifications or approximations of more accurate forms; two of the most accurate being general relativity and relativistic statistical mechanics.

Geometric optics is an approximation to the quantum theory of light, and does not have a superior "classical" form.

When both quantum mechanics and classical mechanics cannot apply, such as at the quantum level with many degrees of freedom, quantum field theory (QFT) is of use. QFT deals with small distances and large speeds with many degrees of freedom as well as the possibility of any change in the number of particles throughout the interaction. When treating large degrees of freedom at the macroscopic level, statistical mechanics becomes useful. Statistical mechanics describes the behavior of large (but countable) numbers of particles and their interactions as a whole at the macroscopic level. Statistical mechanics is mainly used in thermodynamicsfor systems that lie outside the bounds of the assumptions of classical thermodynamics. In the case of high velocity objects approaching the speed of light, classical mechanics is enhanced by special relativity. In case that objects become extremely heavy (i.e. their Schwarzschild radius is not negligibly small for a given application), deviations from Newtonian mechanics become apparent and can be quantified by using the Parameterized post-Newtonian formalism. In that case, General relativity (GR) becomes applicable. However, until now there is no theory of Quantum gravity unifying GR and QFT in the sense that it could be used when objects become extremely small and heavy.

The Newtonian Approximation to Special Relativity

In special relativity, the momentum of a particle is given by

$$\mathbf{p} = \frac{m\mathbf{v}}{\sqrt{1 - v^2 / c^2}},$$

where m is the particle,s rest mass, $\mathbf{v}$ its velocity, v is the modulus of $\mathbf{v}$, and c is the speed of light.

If v is very small compared to c, v^2/c^2 is approximately zero, and so

$$\mathbf{p} \approx m\mathbf{v}.$$

Thus the Newtonian equation $\mathbf{p} = m\mathbf{v}$ is an approximation of the relativistic equation for bodies moving with low speeds compared to the speed of light.

For example, the relativistic cyclotron frequency of a cyclotron, gyrotron, or high voltage magnetron is given by

$$f = f_c \frac{m_0}{m_0 + T / c^2},$$

where f_c is the classical frequency of an electron (or other charged particle) with kinetic energy T and (rest) mass m_0 circling in a magnetic field. The (rest) mass of an electron is 511 keV. So the frequency correction is 1% for a magnetic vacuum tube with a 5.11 kV direct current accelerating voltage.

The Classical Approximation to Quantum Mechanics

The ray approximation of classical mechanics breaks down when the de Broglie wavelength is not much smaller than other dimensions of the system. For non-relativistic particles, this wavelength is

$$\lambda = \frac{h}{p}$$

where h is Planck's constant and p is the momentum.

Again, this happens with electrons before it happens with heavier particles. For example, the electrons used by Clinton Davisson and Lester Germer in 1927, accelerated by 54 V, had a wavelength of 0.167 nm, which was long enough to exhibit a single diffraction side lobe when reflecting from the face of a nickel crystal with atomic spacing of 0.215 nm. With a larger vacuum chamber, it would seem relatively easy to increase the angular resolution from around a radian to a milliradian and see quantum diffraction from the periodic patterns of integrated circuit computer memory.

More practical examples of the failure of classical mechanics on an engineering scale are conduction by quantum tunneling in tunnel diodes and very narrow transistor gates in integrated circuits.

Classical mechanics is the same extreme high frequency approximation as geometric optics. It is more often accurate because it describes particles and bodies with rest mass. These have more momentum and therefore shorter De Broglie wavelengths than massless particles, such as light, with the same kinetic energies.

Quantum Mechanics

Quantum mechanics is the study of the dynamics of particles at its most fundamental level. The state of a particle, such as its position or momentum, is described by a statistical distribution given by its wavefunction. As this name suggests, this formalism gives matter many properties that are classically associated with waves.

It results in what may appear to be some very strange conclusions about the physical world. At the scale of atoms and electrons, many of the equations of classical mechanics, which describe how things move at everyday sizes and speeds, cease to be useful. In classical mechanics, objects exist in a specific place at a specific time. However, in quantum mechanics, objects instead exist in a haze of probability; they have a certain chance of being at point A, another chance of being at point B and so on.

Three Revolutionary Principles

Quantum mechanics (QM) developed over many decades, beginning as a set of controversial mathematical explanations of experiments that the math of classical mechanics could not explain. It began at the turn of the 20th century, around the same time that Albert Einstein published his theory of relativity, a separate mathematical revolution in physics that describes the motion of things at high speeds. Unlike relativity, however, the origins of QM cannot be attributed to any one scientist. Rather, multiple scientists contributed to a foundation of three revolutionary principles that gradually gained acceptance and experimental verification between 1900 and 1930. They are:

Quantized properties: Certain properties, such as position, speed and color, can sometimes only occur in specific, set amounts, much like a dial that "clicks" from number to number. This challenged a fundamental assumption of classical mechanics, which said that such properties should exist on a smooth, continuous spectrum. To describe the idea that some properties "clicked" like a dial with specific settings, scientists coined the word "quantized."

Particles of light: Light can sometimes behave as a particle. This was initially met with harsh criticism, as it ran contrary to 200 years of experiments showing that light behaved as a wave; much like ripples on the surface of a calm lake. Light behaves similarly in that it bounces off walls and bends around corners, and that the crests and troughs of the wave can add up or cancel out. Added wave crests result in brighter light, while waves that cancel out produce darkness. A light source can be thought of as a ball on a stick being rhythmically dipped in the center of a lake. The color emitted corresponds to the distance between the crests, which is determined by the speed of the ball's rhythm.

Waves of matter: Matter can also behave as a wave. This ran counter to the roughly 30 years of experiments showing that matter (such as electrons) exists as particles.

Quantized Properties

In 1900, German physicist Max Planck sought to explain the distribution of colors emitted over the spectrum in the glow of red-hot and white-hot objects, such as light-bulb filaments. When making physical sense of the equation he had derived to describe this distribution, Planck realized it implied that combinations of only certain colors (albeit a great number of them) were emitted, specifically those that were whole-number multiples of some base value. Somehow, colors were quantized! This was unexpected because light was understood to act as a wave, meaning that values of color should be a continuous spectrum. What could be forbidding atoms from producing the colors between these whole-number multiples? This seemed so strange that Planck regarded quantization as nothing more than a mathematical trick. According to Helge Kragh in his 2000 article in Physics World magazine, "Max Planck, the Reluctant Revolutionary," "If a revolution occurred in physics in December 1900, nobody seemed to notice it. Planck was no exception."

Planck's equation also contained a number that would later become very important to future development of QM; today, it's known as "Planck's Constant."

Quantization helped to explain other mysteries of physics. In 1907, Einstein used Planck's hypothesis of quantization to explain why the temperature of a solid changed by different amounts if you put the same amount of heat into the material but changed the starting temperature.

Since the early 1800s, the science of spectroscopy had shown that different elements emit and absorb specific colors of light called "spectral lines." Though spectroscopy was a reliable method for determining the elements contained in objects such as distant stars, scientists were puzzled about *why* each element gave off those specific lines in the first place. In 1888, Johannes Rydberg derived an equation that described the spectral lines emitted by hydrogen, though nobody could explain why the equation worked. This changed in 1913 when Niels Bohr applied Planck,s hypothesis of quantization to Ernest Rutherford,s 1911 "planetary" model of the atom, which postulated that electrons orbited the nucleus the same way that planets orbit the sun. According to

Physics 2000 (a site from the University of Colorado), Bohr proposed that electrons were restricted to "special" orbits around an atom,s nucleus. They could "jump" between special orbits, and the energy produced by the jump caused specific colors of light, observed as spectral lines. Though quantized properties were invented as but a mere mathematical trick, they explained so much that they became the founding principle of QM.

Particles of Light

In 1905, Einstein published a paper, "Concerning an Heuristic Point of View Toward the Emission and Transformation of Light," in which he envisioned light traveling not as a wave, but as some manner of "energy quanta." This packet of energy, Einstein suggested, could "be absorbed or generated only as a whole," specifically when an atom "jumps" between quantized vibration rates. This would also apply, as would be shown a few years later, when an electron "jumps" between quantized orbits. Under this model, Einstein's "energy quanta" contained the energy difference of the jump; when divided by Planck's constant, that energy difference determined the color of light carried by those quanta.

With this new way to envision light, Einstein offered insights into the behavior of nine different phenomena, including the specific colors that Planck described being emitted from a light-bulb filament. It also explained how certain colors of light could eject electrons off metal surfaces, a phenomenon known as the "photoelectric effect." However, Einstein wasn't wholly justified in taking this leap, said Stephen Klassen, an associate professor of physics at the University of Winnipeg. In a 2008 paper, "The Photoelectric Effect: Rehabilitating the Story for the Physics Classroom," Klassen states that Einstein's energy quanta aren't necessary for explaining all of those nine phenomena. Certain mathematical treatments of light as a wave are still capable of describing both the specific colors that Planck described being emitted from a light-bulb filament and the photoelectric effect. Indeed, in Einstein's controversial winning of the 1921 Nobel Prize, the Nobel committee only acknowledged "his discovery of the law of the photoelectric effect," which specifically did not rely on the notion of energy quanta.

Roughly two decades after Einstein's paper, the term "photon" was popularized for describing energy quanta, thanks to the 1923 work of Arthur Compton, who showed that light scattered by an electron beam changed in color. This showed that particles of light (photons) were indeed colliding with particles of matter (electrons), thus confirming Einstein's hypothesis. By now, it was clear that light could behave both as a wave and a particle, placing light's "wave-particle duality" into the foundation of QM.

Waves of Matter

Since the discovery of the electron in 1896, evidence that all matter existed in the form of particles was slowly building. Still, the demonstration of light's wave-particle duality made scientists question whether matter was limited to acting *only* as particles. Perhaps wave-particle duality could ring true for matter as well? The first scientist to make substantial headway with this reasoning was a French physicist named Louis de Broglie. In 1924, de Broglie used the equations of Einstein,s theory of special relativity to show that particles can exhibit wave-like characteristics, and that waves can exhibit particle-like characteristics. Then in 1925, two scientists, working independently and using separate lines of mathematical thinking, applied de Broglie,s reasoning to explain how electrons whizzed around in atoms (a phenomenon that was unexplainable using the equations of classical mechanics). In Germany, physicist Werner Heisenberg (teaming with Max Born and Pascual Jordan) accomplished this by developing "matrix mechanics." Austrian physicist Erwin Schrödinger developed a similar theory called "wave mechanics." Schrödinger showed in 1926 that these two approaches were equivalent.

The Heisenberg-Schrödinger model of the atom, in which each electron acts as a wave (sometimes referred to as a "cloud") around the nucleus of an atom replaced the Rutherford-Bohr model. One stipulation of the new model was that the ends of the wave that forms an electron must meet. In "Quantum Mechanics in Chemistry, 3rd Ed." (W.A. Benjamin, 1981), Melvin Hanna writes, "The imposition of the boundary conditions has restricted the energy to discrete values." A consequence of this stipulation is that only whole numbers of crests and troughs are allowed, which explains why some properties are quantized. In the Heisenberg-Schrödinger model of the atom, electrons obey a "wave function" and occupy "orbitals" rather than orbits. Unlike the circular orbits of the Rutherford-Bohr model, atomic orbitals have a variety of shapes ranging from spheres to dumbbells to daisies.

In 1927, Walter Heitler and Fritz London further developed wave mechanics to show how atomic orbitals could combine to form molecular orbitals, effectively showing why atoms bond to one another to form molecules. This was yet another problem that had been unsolvable using the math of classical mechanics. These insights gave rise to the field of "quantum chemistry."

The Uncertainty Principle

Also in 1927, Heisenberg made another major contribution to quantum physics. He reasoned that since matter acts as waves, some properties, such as an electron's position and speed, are "complementary," meaning there's a limit (related to Planck's constant) to how well the precision of each property can be known. Under what would come to be called "Heisenberg's uncertainty principle," it was reasoned that the more precisely an electron's position is known, the less precisely its speed can be known, and vice versa. This uncertainty principle applies to everyday-size objects as well, but is not noticeable because the lack of precision is extraordinarily tiny. According to Dave Slaven of Morningside College (Sioux City, IA), if a baseball's speed is known to within a precision of 0.1 mph, the maximum precision to which it is possible to know the ball's position is 0.0000000000 0000000000000000008 millimeters.

Onward

The principles of quantization, wave-particle duality and the uncertainty principle ushered in a new era for QM. In 1927, Paul Dirac applied a quantum understanding of electric and magnetic fields to give rise to the study of "quantum field theory" (QFT), which treated particles (such as photons and electrons) as excited states of an underlying physical field. Work in QFT continued for a decade until scientists hit a roadblock: Many equations in QFT stopped making physical sense because they produced results of infinity. After a decade of stagnation, Hans Bethe made a breakthrough in 1947 using a technique called "renormalization." Here, Bethe realized that all infinite results related to two phenomena (specifically "electron self-energy" and "vacuum polarization") such that the observed values of electron mass and electron charge could be used to make all the infinities disappear.

Since the breakthrough of renormalization, QFT has served as the foundation for developing quantum theories about the four fundamental forces of nature: 1) electromagnetism, 2) the weak nuclear force, 3) the strong nuclear force and 4) gravity. The first insight provided by QFT was a quantum description of electromagnetism through "quantum electrodynamics" (QED), which made strides in the late 1940s and early 1950s. Next was a quantum description of the weak nuclear force, which was unified with electromagnetism to build "electroweak theory" (EWT) throughout the 1960s. Finally came a quantum treatment of the strong nuclear force using "quantum chromodynamics" (QCD) in the 1960s and 1970s. The theories of QED, EWT and QCD together form the basis of the Standard Model of particle physics. Unfortunately, QFT has yet to produce a quantum theory of gravity. That quest continues today in the studies of string theory and loop quantum gravity.

Statistical Mechanics

Statistical mechanics is the branch of physics that combines the principles and procedures of statistics with the laws of both classical and quantum mechanics, particularly with respect to the field of thermodynamics. It aims to predict and explain the measurable properties of macroscopic systems on the basis of the properties and behaviour of the microscopic constituents of those systems. Statistical mechanics interprets, for example, thermal energy as the energy of atomic particles in disordered states and temperature as a quantitative measure of how energy is shared among such particles. Statistical mechanics draws heavily on the laws of probability so that it does not concentrate on the behaviour of every individual particle in a macroscopic substance but on the average behaviour of a large number of particles of the same kind.

Depending on how you look at it, statistical mechanics is either the least fundamental or most fundamental of all fields of physics. That is because it is not really science at all. It is pure mathematics.

In other subjects, you learn about "natural laws": Newton's second law, Maxwell's equations, Schrödinger's equation, etc. These laws are not derived from anything else. They were discovered experimentally and then assumed to reflect fundamental aspects of reality. But statistical mechanics does not involve any natural laws of this sort. Instead, it is a set of techniques that can be applied to nearly any physical system, no matter what laws that system obeys. That is why I call it the most fundamental field of physics. New theories may replace old ones, and natural laws may turn out to be merely approximations to deeper laws. But statistical mechanics remains valid through it all, and whatever new laws are discovered, it will almost certainly work just as well with them as it did with the old ones.

It grew out of thermodynamics, an earlier theory that described the behavior of a mysterious substance called "heat" or "caloric". Thermodynamics was a physical theory of the more conventional sort. It involved natural laws discovered by experiment, and made no claims about why those laws happened to hold. Caloric was supposed to be a substance much like other forms of matter. But this view turned out to be incorrect. Heat is actually an emergent phenomenon: a mathematical quantity defined in terms of a more detailed theory (the movement of individual atoms). The "laws" of thermodynamics can be derived from that deeper theory by applying statistical techniques. If you ignore the details of how each atom is moving, you find they collectively behave in a way that resembles a continuous fluid. That is what statistical mechanics is all about: deriving high level descriptions by starting from lower level ones and then averaging out lots of details.

Is it possible other theories could be explained in the same way? That is an open question, and a fascinating one. Many physicists suspect gravity is an emergent phenomenon, that it arises from the collective behavior of some deeper degrees of freedom. Statistical interpretations have also been proposed for quantum mechanics. These are all very speculative, of course, and they could easily turn out to be wrong. But they also could easily turn out to be right. Based on what we know today, it is entirely possible that the very structure of spacetime is a consequence of statistical mechanics.

Principles: Mechanics and Ensembles

In physics, there are two types of mechanics usually examined: classical mechanics and quantum mechanics. For both types of mechanics, the standard mathematical approach is to consider two concepts:

1. The complete state of the mechanical system at a given time, mathematically encoded as a phase point (classical mechanics) or a pure quantum state vector (quantum mechanics).
2. An equation of motion which carries the state forward in time: Hamilton's equations (classical mechanics) or the time-dependent Schrödinger equation (quantum mechanics)

Using these two concepts, the state at any other time, past or future, can in principle be calculated. There is however a disconnection between these laws and everyday life experiences, as we do not find it necessary (nor even theoretically possible) to know exactly at a microscopic level the simultaneous positions and velocities of each molecule while carrying out processes at the human scale (for example, when performing a chemical reaction). Statistical mechanics fills this disconnection between the laws of mechanics and the practical experience of incomplete knowledge, by adding some uncertainty about which state the system is in.

Whereas ordinary mechanics only considers the behaviour of a single state, statistical mechanics introduces the statistical ensemble, which is a large collection of virtual, independent copies of the system in various states. The statistical ensemble is a probability distribution over all possible states of the system. In classical statistical mechanics, the ensemble is a probability distribution over phase points (as opposed to a single phase point in ordinary mechanics), usually represented as a distribution in a phase space with canonical coordinates. In quantum statistical mechanics, the ensemble is a probability distribution over pure states, and can be compactly summarized as a density matrix.

As is usual for probabilities, the ensemble can be interpreted in different ways:

- An ensemble can be taken to represent the various possible states that a *single system* could be in (epistemic probability, a form of knowledge), or
- The members of the ensemble can be understood as the states of the systems in experiments repeated on independent systems which have been prepared in a similar but imperfectly controlled manner (empirical probability), in the limit of an infinite number of trials.

These two meanings are equivalent for many purposes, and will be used interchangeably in this topic.

However the probability is interpreted, each state in the ensemble evolves over time according to the equation of motion. Thus, the ensemble itself (the probability distribution over states) also evolves, as the virtual systems in the ensemble continually leave one state and enter another. The ensemble evolution is given by the Liouville equation (classical mechanics) or the von Neumann equation (quantum mechanics). These equations are simply derived by the application of the mechanical equation of motion separately to each virtual system contained in the ensemble, with the probability of the virtual system being conserved over time as it evolves from state to state.

One special class of ensemble is those ensembles that do not evolve over time. These ensembles are known as *equilibrium ensembles* and their condition is known as *statistical equilibrium*. Statistical equilibrium occurs if, for each state in the ensemble, the ensemble also contains all of its future and past states with probabilities equal to the probability of being in that state. The study of equilibrium ensembles of isolated systems is the focus of statistical thermodynamics. Non-equilibrium statistical mechanics addresses the more general case of ensembles that change over time, and/or ensembles of non-isolated systems.

Statistical Thermodynamics

The primary goal of statistical thermodynamics (also known as equilibrium statistical mechanics) is to derive the classical thermodynamics of materials in terms of the properties of their constituent particles and the interactions between them. In other words, statistical thermodynamics provides a connection between the macroscopic properties of materials in thermodynamic equilibrium, and the microscopic behaviours and motions occurring inside the material.

Whereas statistical mechanics proper involves dynamics, here the attention is focussed on *statistical equilibrium* (steady state). Statistical equilibrium does not mean that the particles have stopped moving (mechanical equilibrium), rather, only that the ensemble is not evolving.

Fundamental Postulate

A sufficient (but not necessary) condition for statistical equilibrium with an isolated system is that the probability distribution is a function only of conserved properties (total energy, total particle numbers, etc.). There are many different equilibrium ensembles that can be considered, and only some of them correspond to thermodynamics. Additional postulates are necessary to motivate why the ensemble for a given system should have one form or another.

A common approach found in many textbooks is to take the *equal a priori probability postulate*. This postulate states that

For an isolated system with an exactly known energy and exactly known composition, the system can be found with equal probability *in any microstate consistent with that knowledge.*

The equal a priori probability postulate therefore provides a motivation for the microcanonical ensemble described below. There are various arguments in favour of the equal a priori probability postulate:

- Ergodic hypothesis: An ergodic system is one that evolves over time to explore "all accessible" states: all those with the same energy and composition. In an ergodic system, the microcanonical ensemble is the only possible equilibrium ensemble with fixed energy. This approach has limited applicability, since most systems are not ergodic.

- Principle of indifference: In the absence of any further information, we can only assign equal probabilities to each compatible situation.
- Maximum information entropy: A more elaborate version of the principle of indifference states that the correct ensemble is the ensemble that is compatible with the known information and that has the largest Gibbs entropy (information entropy).

Other fundamental postulates for statistical mechanics have also been proposed.

Three Thermodynamic Ensembles

There are three equilibrium ensembles with a simple form that can be defined for any isolated systembounded inside a finite volume. These are the most often discussed ensembles in statistical thermodynamics. In the macroscopic limit they all correspond to classical thermodynamics.

Microcanonical Ensemble

Microcanonical ensemble describes a system with a precisely given energy and fixed composition (precise number of particles). The microcanonical ensemble contains with equal probability each possible state that is consistent with that energy and composition.

Canonical Ensemble

Canonical ensemble describes a system of fixed composition that is in thermal equilibrium with a heat bath of a precise temperature. The canonical ensemble contains states of varying energy but identical composition; the different states in the ensemble are accorded different probabilities depending on their total energy.

Grand Canonical Ensemble

Grand canonical ensemble describes a system with non-fixed composition (uncertain particle numbers) that is in thermal and chemical equilibrium with a thermodynamic reservoir. The reservoir has a precise temperature, and precise chemical potentials for various types of particle. The grand canonical ensemble contains states of varying energy and varying numbers of particles; the different states in the ensemble are accorded different probabilities depending on their total energy and total particle numbers.

For systems containing many particles (the thermodynamic limit), all three of the ensembles listed above tend to give identical behaviour. It is then simply a matter of mathematical convenience which ensemble is used. The Gibbs theorem about equivalence of ensembles was developed into the theory of concentration of measure phenomenon, which has applications in many areas of science, from functional analysis to methods of artificial intelligence and big data technology.

Important cases where the thermodynamic ensembles *do not* give identical results include:

- Microscopic systems.
- Large systems at a phase transition.
- Large systems with long-range interactions.

In these cases the correct thermodynamic ensemble must be chosen as there are observable differences between these ensembles not just in the size of fluctuations, but also in average quantities such as the distribution of particles. The correct ensemble is that which corresponds to the way the system has been prepared and characterized—in other words, the ensemble that reflects the knowledge about that system.

	Thermodynamic ensembles		
	Microcanonical	**Canonical**	**Grand canonical**
Fixed variables	N, E, V	N, T, V	μ, T, V
Microscopic features	• Number of microstates • W	• Canonical partition function • $z = \sum_k e^{-E_k / k_B T}$	• Grand partition function • $\mathcal{Z} = \sum_k e^{-(E_k - \mu N_k)/ K_B T}$
Macroscopic function	• Boltzmann entropy • $S = k_B \log W$	• Helmholtz free energy • $F = -k_B T \log Z$	• Grand potential • $\Omega = -k_B T \log \mathcal{Z}$

Calculation Methods

Once the characteristic state function for an ensemble has been calculated for a given system, that system is 'solved' (macroscopic observables can be extracted from the characteristic state function). Calculating the characteristic state function of a thermodynamic ensemble is not necessarily a simple task, however, since it involves considering every possible state of the system. While some hypothetical systems have been exactly solved, the most general (and realistic) case is too complex for an exact solution. Various approaches exist to approximate the true ensemble and allow calculation of average quantities.

Exact

There are some cases which allow exact solutions.

- For very small microscopic systems, the ensembles can be directly computed by simply enumerating over all possible states of the system (using exact diagonalization in quantum mechanics, or integral over all phase space in classical mechanics).
- Some large systems consist of many separable microscopic systems, and each of the subsystems can be analysed independently. Notably, idealized gases of non-interacting particles have this property, allowing exact derivations of Maxwell–Boltzmann statistics, Fermi–Dirac statistics, and Bose–Einstein statistics.
- A few large systems with interaction have been solved. By the use of subtle mathematical

techniques, exact solutions have been found for a few toy models. Some examples include the Bethe ansatz, square-lattice Ising model in zero field, hard hexagon model.

Monte Carlo

One approximate approach that is particularly well suited to computers is the Monte Carlo method, which examines just a few of the possible states of the system, with the states chosen randomly (with a fair weight). As long as these states form a representative sample of the whole set of states of the system, the approximate characteristic function is obtained. As more and more random samples are included, the errors are reduced to an arbitrarily low level.

- The Metropolis–Hastings algorithm is a classic Monte Carlo method which was initially used to sample the canonical ensemble.
- Path integral Monte Carlo, also used to sample the canonical ensemble.

Other

- For rarefied non-ideal gases, approaches such as the cluster expansion use perturbation theory to include the effect of weak interactions, leading to a virial expansion.
- For dense fluids, another approximate approach is based on reduced distribution functions, in particular the radial distribution function.
- Molecular dynamics computer simulations can be used to calculate microcanonical ensembleaverages, in ergodic systems. With the inclusion of a connection to a stochastic heat bath, they can also model canonical and grand canonical conditions.
- Mixed methods involving non-equilibrium statistical mechanical results may be useful.

Non-equilibrium Statistical Mechanics

There are many physical phenomena of interest that involve quasi-thermodynamic processes out of equilibrium, for example:

- Heat transport by the internal motions in a material, driven by a temperature imbalance,
- Electric currents carried by the motion of charges in a conductor, driven by a voltage imbalance,
- Spontaneous chemical reactions driven by a decrease in free energy,
- Friction, dissipation, quantum decoherence,
- Systems being pumped by external forces (optical pumping, etc.), and
- Irreversible processes in general.

All of these processes occur over time with characteristic rates, and these rates are of importance for engineering. The field of non-equilibrium statistical mechanics is concerned with understanding these non-equilibrium processes at the microscopic level. (Statistical thermodynamics can only be used to calculate the final result, after the external imbalances have been removed and the ensemble has settled back down to equilibrium).

In principle, non-equilibrium statistical mechanics could be mathematically exact: ensembles for an isolated system evolve over time according to deterministic equations such as Liouville's equation or its quantum equivalent, the von Neumann equation. These equations are the result of applying the mechanical equations of motion independently to each state in the ensemble. Unfortunately, these ensemble evolution equations inherit much of the complexity of the underlying mechanical motion, and so exact solutions are very difficult to obtain. Moreover, the ensemble evolution equations are fully reversible and do not destroy information (the ensemble's Gibbs entropy is preserved). In order to make headway in modelling irreversible processes, it is necessary to consider additional factors besides probability and reversible mechanics.

Non-equilibrium mechanics is therefore an active area of theoretical research as the range of validity of these additional assumptions continues to be explored.

Stochastic Methods

One approach to non-equilibrium statistical mechanics is to incorporate stochastic (random) behaviour into the system. Stochastic behaviour destroys information contained in the ensemble. While this is technically inaccurate (aside from hypothetical situations involving black holes, a system cannot in itself cause loss of information), the randomness is added to reflect that information of interest becomes converted over time into subtle correlations within the system, or to correlations between the system and environment. These correlations appear as chaotic or pseudorandom influences on the variables of interest. By replacing these correlations with randomness proper, the calculations can be made much easier.

- *Boltzmann transport equation*: An early form of stochastic mechanics appeared even before the term "statistical mechanics" had been coined, in studies of kinetic theory. James Clerk Maxwell had demonstrated that molecular collisions would lead to apparently chaotic motion inside a gas. Ludwig Boltzmann subsequently showed that, by taking this molecular chaos for granted as a complete randomization, the motions of particles in a gas would follow a simple Boltzmann transport equationthat would rapidly restore a gas to an equilibrium state.

The Boltzmann transport equation and related approaches are important tools in non-equilibrium statistical mechanics due to their extreme simplicity. These approximations work well in systems where the "interesting" information is immediately (after just one collision) scrambled up into subtle correlations, which essentially restricts them to rarefied gases. The Boltzmann transport equation has been found to be very useful in simulations of electron transport in lightly doped semiconductors (in transistors), where the electrons are indeed analogous to a rarefied gas.

A quantum technique related in theme is the random phase approximation.

- *BBGKY hierarchy*: In liquids and dense gases, it is not valid to immediately discard the correlations between particles after one collision. The BBGKY hierarchy (Bogoliubov–Born–Green–Kirkwood–Yvon hierarchy) gives a method for deriving Boltzmann-type equations but also extending them beyond the dilute gas case, to include correlations after a few collisions.

- *Keldysh formalism* (a.k.a. NEGF—non-equilibrium Green functions): A quantum approach to including stochastic dynamics is found in the Keldysh formalism. This approach often used in electronic quantum transport calculations.

Near-equilibrium Methods

Another important class of non-equilibrium statistical mechanical models deals with systems that are only very slightly perturbed from equilibrium. With very small perturbations, the response can be analysed in linear response theory. A remarkable result, as formalized by the fluctuation-dissipation theorem, is that the response of a system when near equilibrium is precisely related to the fluctuations that occur when the system is in total equilibrium. Essentially, a system that is slightly away from equilibrium—whether put there by external forces or by fluctuations—relaxes towards equilibrium in the same way, since the system cannot tell the difference or "know" how it came to be away from equilibrium.

This provides an indirect avenue for obtaining numbers such as ohmic conductivity and thermal conductivity by extracting results from equilibrium statistical mechanics. Since equilibrium statistical mechanics is mathematically well defined and (in some cases) more amenable for calculations, the fluctuation-dissipation connection can be a convenient shortcut for calculations in near-equilibrium statistical mechanics.

A few of the theoretical tools used to make this connection include:

- Fluctuation–dissipation theorem
- Onsager reciprocal relations
- Green–Kubo relations
- Landauer–Büttiker formalism
- Mori–Zwanzig formalism

Hybrid Methods

An advanced approach uses a combination of stochastic methods and linear response theory. As an example, one approach to compute quantum coherence effects (weak localization, conductance fluctuations) in the conductance of an electronic system is the use of the Green-Kubo relations, with the inclusion of stochastic dephasing by interactions between various electrons by use of the Keldysh method.

Applications Outside Thermodynamics

The ensemble formalism also can be used to analyze general mechanical systems with uncertainty in knowledge about the state of a system. Ensembles are also used in:

- Propagation of uncertainty over time,

- Regression analysis of gravitational orbits,
- Ensemble forecasting of weather,
- Dynamics of neural networks,
- Bounded-rational potential games in game theory and economics.

Thermodynamics

Thermodynamics is the branch of physics that deals with the relationships between heat and other forms of energy. In particular, it describes how thermal energy is converted to and from other forms of energy and how it affects matter.

Thermal energy is the energy a substance or system has due to its temperature, i.e., the energy of moving or vibrating molecules, according to the Energy Education website of the Texas Education Agency. Thermodynamics involves measuring this energy, which can be "exceedingly complicated," according to David McKee, a professor of physics at Missouri Southern State University. "The systems that we study in thermodynamics consist of very large numbers of atoms or molecules interacting in complicated ways. But, if these systems meet the right criteria, which we call equilibrium, they can be described with a very small number of measurements or numbers. Often this is idealized as the mass of the system, the pressure of the system, and the volume of the system, or some other equivalent set of numbers. Three numbers describe 10^{26} or 10^{30} nominal independent variables."

Heat

Thermodynamics, then, is concerned with several properties of matter; foremost among these is heat. Heat is energy transferred between substances or systems due to a temperature difference between them, according to Energy Education. As a form of energy, heat is conserved, i.e., it cannot be created or destroyed. It can, however, be transferred from one place to another. Heat can also be converted to and from other forms of energy. For example, a steam turbine can convert heat to kinetic energy to run a generator that converts kinetic energy to electrical energy. A light bulb can convert this electrical energy to electromagnetic radiation (light), which, when absorbed by a surface, is converted back into heat.

Temperature

The amount of heat transferred by a substance depends on the speed and number of atoms or molecules in motion, according to Energy Education. The faster the atoms or molecules move, the higher the temperature, and the more atoms or molecules that are in motion, the greater the quantity of heat they transfer.

Temperature is "a measure of the average kinetic energy of the particles in a sample of matter, expressed in terms of units or degrees designated on a standard scale," according to the American Heritage Dictionary. The most commonly used temperature scale is Celsius, which is based on the freezing and boiling points of water, assigning respective values of 0 degrees C and 100 degrees C.

The Fahrenheit scale is also based on the freezing and boiling points of water which have assigned values of 32 F and 212 F, respectively.

Scientists worldwide, however, use the Kelvin (K with no degree sign) scale, named after William Thomson, 1st Baron Kelvin, because it works in calculations. This scale uses the same increment as the Celsius scale, i.e., a temperature change of 1 C is equal to 1 K. However, the Kelvin scale starts at absolute zero, the temperature at which there is a total absence of heat energy and all molecular motion stops. A temperature of 0 K is equal to minus 459.67 F or minus 273.15 C.

Specific Heat

The amount of heat required to increase the temperature of a certain mass of a substance by a certain amount is called specific heat, or specific heat capacity, according to Wolfram Research. The conventional unit for this is calories per gram per kelvin. The calorie is defined as the amount of heat energy required to raise the temperature of 1 gram of water at 4 C by 1 degree.

The specific heat of a metal depends almost entirely on the number of atoms in the sample, not its mass. For instance, a kilogram of aluminum can absorb about seven times more heat than a kilogram of lead. However, lead atoms can absorb only about 8 percent more heat than an equal number of aluminum atoms. A given mass of water, however, can absorb nearly five times as much heat as an equal mass of aluminum. The specific heat of a gas is more complex and depends on whether it is measured at constant pressure or constant volume.

Thermal Conductivity

Thermal conductivity (k) is "the rate at which heat passes through a specified material, expressed as the amount of heat that flows per unit time through a unit area with a temperature gradient of one degree per unit distance," according to the Oxford Dictionary. The unit for k is watts (W) per meter (m) per kelvin (K). Values of k for metals such as copper and silver are relatively high at 401 and 428 W/m·K, respectively. This property makes these materials useful for automobile radiators and cooling fins for computer chips because they can carry away heat quickly and exchange it with the environment. The highest value of k for any natural substance is diamond at 2,200 W/m·K.

Other materials are useful because they are extremely poor conductors of heat; this property is referred to as thermal resistance, or R-value, which describes the rate at which heat is transmitted through the material. These materials, such as rock wool, goose down and Styrofoam, are used for insulation in exterior building walls, winter coats and thermal coffee mugs. R-value is given in units of square feet times degrees Fahrenheit times hours per British thermal unit ($ft^2 \cdot {}^\circ F \cdot h/Btu$) for a 1-inch-thick slab.

Newton's Law of Cooling

In 1701, Sir Isaac Newton first stated his Law of Cooling in a short article titled "Scala graduum Caloris" ("A Scale of the Degrees of Heat") in the Philosophical Transactions of the Royal Society. Newton's statement of the law translates from the original Latin as, "the excess of the degrees of the heat .. were in geometrical progression when the times are in an arithmetical progression."

Worcester Polytechnic Institute gives a more modern version of the law as "the rate of change of temperature is proportional to the difference between the temperature of the object and that of the surrounding environment."

This results in an exponential decay in the temperature difference. For example, if a warm object is placed in a cold bath, within a certain length of time, the difference in their temperatures will decrease by half. Then in that same length of time, the remaining difference will again decrease by half. This repeated halving of the temperature difference will continue at equal time intervals until it becomes too small to measure.

Heat Transfer

Heat can be transferred from one body to another or between a body and the environment by three different means: conduction, convection and radiation. Conduction is the transfer of energy *through* a solid material. Conduction between bodies occurs when they are in direct contact, and molecules transfer their energy across the interface.

Convection is the transfer of heat to or from a fluid medium. Molecules in a gas or liquid in contact with a solid body transmit or absorb heat to or from that body and then move away, allowing other molecules to move into place and repeat the process. Efficiency can be improved by increasing the surface area to be heated or cooled, as with a radiator, and by forcing the fluid to move over the surface, as with a fan.

Radiation is the emission of electromagnetic (EM) energy, particularly infrared photons that carry heat energy. All matter emits and absorbs some EM radiation, the net amount of which determines whether this causes a loss or gain in heat.

The Carnot Cycle

In 1824, Nicolas Léonard Sadi Carnot proposed a model for a heat engine based on what has come to be known as the Carnot cycle. The cycle exploits the relationships among pressure, volume and temperature of gasses and how an input of energy can change form and do work outside the system.

Compressing a gas increases its temperature so it becomes hotter than its environment. Heat can then be removed from the hot gas using a heat exchanger. Then, allowing it to expand causes it to cool. This is the basic principle behind heat pumps used for heating, air conditioning and refrigeration.

Conversely, heating a gas increases its pressure, causing it to expand. The expansive pressure can then be used to drive a piston, thus converting heat energy into kinetic energy. This is the basic principle behind heat engines.

Entropy

All thermodynamic systems generate waste heat. This waste results in an increase in entropy, which for a closed system is "a quantitative measure of the amount of thermal energy not available to do work". Entropy in any closed system *always* increases; it *never* decreases. Additionally, moving parts produce waste heat due to friction, and radiative heat inevitably leaks from the system.

This makes so-called perpetual motion machines impossible. Siabal Mitra, a professor of physics at Missouri State University, explains, "You cannot build an engine that is 100 percent efficient, which means you cannot build a perpetual motion machine. However, there are a lot of folks out there who still don't believe it, and there are people who are still trying to build perpetual motion machines."

Entropy is also defined as "a measure of the disorder or randomness in a closed system," which also inexorably increases. You can mix hot and cold water, but because a large cup of warm water is more disordered than two smaller cups containing hot and cold water, you can never separate it back into hot and cold without adding energy to the system. Put another way, you can't unscramble an egg or remove cream from your coffee. While some processes appear to be completely reversible, in practice, none actually are. Entropy, therefore, provides us with an arrow of time: forward is the direction of increasing entropy.

The Four Laws of Thermodynamics

The fundamental principles of thermodynamics were originally expressed in three laws. Later, it was determined that a more fundamental law had been neglected, apparently because it had seemed so obvious that it did not need to be stated explicitly. To form a complete set of rules, scientists decided this most fundamental law needed to be included. The problem, though, was that the first three laws had already been established and were well known by their assigned numbers. When faced with the prospect of renumbering the existing laws, which would cause considerable confusion, or placing the pre-eminent law at the end of the list, which would make no logical sense, a British physicist, Ralph H. Fowler, came up with an alternative that solved the dilemma: he called the new law the "Zeroth Law." In brief, these laws are:

- The Zeroth Law states that if two bodies are in thermal equilibrium with some third body, then they are also in equilibrium with each other. This establishes temperature as a fundamental and measurable property of matter.
- The First Law states that the total increase in the energy of a system is equal to the increase in thermal energy plus the work done on the system. This states that heat is a form of energy and is therefore subject to the principle of conservation.
- The Second Law states that heat energy cannot be transferred from a body at a lower temperature to a body at a higher temperature without the addition of energy. This is why it costs money to run an air conditioner.
- The Third Law states that the entropy of a pure crystal at absolute zero is zero. As explained above, entropy is sometimes called "waste energy," i.e., energy that is unable to do work, and since there is no heat energy whatsoever at absolute zero, there can be no waste energy. Entropy is also a measure of the disorder in a system, and while a perfect crystal is by definition perfectly ordered, any positive value of temperature means there is motion within the crystal, which causes disorder. For these reasons, there can be no physical system with lower entropy, so entropy always has a positive value.

The science of thermodynamics has been developed over centuries, and its principles apply to nearly every device ever invented. Its importance in modern technology cannot be overstated.

Magnetism

Magnetism is one aspect of the combined electromagnetic force. It refers to physical phenomena arising from the force caused by magnets, objects that produce fields that attract or repel other objects.

A magnetic field exerts a force on particles in the field due to the Lorentz force, according to Georgia State University's HyperPhysics website. The motion of electrically charged particles gives rise to magnetism. The force acting on an electrically charged particle in a magnetic field depends on the magnitude of the charge, the velocity of the particle, and the strength of the magnetic field.

All materials experience magnetism, some more strongly than others. Permanent magnets, made from materials such as iron, experience the strongest effects, known as ferromagnetism. With rare exception, this is the only form of magnetism strong enough to be felt by people.

The Source of Magnetism

All matter is composed of atoms, and atoms are composed of protons, neutrons and electrons. The protons and neutrons are located in the atom's nucleus and the electrons are in constant motion around the nucleus. Electrons carry a negative electrical charge and produce a magnetic field as they move through space. A magnetic field is produced whenever an electrical charge is in motion. The strength of this field is called the magnetic moment.

This may be hard to visualize on a subatomic scale but consider electric current flowing through a conductor. When the electrons (electric current) are flowing through the conductor, a magnetic field forms around the conductor. The magnetic field can be detected using a compass. The magnetic field will place a force on the compass needle, which is another example of a dipole.

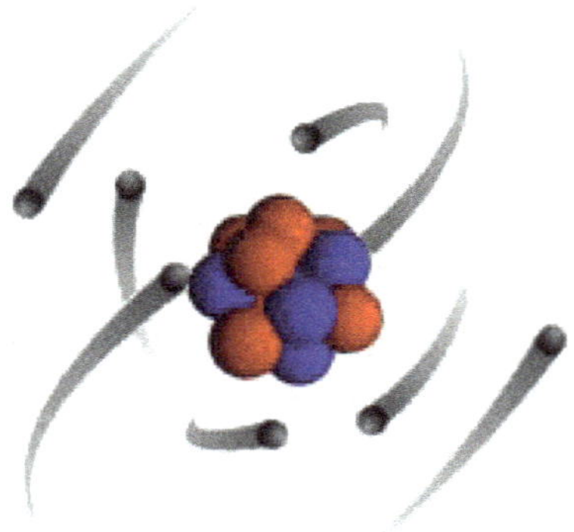

Since all matter is comprised of atoms, all materials are affected in some way by a magnetic field. However, not all materials react the same way.

Materials

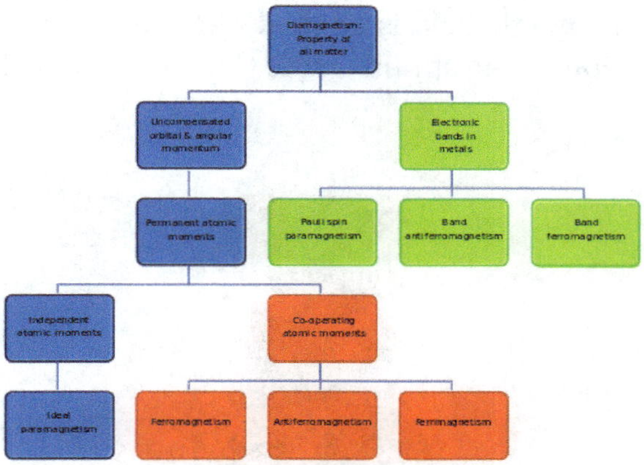

Hierarchy of types of magnetism.

Diamagnetism

Diamagnetism appears in all materials, and is the tendency of a material to oppose an applied magnetic field, and therefore, to be repelled by a magnetic field. However, in a material with paramagnetic properties (that is, with a tendency to enhance an external magnetic field), the paramagnetic behavior dominates. Thus, despite its universal occurrence, diamagnetic behavior is observed only in a purely diamagnetic material. In a diamagnetic material, there are no unpaired electrons, so the intrinsic electron magnetic moments cannot produce any bulk effect. In these cases, the magnetization arises from the electrons, orbital motions, which can be understood classically as follows:

When a material is put in a magnetic field, the electrons circling the nucleus will experience, in addition to their Coulomb attraction to the nucleus, a Lorentz force from the magnetic field. Depending on which direction the electron is orbiting, this force may increase the centripetal force on the electrons, pulling them in towards the nucleus, or it may decrease the force, pulling them away from the nucleus. This effect systematically increases the orbital magnetic moments that were aligned opposite the field, and decreases the ones aligned parallel to the field (in accordance with Lenz's law). This results in a small bulk magnetic moment, with an opposite direction to the applied field.

Note that this description is meant only as a heuristic; the Bohr-van Leeuwen theorem shows that diamagnetism is impossible according to classical physics, and that a proper understanding requires a quantum-mechanical description.

Note that all materials undergo this orbital response. However, in paramagnetic and ferromagnetic substances, the diamagnetic effect is overwhelmed by the much stronger effects caused by the unpaired electrons.

Paramagnetism

In a paramagnetic material there are *unpaired electrons*; i.e., atomic or molecular orbitals with exactly one electron in them. While paired electrons are required by the Pauli exclusion principle to have their intrinsic ('spin,) magnetic moments pointing in opposite directions, causing their magnetic fields to cancel out, an unpaired electron is free to align its magnetic moment in any direction. When an external magnetic field is applied, these magnetic moments will tend to align themselves in the same direction as the applied field, thus reinforcing it.

Ferromagnetism

Tip of permanent magnet with coins demonstrating ferromagnetism

A ferromagnet, like a paramagnetic substance, has unpaired electrons. However, in addition to the electrons' intrinsic magnetic moment's tendency to be parallel to an applied field, there is also in these materials a tendency for these magnetic moments to orient parallel to each other to maintain a lowered-energy state. Thus, even in the absence of an applied field, the magnetic moments of the electrons in the material spontaneously line up parallel to one another.

Every ferromagnetic substance has its own individual temperature, called the Curie temperature, or Curie point, above which it loses its ferromagnetic properties. This is because the thermal tendency to disorder overwhelms the energy-lowering due to ferromagnetic order.

Ferromagnetism only occurs in a few substances; the common ones are iron, nickel, cobalt, their alloys, and some alloys of rare-earth metals.

Magnetic Domains

The magnetic moments of atoms in a ferromagnetic material cause them to behave something like tiny permanent magnets. They stick together and align themselves into small regions of more or less uniform alignment called magnetic domains or Weiss domains. Magnetic domains can be observed with a magnetic force microscope to reveal magnetic domain boundaries that resemble white lines in the sketch. There are many scientific experiments that can physically show magnetic fields.

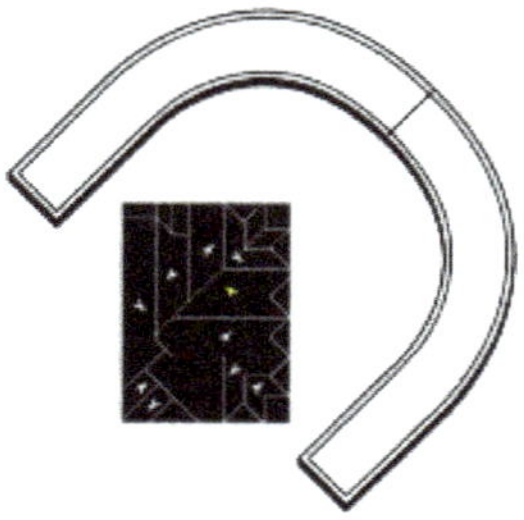

Magnetic domains boundaries (white lines) in ferromagnetic material (black rectangle)

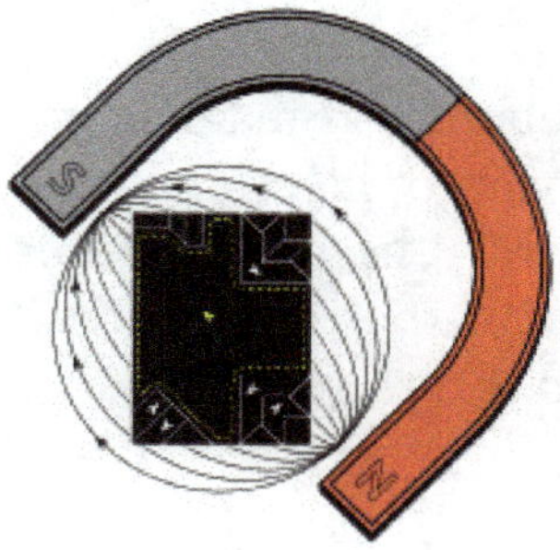

Effect of a magnet on the domains

When a domain contains too many molecules, it becomes unstable and divides into two domains aligned in opposite directions, so that they stick together more stably, as shown at the right.

When exposed to a magnetic field, the domain boundaries move, so that the domains aligned with the magnetic field grow and dominate the structure (dotted yellow area), as shown at the left. When the magnetizing field is removed, the domains may not return to an unmagnetized state. This results in the ferromagnetic material's being magnetized, forming a permanent magnet.

When magnetized strongly enough that the prevailing domain overruns all others to result in only one single domain, the material is magnetically saturated. When a magnetized ferromagnetic material is heated to the Curie point temperature, the molecules are agitated to the point that the magnetic domains lose the organization, and the magnetic properties they cause cease. When the material is cooled, this domain alignment structure spontaneously returns, in a manner roughly analogous to how a liquid can freeze into a crystalline solid.

Antiferromagnetism

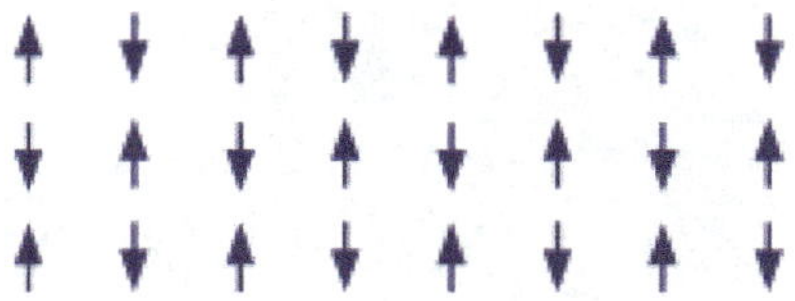

Antiferromagnetic ordering

In an antiferromagnet, unlike a ferromagnet, there is a tendency for the intrinsic magnetic moments of neighboring valence electrons to point in *opposite* directions. When all atoms are arranged in a substance so that each neighbor is anti-parallel, the substance is antiferromagnetic. Antiferromagnets have a zero net magnetic moment, meaning that no field is produced by them. Antiferromagnets are less common compared to the other types of behaviors and are mostly observed at low temperatures. In varying temperatures, antiferromagnets can be seen to exhibit diamagnetic and ferromagnetic properties.

In some materials, neighboring electrons prefer to point in opposite directions, but there is no geometrical arrangement in which *each* pair of neighbors is anti-aligned. This is called a spin glass and is an example of geometrical frustration.

Ferrimagnetism

Like ferromagnetism, ferrimagnets retain their magnetization in the absence of a field. However, like antiferromagnets, neighboring pairs of electron spins tend to point in opposite directions. These two properties are not contradictory, because in the optimal geometrical arrangement, there is more magnetic moment from the sublattice of electrons that point in one direction, than from the sublattice that points in the opposite direction.

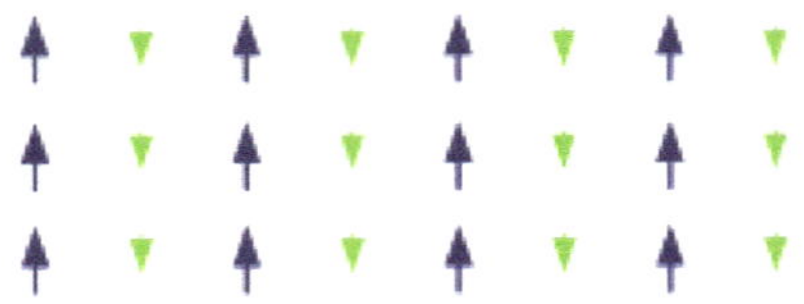

Ferrimagnetic ordering

Most ferrites are ferrimagnetic. The first discovered magnetic substance, magnetite, is a ferrite and was originally believed to be a ferromagnet; Louis Néel disproved this, however, after discovering ferrimagnetism.

Superparamagnetism

When a ferromagnet or ferrimagnet is sufficiently small, it acts like a single magnetic spin that is subject to Brownian motion. Its response to a magnetic field is qualitatively similar to the response of a paramagnet, but much larger.

Other Types of Magnetism

- Metamagnetism
- Molecule-based magnet
- Spin glass

Electromagnetism

Electromagnetism is the science of charge and of the forces and fields associated with charge. Electricity and magnetism are two aspects of electromagnetism.

Electricity and magnetism were long thought to be separate forces. It was not until the 19th century that they were finally treated as interrelated phenomena. In 1905 Albert Einstein's special theory of relativity established beyond a doubt that both are aspects of one common phenomenon. At a practical level, however, electric and magnetic forces behave quite differently and are described by different equations. Electric forces are produced by electric charges either at rest or in motion. Magnetic forces, on the other hand, are produced only by moving charges and act solely on charges in motion.

Electric phenomena occur even in neutral matter because the forces act on the individual charged constituents. The electric force in particular is responsible for most of the physical and chemical properties of atoms and molecules. It is enormously strong compared with gravity. For example, the absence of only one electron out of every billion molecules in two 70-kilogram (154-pound) persons standing two metres (two yards) apart would repel them with a 30,000-ton force. On a more familiar scale, electric phenomena are responsible for the lightning and thunder accompanying certain storms.

Electric and magnetic forces can be detected in regions called electric and magnetic fields. These fields are fundamental in nature and can exist in space far from the charge or current that generated them. Remarkably, electric fields can produce magnetic fields and vice versa, independent of any external charge. A changing magnetic field produces an electric field, as the English physicist Michael Faraday discovered in work that forms the basis of electric power generation. Conversely, a changing electric field produces a magnetic field, as the Scottish physicist James Clerk Maxwell deduced. The mathematical equations formulated by Maxwell incorporated light and wave phenomena into electromagnetism. He showed that electric and magnetic fields travel together through space as waves of electromagnetic radiation, with the changing fields mutually sustaining each other. Examples of electromagnetic waves traveling through space independent of matter are radio and television waves, microwaves, infrared rays, visible light, ultraviolet light, X-rays, and gamma rays. All of these waves travel at the same speed—namely, the velocity of light (roughly 300,000 kilometres, or 186,000 miles, per second). They differ from each other only in the frequency at which their electric and magnetic fields oscillate.

Maxwell's equations still provide a complete and elegant description of electromagnetism down to, but not including, the subatomic scale. The interpretation of his work, however, was broadened in the 20th century. Einstein's special relativity theory merged electric and magnetic fields into one common field and limited the velocity of all matter to the velocity of electromagnetic radiation. During the late 1960s, physicists discovered that other forces in nature have fields with a mathematical structure similar to that of the electromagnetic field. These other forces are the strong force, responsible for the energy released in nuclear fusion, and the weak force, observed in the radioactive decay of unstable atomic nuclei. In particular, the weak and electromagnetic forces have been combined into a common force called the electroweak force. The goal of many physicists to unite all of the fundamental forces, including gravity, into one grand unified theory has not been attained to date.

An important aspect of electromagnetism is the science of electricity, which is concerned with the behaviour of aggregates of charge, including the distribution of charge within matter and the motion of charge from place to place. Different types of materials are classified as either conductors or insulators on the basis of whether charges can move freely through their constituent matter. Electric current is the measure of the flow of charges; the laws governing currents in matter are important in technology, particularly in the production, distribution, and control of energy.

The concept of voltage, like those of charge and current, is fundamental to the science of electricity. Voltage is a measure of the propensity of charge to flow from one place to another; positive charges generally tend to move from a region of high voltage to a region of lower voltage. A common problem in electricity is determining the relationship between voltage and current or charge in a given physical situation.

Fundamentals

Everyday modern life is pervaded by electromagnetic phenomena. When a lightbulb is switched on, a current flows through a thin filament in the bulb, and the current heats the filament to such a high temperature that it glows, illuminating its surroundings. Electric clocks and connections link simple devices of this kind into complex systems such as traffic lights that are timed and synchronized with the speed of vehicular flow. Radio and television sets receive information carried by electromagnetic waves traveling through space at the speed of light. To start an automobile, currents in an electric starter motor generate magnetic fields that rotate the motor shaft and drive engine pistons to compress an explosive mixture of gasoline and air; the spark initiating the combustion is an electric discharge, which makes up a momentary current flow.

Coulomb's Law

Many of these devices and phenomena are complex, but they derive from the same fundamental laws of electromagnetism. One of the most important of these is Coulomb's law, which describes the electric force between charged objects. Formulated by the 18th-century French physicist Charles-Augustin de Coulomb, it is analogous to Newton's law for the gravitational force. Both gravitational and electric forces decrease with the square of the distance between the objects, and both forces act along a line between them. In Coulomb's law, however, the magnitude and sign of the electric force are determined by the charge, rather than the mass, of an object. Thus, charge determines how electromagnetism influences the motion of charged objects. (Charge is a basic property of matter. Every constituent of matter has an electric charge with a value that can be positive, negative, or zero. For example, electrons are negatively charged, and atomic nuclei are positively charged. Most bulk matter has an equal amount of positive and negative charge and thus has zero net charge.)

According to Coulomb, the electric force for charges at rest has the following properties:

(1) Like charges repel each other, and unlike charges attract. Thus, two negative charges repel one another, while a positive charge attracts a negative charge.

(2) The attraction or repulsion acts along the line between the two charges.

(3) The size of the force varies inversely as the square of the distance between the two charges. Therefore, if the distance between the two charges is doubled, the attraction or repulsion becomes weaker, decreasing to one-fourth of the original value. If the charges come 10 times closer, the size of the force increases by a factor of 100.

(4) The size of the force is proportional to the value of each charge. The unit used to measure charge is the coulomb (C). If there were two positive charges, one of 0.1 coulomb and the second of 0.2 coulomb, they would repel each other with a force that depends on the product 0.2×0.1. If each of the charges were reduced by one-half, the repulsion would be reduced to one-quarter of its former value.

Static cling is a practical example of the Coulomb force. In static cling, garments made of synthetic material collect a charge, especially in dry winter air. A plastic or rubber comb passed quickly through hair also becomes charged and will pick up bits of paper. The synthetic fabric and the comb are insulators; charge on these objects cannot move easily from one part of the object to another. Similarly, an office copy machine uses electric force to attract particles of ink to paper.

Principle of Charge Conservation

Like Coulomb's law, the principle of charge conservation is a fundamental law of nature. According to this principle, the charge of an isolated system cannot change. If an additional positively charged particle appears within a system, a particle with a negative charge of the same magnitude will be created at the same time; thus, the principle of conservation of charge is maintained. In nature, a pair of oppositely charged particles is created when high-energy radiation interacts with matter; an electron and a positron are created in a process known as pair production.

The smallest subdivision of the amount of charge that a particle can have is the charge of one proton, $+1.602 \times 10^{-19}$ coulomb. The electron has a charge of the same magnitude but opposite sign—i.e., -1.602×10^{-19} coulomb. An ordinary flashlight battery delivers a current that provides a total charge flow of approximately 5,000 coulomb, which corresponds to more than 10^{22} electrons, before it is exhausted.

Electric current is a measure of the flow of charge, as, for example, charge flowing through a wire. The size of the current is measured in amperes and symbolized by i. An ampere of current represents the passage of one coulomb of charge per second, or 6.2 billion billion electrons (6.2×10^{18} electrons) per second. A current is positive when it is in the direction of the flow of positive charges; its direction is opposite to the flow of negative charges.

Electric Fields and Forces

The force and conservation laws are only two aspects of electromagnetism, however. Electric and magnetic forces are caused by electromagnetic fields. The term *field* denotes a property of space, so that the field quantity has a numerical value at each point of space. These values may also vary with time. The value of the electric or magnetic field is a vector—i.e., a quantity having both magnitude and direction. The value of the electric field at a point in space, for example, equals the force that would be exerted on a unit charge at that position in space.

Every charged object sets up an electric field in the surrounding space. A second charge "feels" the presence of this field. The second charge is either attracted toward the initial charge or repelled from it, depending on the signs of the charges. Of course, since the second charge also has an electric field, the first charge feels its presence and is either attracted or repelled by the second charge too.

The electric field from a charge is directed away from the charge when the charge is positive and toward the charge when it is negative. The arrows point in the direction of the electric field, and the length of the arrows indicates the strength of the field at the midpoint of the arrows.

If a positive charge were placed in the electric field, it would feel a force in the direction of the field. A negative charge would feel a force in the direction opposite the direction of the field.

In calculations, it is often more convenient to deal directly with the electric field than with the charges. Frequently, more is known about the field than about the distribution of charges in space. For example, the distribution of charges in conductors is generally unknown because the charges move freely within the conductor. In static situations, however, the electric field in a conductor in equilibrium has a definite value, zero, because any force on the charges inside the conductor redistributes them until the field vanishes. The unit of electric field is newtons per coulomb, or volts per metre.

The electric potential is another useful field. It provides an alternative to the electric field in electrostatics problems. The potential is easier to use, however, because it is a single number, a scalar, instead of a vector. The difference in potential between two places measures the degree to which charges are influenced to move from one place to another. If the potential is the same at two places (i.e., if the places have the same voltage), charges will not be influenced to move from one place to the other. The potential on an object or at some point in space is measured in volts; it equals the electrostatic energy that a unit charge would have at that position. In a typical 12-volt car battery, the battery terminal that is marked with a + sign is at a potential 12 volts greater than the potential of the terminal marked with the – sign. When a wire, such as the filament of a car headlight, is connected between the + and the – terminals of the battery, charges move through the filament as an electric current and heat the filament, and the hot filament radiates light.

Magnetic Fields and Forces

The magnetic force influences only those charges that are already in motion. It is transmitted by the magnetic field. Both magnetic fields and magnetic forces are more complicated than electric fields and electric forces. The magnetic field does not point along the direction of the source of the field; instead, it points in a perpendicular direction. In addition, the magnetic force acts in a direction that is perpendicular to the direction of the field. In comparison, both the electric force and the electric field point directly toward or away from the charge.

The magnetic field is produced by a current of charge in a wire. Certain materials, such as copper, silver, and aluminum, are conductors that allow charge to flow freely from place to place. If an external influence establishes a current in a conductor, the current generates a magnetic field. For a long straight wire, the magnetic field has a direction that encircles the wire on a plane perpendicular to the wire. The strength of the magnetic field decreases with distance from the wire. The arrows in figure represent the size and direction of the magnetic field for a current moving in the direction indicated. Figure A shows an end view with the current coming toward the reader, while figure B provides a three-dimensional view of the magnetic field at one position along the wire.

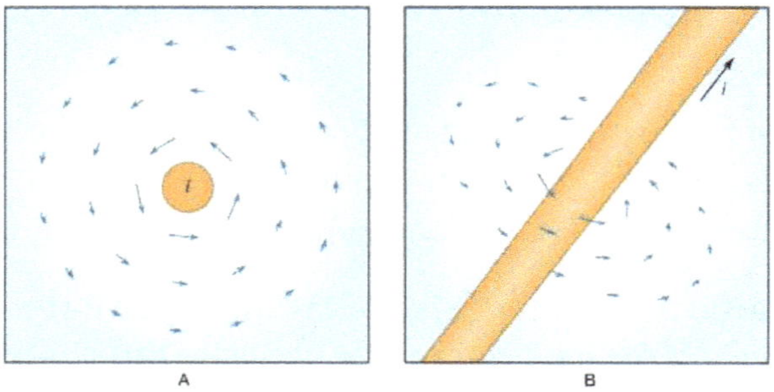

Magnetic field
Magnetic field of a long wire. (A) An end view, with the current flowing toward the viewer. (B) A three-dimensional view.

In subsequent figures, continuous lines will be used to represent the direction of electric and magnetic fields. These lines emphasize the important fact that electric fields begin on positive charges and end on negative charges, while magnetic fields do not have beginnings or ends and close on themselves. The magnetic field shown in figure is unusually simple. Highly complex and useful magnetic fields can be generated by the proper choice of conductors to carry electric currents. Under development are thermonuclear fusion reactors for obtaining energy from the fusion of light nuclei in the form of very hot plasmas of hydrogen isotopes. The plasmas have to be confined by magnetic fields (dubbed "magnetic bottles") as no material container can withstand such high temperatures. Charged particles are also confined by magnetic fields in nature. Large numbers of charged particles, mostly protons and electrons, are trapped in huge bands around Earth by its magnetic field. These bands are known as the Van Allen radiation belts. Disturbance of Earth's confining magnetic field produces spectacular displays, the so-called northern lights, in which trapped charged particles are freed and crash through the atmosphere to Earth.

Interaction of a Magnetic Field with a Charge

How does the magnetic field interact with a charged object? If the charge is at rest, there is no interaction. If the charge moves, however, it is subjected to a force, the size of which increases in direct proportion with the velocity of the charge. The force has a direction that is perpendicular both to the direction of motion of the charge and to the direction of the magnetic field. There are two possible precisely opposite directions for such a force for a given direction of motion. This apparent ambiguity is resolved by the fact that one of the two directions applies to the force on a moving positive charge while the other direction applies to the force on a moving negative charge. Figure below illustrates the directions of the magnetic force on positive charges and on negative charges as they move in a magnetic field that is perpendicular to the motion.

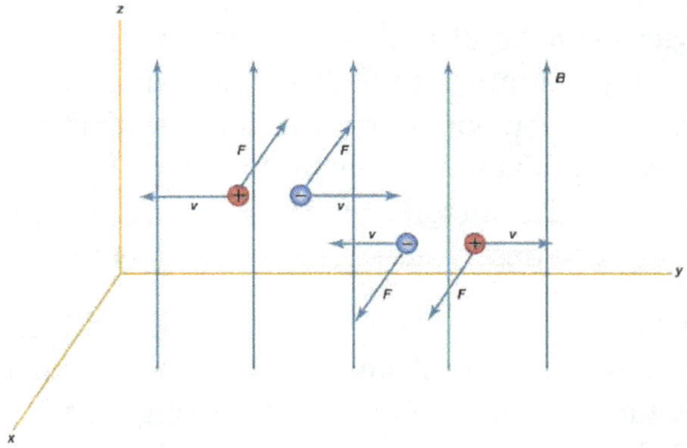

Magnetic force

Magnetic force on moving charges. The magnetic force *F* is proportional to the charge and to the magnitude of velocity v times the magnetic field *B*.

Depending on the initial orientation of the particle velocity to the magnetic field, charges having a constant speed in a uniform magnetic field will follow a circular or helical path.

Electric currents in wires are not the only source of magnetic fields. Naturally occurring minerals exhibit magnetic properties and have magnetic fields. These magnetic fields result from the motion of electrons in the atoms of the material. They also result from a property of electrons called the magnetic dipole moment, which is related to the intrinsic spin of individual electrons. In most materials, little or no field is observed outside the matter because of the random orientation of the various constituent atoms. In some materials such as iron, however, atoms within certain distances tend to become aligned in one particular direction.

Magnets have numerous applications, ranging from use as toys and paper holders on home refrigerators to essential components in electric generators and machines that can accelerate particles to speeds approaching that of light. The practical application of magnetism in technology is greatly enhanced by using iron and other ferromagnetic materials with electric currents in devices like motors. These materials amplify the magnetic field produced by the currents and thereby create more powerful fields.

While electric and magnetic effects are well separated in many phenomena and applications, they are coupled closely together when there are rapid time fluctuations. Faraday's law of induction describes how a time-varying magnetic field produces an electric field. Important practical applications include the electric generator and transformer. In a generator, the physical motion of a magnetic field produces electricity for power. In a transformer, electric power is converted from one voltage level to another by the magnetic field of one circuit inducing an electric current in another circuit.

The existence of electromagnetic waves depends on the interaction between electric and magnetic fields. Maxwell postulated that a time-varying electric field produces a magnetic field. His theory predicted the existence of electromagnetic waves in which each time-varying field produces the other field. For example, radio waves are generated by electronic circuits known as oscillators that cause rapidly oscillating currents to flow in antennas; the rapidly varying magnetic field has an associated varying electric field. The result is the emission of radio waves into space.

Many electromagnetic devices can be described by circuits consisting of conductors and other elements. These circuits may operate with a steady flow of current, as in a flashlight, or with time-varying currents. Important elements in circuits include sources of power called electromotive forces; resistors, which control the flow of current for a given voltage; capacitors, which store charge and energy temporarily; and inductors, which also store electrical energy for a limited time. Circuits with these elements can be described entirely with algebra.

Two mathematical quantities associated with vector fields, like the electric field $\boldsymbol{E}$ and the magnetic field $\boldsymbol{B}$, are useful for describing electromagnetic phenomena. They are the flux of such a field through a surface and the line integral of the field along a path. The flux of a field through a surface measures how much of the field penetrates through the surface; for every small section of the surface, the flux is proportional to the area of that section and depends also on the relative orientation of the section and the field. The line integral of a field along a path measures the degree to which the field is aligned with the path; for every small section of path, it is proportional to the length of that section and is also dependent on the alignment of the field with that section of path. When the field is perpendicular to the path, there is no contribution to the line integral. The fluxes of $\boldsymbol{E}$ and $\boldsymbol{B}$ through a surface and the line integrals of these fields along a path play an important role in electromagnetic theory. As examples, the flux of the electric field $\boldsymbol{E}$ through a closed surface measures the amount of charge contained within the surface; the flux of the magnetic field $\boldsymbol{B}$ through a closed surface is always zero because there are no magnetic monopoles (magnetic charges consisting of a single pole) to act as sources of the magnetic field in the way that charge is a source of the electric field.

Optics

Optics is the scientific study of light. It is a branch of physics that covers the behavior and properties of light.

There are two major branches of optics, physical and geometrical. Physical optics deals primarily with the nature and properties of light itself. Geometrical optics has to do with the principles that govern the image-forming properties of lenses, mirrors, and other devices that make use of light. It also includes optical data processing, which involves the manipulation of the information content of an image formed by coherent optical systems.

Originally, the term optics was used only in relation to the eye and vision. Later, as lenses and other devices for aiding vision began to be developed, these were naturally called optical instruments, and the meaning of the term optics eventually became broadened to cover any application of light, even though the ultimate receiver is not the eye but a physical detector, such as a photographic plate or a television camera. In the 20th century optical methods came to be applied extensively to regions of the electromagnetic radiation spectrum not visible to the eye, such as X-rays, ultraviolet, infrared, and microwave radio waves, and to this extent these regions are now often included in the general field of optics.

Classical Optics

Classical optics is divided into two main branches: geometrical (or ray) optics and physical (or wave) optics. In geometrical optics, light is considered to travel in straight lines, while in physical optics, light is considered as an electromagnetic wave.

Geometrical optics can be viewed as an approximation of physical optics that applies when the wavelength of the light used is much smaller than the size of the optical elements in the system being modelled.

Geometrical Optics

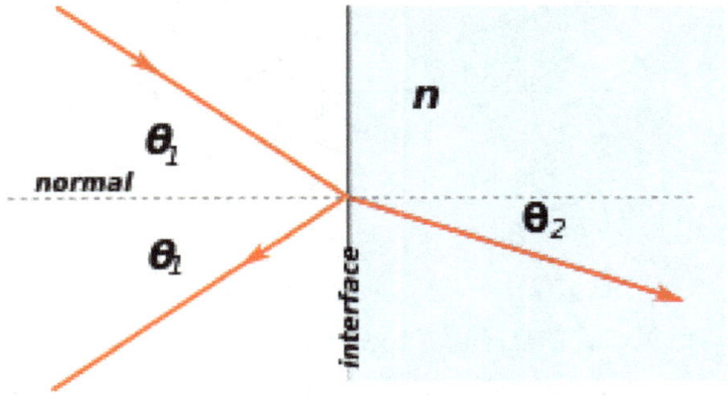

Geometry of reflection and refraction of light rays

Geometrical optics, or *ray optics*, describes the propagation of light in terms of "rays" which travel in straight lines, and whose paths are governed by the laws of reflection and refraction at interfaces between different media. These laws were discovered empirically as far back as 984 AD and have been used in the design of optical components and instruments from then until the present day. They can be summarised as follows:

When a ray of light hits the boundary between two transparent materials, it is divided into a reflected and a refracted ray.

> The law of reflection says that the reflected ray lies in the plane of incidence, and the angle of reflection equals the angle of incidence.
>
> The law of refraction says that the refracted ray lies in the plane of incidence, and the sine of the angle of refraction divided by the sine of the angle of incidence is a constant:
>
> $$\frac{\sin \theta_1}{\sin \theta_2} = n,$$

where n is a constant for any two materials and a given colour of light. If the first material is air or vacuum, n is the refractive index of the second material.

The laws of reflection and refraction can be derived from Fermat's principle which states that *the path taken between two points by a ray of light is the path that can be traversed in the least time.*

Approximations

Geometric optics is often simplified by making the paraxial approximation, or "small angle approximation". The mathematical behaviour then becomes linear, allowing optical components and systems to be described by simple matrices. This leads to the techniques of Gaussian optics and *paraxial ray tracing*, which are used to find basic properties of optical systems, such as approximate image and object positions and magnifications.

Reflections

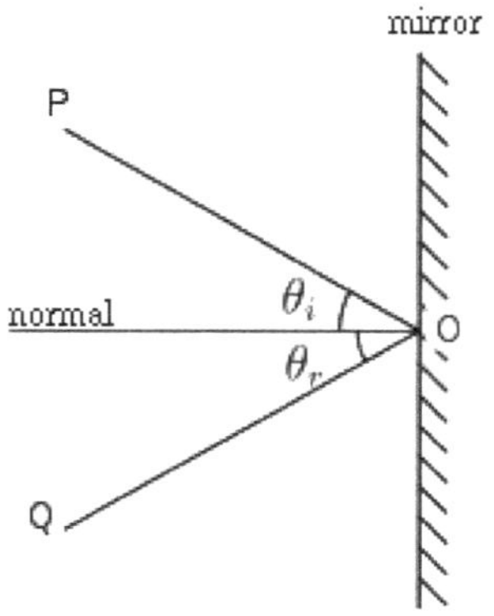

Diagram of specular reflection

Reflections can be divided into two types: specular reflection and diffuse reflection. Specular reflection describes the gloss of surfaces such as mirrors, which reflect light in a simple, predictable way. This allows for production of reflected images that can be associated with an actual (real) or extrapolated (virtual) location in space. Diffuse reflection describes non-glossy materials, such as paper or rock. The reflections from these surfaces can only be described statistically, with the exact distribution of the reflected light depending on the microscopic structure of the material. Many diffuse reflectors are described or can be approximated by Lambert's cosine law, which describes surfaces that have equal luminance when viewed from any angle. Glossy surfaces can give both specular and diffuse reflection.

In specular reflection, the direction of the reflected ray is determined by the angle the incident ray makes with the surface normal, a line perpendicular to the surface at the point where the ray hits. The incident and reflected rays and the normal lie in a single plane, and the angle between the reflected ray and the surface normal is the same as that between the incident ray and the normal. This is known as the Law of Reflection.

For flat mirrors, the law of reflection implies that images of objects are upright and the same distance behind the mirror as the objects are in front of the mirror. The image size is the same as the object size. The law also implies that mirror images are parity inverted, which we perceive as a left-right inversion. Images formed from reflection in two (or any even number of) mirrors are not parity inverted. Corner reflectors retroreflect light, producing reflected rays that travel back in the direction from which the incident rays came.

Mirrors with curved surfaces can be modelled by ray tracing and using the law of reflection at each point on the surface. For mirrors with parabolic surfaces, parallel rays incident on the mirror produce reflected rays that converge at a common focus. Other curved surfaces may also focus light, but with aberrations due to the diverging shape causing the focus to be smeared out in space. In particular, spherical mirrors exhibit spherical aberration. Curved mirrors can form images with magnification greater than or less than one, and the magnification can be negative, indicating that the image is inverted. An upright image formed by reflection in a mirror is always virtual, while an inverted image is real and can be projected onto a screen.

Refractions

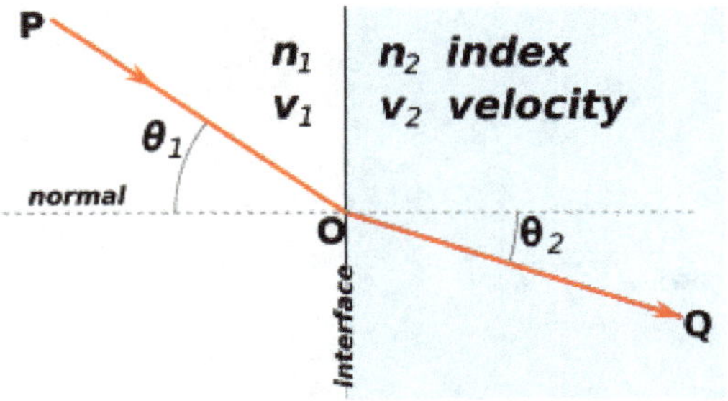

Illustration of Snell's Law for the case $n_1 < n_2$, such as air/water interface

Refraction occurs when light travels through an area of space that has a changing index of refraction; this principle allows for lenses and the focusing of light. The simplest case of refraction occurs when there is an interface between a uniform medium with index of refraction n_1 and another medium with index of refraction n_2. In such situations, Snell's Law describes the resulting deflection of the light ray:

$$n_1 \sin\theta_1 = n_2 \sin\theta_2$$

where θ_1 and θ_2 are the angles between the normal (to the interface) and the incident and refracted waves, respectively.

The index of refraction of a medium is related to the speed, v, of light in that medium by

$$n = c / v,$$

where c is the speed of light in vacuum.

Snell's Law can be used to predict the deflection of light rays as they pass through linear media as long as the indexes of refraction and the geometry of the media are known. For example, the propagation of light through a prism results in the light ray being deflected depending on the shape and orientation of the prism. In most materials, the index of refraction varies with the frequency of the light. Taking this into account, Snell's Law can be used to predict how a prism will disperse light into a spectrum. The discovery of this phenomenon when passing light through a prism is famously attributed to Isaac Newton.

Some media have an index of refraction which varies gradually with position and, thus, light rays in the medium are curved. This effect is responsible for mirages seen on hot days: a change in index of refraction air with height causes light rays to bend, creating the appearance of specular reflections in the distance (as if on the surface of a pool of water). Optical materials with varying index of refraction are called gradient-index (GRIN) materials. Such materials are used to make gradient-index optics.

For light rays travelling from a material with a high index of refraction to a material with a low index of refraction, Snell's law predicts that there is no θ_2 when θ_1 is large. In this case, no transmission occurs; all the light is reflected. This phenomenon is called total internal reflection and allows for fibre optics technology. As light travels down an optical fibre, it undergoes total internal reflection allowing for essentially no light to be lost over the length of the cable.

Lenses

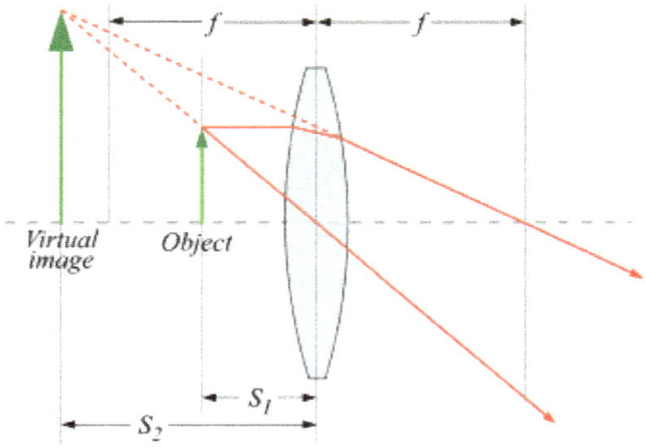

A ray tracing diagram for a converging lens.

A device which produces converging or diverging light rays due to refraction is known as a *lens*. Lenses are characterized by their focal length: a converging lens has positive focal length, while a diverging lens has negative focal length. Smaller focal length indicates that the lens has a stronger converging or diverging effect. The focal length of a simple lens in air is given by the lensmaker's equation.

Ray tracing can be used to show how images are formed by a lens. For a thin lens in air, the location of the image is given by the simple equation

$$\frac{1}{S_1}+\frac{1}{S_2}=\frac{1}{f},$$

where S_1 is the distance from the object to the lens, S_2 is the distance from the lens to the image, and f is the focal length of the lens. In the sign convention used here, the object and image distances are positive if the object and image are on opposite sides of the lens.

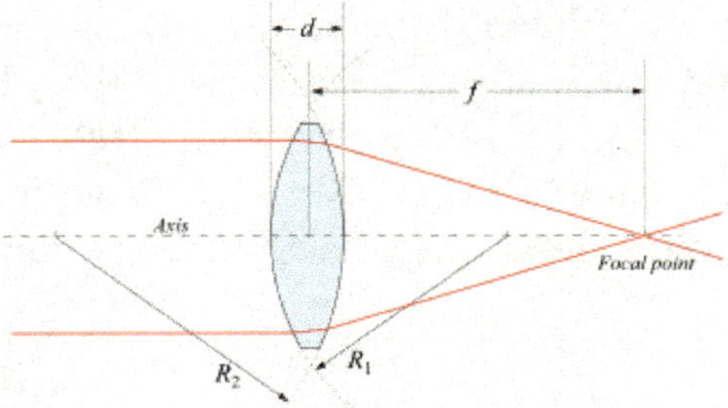

Positive (converging) lens

Incoming parallel rays are focused by a converging lens onto a spot one focal length from the lens, on the far side of the lens. This is called the rear focal point of the lens. Rays from an object at finite distance are focused further from the lens than the focal distance; the closer the object is to the lens, the further the image is from the lens.

With diverging lenses, incoming parallel rays diverge after going through the lens, in such a way that they seem to have originated at a spot one focal length in front of the lens. This is the lens's front focal point. Rays from an object at finite distance are associated with a virtual image that is closer to the lens than the focal point, and on the same side of the lens as the object. The closer the object is to the lens, the closer the virtual image is to the lens. As with mirrors, upright images produced by a single lens are virtual, while inverted images are real.

Lenses suffer from aberrations that distort images. *Monochromatic aberrations* occur because the geometry of the lens does not perfectly direct rays from each object point to a single point on the image, while chromatic aberration occurs because the index of refraction of the lens varies with the wavelength of the light.

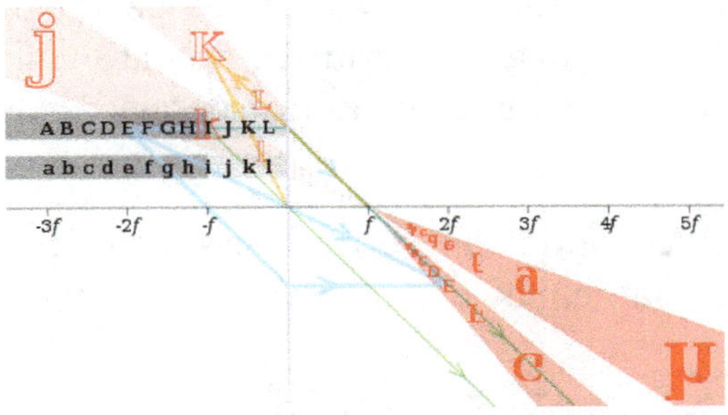

Images of black letters in a thin convex lens of focal length f are shown in red. Selected rays are shown for letters **E**, **I** and **K** in blue, green and orange, respectively. Note that **E** (at $2f$) has an equal-size, real and inverted image; **I** (at f) has its image at infinity; and **K** (at $f/2$) has a double-size, virtual and upright image.

Physical Optics

In physical optics, light is considered to propagate as a wave. This model predicts phenomena such as interference and diffraction, which are not explained by geometric optics. The speed of light waves in air is approximately 3.0×10^8 m/s (exactly 299,792,458 m/s in vacuum). The wavelength of visible light waves varies between 400 and 700 nm, but the term "light" is also often applied to infrared (0.7–300 μm) and ultraviolet radiation (10–400 nm).

The wave model can be used to make predictions about how an optical system will behave without requiring an explanation of what is "waving" in what medium. Until the middle of the 19th century, most physicists believed in an "ethereal" medium in which the light disturbance propagated. The existence of electromagnetic waves was predicted in 1865 by Maxwell's equations. These waves propagate at the speed of light and have varying electric and magnetic fields which are orthogonal to one another, and also to the direction of propagation of the waves. Light waves are now generally treated as electromagnetic waves except when quantum mechanical effects have to be considered.

Modelling and Design of Optical Systems using Physical Optics

Many simplified approximations are available for analysing and designing optical systems. Most of these use a single scalar quantity to represent the electric field of the light wave, rather than using a vector model with orthogonal electric and magnetic vectors. The Huygens–Fresnel equation is one such model. This was derived empirically by Fresnel in 1815, based on Huygens' hypothesis that each point on a wavefront generates a secondary spherical wavefront, which Fresnel combined with the principle of superposition of waves. The Kirchhoff diffraction equation, which is derived using Maxwell's equations, puts the Huygens-Fresnel equation on a firmer physical foundation. Examples of the application of Huygens–Fresnel principle can be found in the sections on diffraction and Fraunhofer diffraction.

More rigorous models, involving the modelling of both electric and magnetic fields of the light wave, are required when dealing with the detailed interaction of light with materials where the interaction depends on their electric and magnetic properties. For instance, the behaviour of a light wave interacting with a metal surface is quite different from what happens when it interacts with a dielectric material. A vector model must also be used to model polarised light.

Numerical modeling techniques such as the finite element method, the boundary element method and the transmission-line matrix method can be used to model the propagation of light in systems which cannot be solved analytically. Such models are computationally demanding and are normally only used to solve small-scale problems that require accuracy beyond that which can be achieved with analytical solutions.

All of the results from geometrical optics can be recovered using the techniques of Fourier optics which apply many of the same mathematical and analytical techniques used in acoustic engineering and signal processing.

Gaussian beam propagation is a simple paraxial physical optics model for the propagation of coherent radiation such as laser beams. This technique partially accounts for diffraction, allowing accurate calculations of the rate at which a laser beam expands with distance, and the minimum size to which the beam can be focused. Gaussian beam propagation thus bridges the gap between geometric and physical optics.

Superposition and Interference

In the absence of nonlinear effects, the superposition principle can be used to predict the shape of interacting waveforms through the simple addition of the disturbances. This interaction of waves to produce a resulting pattern is generally termed "interference" and can result in a variety of outcomes. If two waves of the same wavelength and frequency are *in phase*, both the wave crests and wave troughs align. This results in constructive interference and an increase in the amplitude of the wave, which for light is associated with a brightening of the waveform in that location. Alternatively, if the two waves of the same wavelength and frequency are out of phase, then the wave crests will align with wave troughs and vice versa. This results in destructive interference and a decrease in the amplitude of the wave, which for light is associated with a dimming of the waveform at that location.

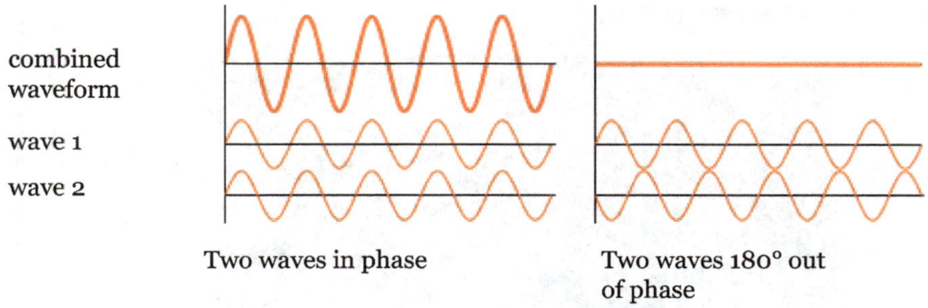

When oil or fuel is spilled, colourful patterns are formed by thin-film interference.

Since the Huygens–Fresnel principle states that every point of a wavefront is associated with the production of a new disturbance, it is possible for a wavefront to interfere with itself constructively or destructively at different locations producing bright and dark fringes in regular and predictable patterns. Interferometry is the science of measuring these patterns, usually as a means of making precise determinations of distances or angular resolutions. The Michelson interferometer was a famous instrument which used interference effects to accurately measure the speed of light.

The appearance of thin films and coatings is directly affected by interference effects. Antireflective coatings use destructive interference to reduce the reflectivity of the surfaces they coat, and can be used to minimise glare and unwanted reflections. The simplest case is a single layer with thickness one-fourth the wavelength of incident light. The reflected wave from the top of the film and the reflected wave from the film/material interface are then exactly 180° out of phase, causing destructive interference. The waves are only exactly out of phase for one wavelength, which would typically be chosen to be near the centre of the visible spectrum, around 550 nm. More complex designs using multiple layers can achieve low reflectivity over a broad band, or extremely low reflectivity at a single wavelength.

Constructive interference in thin films can create strong reflection of light in a range of wavelengths, which can be narrow or broad depending on the design of the coating. These films are used to make dielectric mirrors, interference filters, heat reflectors, and filters for colour separation in colour television cameras. This interference effect is also what causes the colourful rainbow patterns seen in oil slicks.

Diffraction and Optical Resolution

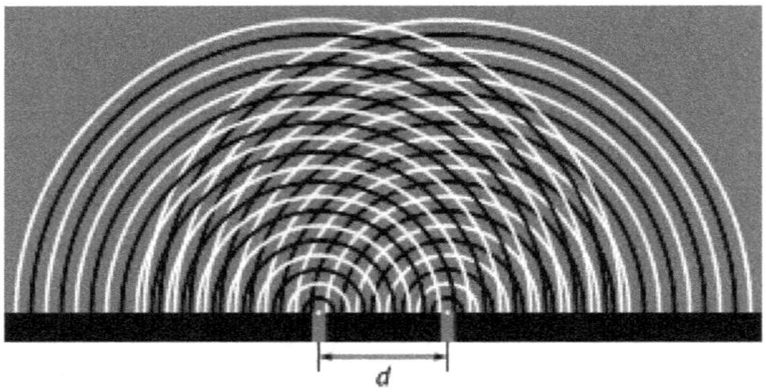

Diffraction on two slits separated by distance d. The bright fringes occur along lines where black lines intersect with black lines and white lines intersect with white lines. These fringes are separated by angle θ and are numbered as order n.

Diffraction is the process by which light interference is most commonly observed. The effect was first described in 1665 by Francesco Maria Grimaldi, who also coined the term from the Latin *diffringere*, 'to break into pieces'. Later that century, Robert Hooke and Isaac Newton also described phenomena now known to be diffraction in Newton's rings while James Gregory recorded his observations of diffraction patterns from bird feathers.

The first physical optics model of diffraction that relied on the Huygens–Fresnel principle was developed in 1803 by Thomas Young in his interference experiments with the interference patterns of two closely spaced slits. Young showed that his results could only be explained if the two slits acted as two unique sources of waves rather than corpuscles. In 1815 and 1818, Augustin-Jean Fresnel firmly established the mathematics of how wave interference can account for diffraction.

The simplest physical models of diffraction use equations that describe the angular separation of light and dark fringes due to light of a particular wavelength (λ). In general, the equation takes the form

$$m\lambda = d \sin\theta$$

where d is the separation between two wavefront sources (in the case of Young's experiments, it was two slits), θ is the angular separation between the central fringe and the m th order fringe, where the central maximum is $m = 0$.

This equation is modified slightly to take into account a variety of situations such as diffraction through a single gap, diffraction through multiple slits, or diffraction through a diffraction grating that contains a large number of slits at equal spacing. More complicated models of diffraction require working with the mathematics of Fresnel or Fraunhofer diffraction.

X-ray diffraction makes use of the fact that atoms in a crystal have regular spacing at distances that are on the order of one angstrom. To see diffraction patterns, x-rays with similar wavelengths to that spacing are passed through the crystal. Since crystals are three-dimensional objects rather than two-dimensional gratings, the associated diffraction pattern varies in two directions according to Bragg reflection, with the associated bright spots occurring in unique patterns and d being twice the spacing between atoms.

Diffraction effects limit the ability for an optical detector to optically resolve separate light sources. In general, light that is passing through an aperture will experience diffraction and the best images that can be created (as described in diffraction-limited optics) appear as a central spot with surrounding bright rings, separated by dark nulls; this pattern is known as an Airy pattern, and the central bright lobe as an Airy disk. The size of such a disk is given by

$$sin\,\theta = 1.22\frac{\lambda}{D}$$

where θ is the angular resolution, λ is the wavelength of the light, and D is the diameter of the lens aperture. If the angular separation of the two points is significantly less than the Airy disk angular radius, then the two points cannot be resolved in the image, but if their angular separation is much greater than this, distinct images of the two points are formed and they can therefore be resolved. Rayleigh defined the somewhat arbitrary "Rayleigh criterion" that two points whose angular separation is equal to the Airy disk radius (measured to first null, that is, to the first place where no light is seen) can be considered to be resolved. It can be seen that the greater the diameter of the lens or its aperture, the finer the resolution. Interferometry, with its ability to mimic extremely large baseline apertures, allows for the greatest angular resolution possible.

For astronomical imaging, the atmosphere prevents optimal resolution from being achieved in the visible spectrum due to the atmospheric scattering and dispersion which cause stars to twinkle. Astronomers refer to this effect as the quality of astronomical seeing. Techniques known as adaptive optics have been used to eliminate the atmospheric disruption of images and achieve results that approach the diffraction limit.

Dispersion and Scattering

Refractive processes take place in the physical optics limit, where the wavelength of light is similar to other distances, as a kind of scattering. The simplest type of scattering is Thomson scattering which occurs when electromagnetic waves are deflected by single particles. In the limit of Thomson scattering, in which the wavelike nature of light is evident, light is dispersed independent of the frequency, in contrast to Compton scattering which is frequency-dependent and strictly a quantum mechanical process, involving the nature of light as particles. In a statistical sense, elastic scattering of light by numerous particles much smaller than the wavelength of the light is a process known as Rayleigh scattering while the similar process for scattering by particles that are similar or larger in wavelength is known as Mie scattering with the Tyndall effect being a commonly observed result. A small proportion of light scattering from atoms or molecules may undergo Raman scattering, wherein the frequency changes due to excitation of the atoms and molecules. Brillouin scattering occurs when the frequency of light changes due to local changes with time and movements of a dense material.

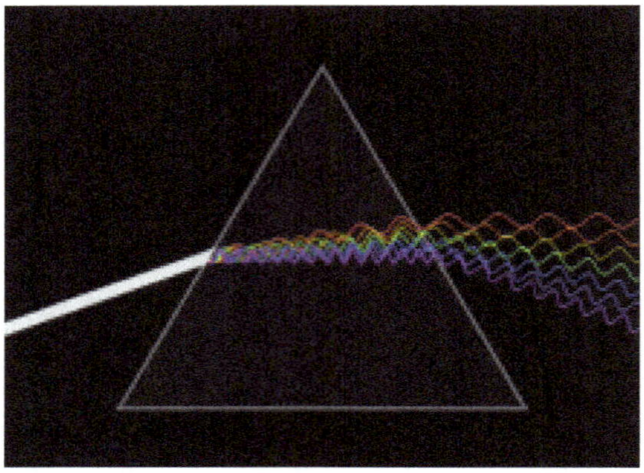

Dispersion of light through a prism. High frequency (blue) light is deflected the most, and low frequency (red) the least.

Dispersion occurs when different frequencies of light have different phase velocities, due either to material properties (*material dispersion*) or to the geometry of an optical waveguide (*waveguide dispersion*). The most familiar form of dispersion is a decrease in index of refraction with increasing wavelength, which is seen in most transparent materials. This is called "normal dispersion". It occurs in all dielectric materials, in wavelength ranges where the material does not absorb light. In wavelength ranges where a medium has significant absorption, the index of refraction can increase with wavelength. This is called "anomalous dispersion".

The separation of colours by a prism is an example of normal dispersion. At the surfaces of the prism, Snell's law predicts that light incident at an angle θ to the normal will be refracted at an angle arcsin(sin (θ) / *n*). Thus, blue light, with its higher refractive index, is bent more strongly than red light, resulting in the well-known rainbow pattern.

Material dispersion is often characterised by the Abbe number, which gives a simple measure of dispersion based on the index of refraction at three specific wavelengths. Waveguide dispersion is dependent on the propagation constant. Both kinds of dispersion cause changes in the group characteristics of the wave, the features of the wave packet that change with the same frequency as the amplitude of the electromagnetic wave. "Group velocity dispersion" manifests as a spreading-out of the signal "envelope" of the radiation and can be quantified with a group dispersion delay parameter:

$$D = \frac{1}{v_g^2}\frac{dv_g}{d\lambda}$$

where v_g is the group velocity. For a uniform medium, the group velocity is

$$v_g = c\left(n - \lambda\frac{dn}{d\lambda}\right)^{-1}$$

where n is the index of refraction and c is the speed of light in a vacuum. This gives a simpler form for the dispersion delay parameter:

$$D = -\frac{\lambda}{c}\frac{d^2n}{d\lambda^2}.$$

If D is less than zero, the medium is said to have *positive dispersion* or normal dispersion. If D is greater than zero, the medium has *negative dispersion*. If a light pulse is propagated through a normally dispersive medium, the result is the higher frequency components slow down more than the lower frequency components. The pulse therefore becomes *positively chirped*, or *up-chirped*, increasing in frequency with time. This causes the spectrum coming out of a prism to appear with red light the least refracted and blue/violet light the most refracted. Conversely, if a pulse travels through an anomalously (negatively) dispersive medium, high frequency components travel faster than the lower ones, and the pulse becomes *negatively chirped*, or *down-chirped*, decreasing in frequency with time.

The result of group velocity dispersion, whether negative or positive, is ultimately temporal spreading of the pulse. This makes dispersion management extremely important in optical communications systems based on optical fibres, since if dispersion is too high, a group of pulses representing information will each spread in time and merge, making it impossible to extract the signal.

Polarization

Polarization is a general property of waves that describes the orientation of their oscillations. For transverse waves such as many electromagnetic waves, it describes the orientation of the oscillations in the plane perpendicular to the wave's direction of travel. The oscillations may be oriented in a single direction (linear polarization), or the oscillation direction may rotate as the wave travels (circular or elliptical polarization). Circularly polarised waves can rotate rightward or leftward in the direction of travel, and which of those two rotations is present in a wave is called the wave's chirality.

The typical way to consider polarization is to keep track of the orientation of the electric field vector as the electromagnetic wave propagates. The electric field vector of a plane wave may be arbitrarily divided into two perpendicular components labeled x and y (with **z** indicating the direction of travel). The shape traced out in the x-y plane by the electric field vector is a Lissajous figure that describes the *polarization state*. The following figures show some examples of the evolution of the electric field vector (blue), with time (the vertical axes), at a particular point in space, along with its x and y components (red/left and green/right), and the path traced by the vector in the plane (purple): The same evolution would occur when looking at the electric field at a particular time while evolving the point in space, along the direction opposite to propagation.

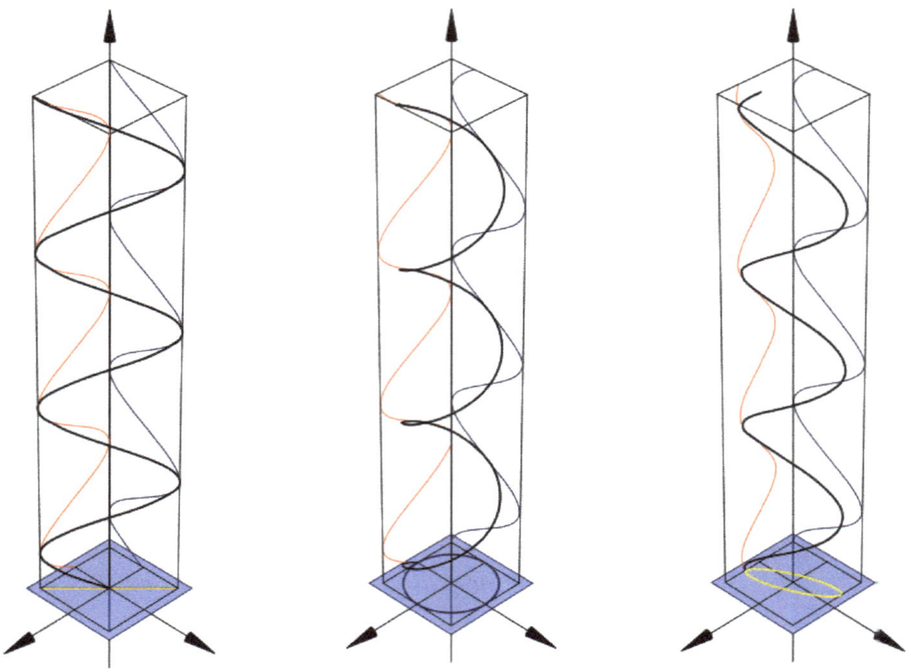

In the left most figure above, the x and y components of the light wave are in phase. In this case, the ratio of their strengths is constant, so the direction of the electric vector (the vector sum of these two components) is constant. Since the tip of the vector traces out a single line in the plane, this special case is called linear polarization. The direction of this line depends on the relative amplitudes of the two components.

In the middle figure, the two orthogonal components have the same amplitudes and are 90° out of phase. In this case, one component is zero when the other component is at maximum or minimum amplitude. There are two possible phase relationships that satisfy this requirement: the x component can be 90° ahead of the y component or it can be 90° behind the y component. In this special case, the electric vector traces out a circle in the plane, so this polarization is called circular polarization. The rotation direction in the circle depends on which of the two phase relationships exists and corresponds to *right-hand circular polarization* and *left-hand circular polarization*.

In all other cases, where the two components either do not have the same amplitudes and/or their phase difference is neither zero nor a multiple of 90°, the polarization is called elliptical polarization because the electric vector traces out an ellipse in the plane (the *polarization ellipse*). This is shown in the above figure on the right. Detailed mathematics of polarization is done using Jones calculus and is characterised by the Stokes parameters.

Changing Polarization

Media that have different indexes of refraction for different polarization modes are called *birefringent*. Well known manifestations of this effect appear in optical wave plates/retarders (linear modes) and in Faraday rotation/optical rotation (circular modes). If the path length in the birefringent medium is sufficient, plane waves will exit the material with a significantly different propagation direction, due to refraction. For example, this is the case with macroscopic crystals of calcite, which present the viewer with two offset, orthogonally polarised images of whatever is viewed through them. It was this effect that provided the first discovery of polarization, by Erasmus Bartholinus in 1669. In addition, the phase shift, and thus the change in polarization state, is usually frequency dependent, which, in combination with dichroism, often gives rise to bright colours and rainbow-like effects. In mineralogy, such properties, known as pleochroism, are frequently exploited for the purpose of identifying minerals using polarization microscopes. Additionally, many plastics that are not normally birefringent will become so when subject to mechanical stress, a phenomenon which is the basis of photoelasticity. Non-birefringent methods, to rotate the linear polarization of light beams, include the use of prismatic polarization rotators which use total internal reflection in a prism set designed for efficient collinear transmission.

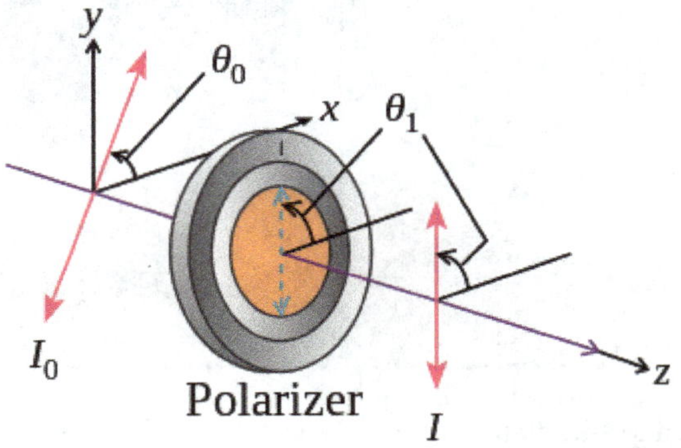

A polariser changing the orientation of linearly polarised light.
In this picture, $\theta_1 - \theta_o = \theta_i$.

Media that reduce the amplitude of certain polarization modes are called *dichroic*, with devices that block nearly all of the radiation in one mode known as *polarizing filters* or simply "polarisers". Malus' law, which is named after Étienne-Louis Malus, says that when a perfect polariser is placed in a linear polarised beam of light, the intensity, *I*, of the light that passes through is given by

$$I = I_0 \cos^2 \theta_i, \quad ,$$

where

I_0 is the initial intensity,

and θ_i is the angle between the light's initial polarization direction and the axis of the polariser.

A beam of unpolarised light can be thought of as containing a uniform mixture of linear polarizations at all possible angles. Since the average value of $\cos^2 \theta$ is 1/2, the transmission coefficient becomes

$$\frac{I}{I_0} = \frac{1}{2}$$

In practice, some light is lost in the polariser and the actual transmission of unpolarised light will be somewhat lower than this, around 38% for Polaroid-type polarisers but considerably higher (>49.9%) for some birefringent prism types.

In addition to birefringence and dichroism in extended media, polarization effects can also occur at the (reflective) interface between two materials of different refractive index. These effects are treated by the Fresnel equations. Part of the wave is transmitted and part is reflected, with the ratio depending on angle of incidence and the angle of refraction. In this way, physical optics recovers Brewster's angle. When light reflects from a thin film on a surface, interference between the reflections from the film's surfaces can produce polarization in the reflected and transmitted light.

Natural Light

The effects of a polarising filter on the sky in a photograph. Left picture is taken without polariser. For the right picture, filter was adjusted to eliminate certain polarizations of the scattered blue light from the sky.

Most sources of electromagnetic radiation contain a large number of atoms or molecules that emit light. The orientation of the electric fields produced by these emitters may not be correlated, in which case the light is said to be *unpolarised*. If there is partial correlation between the emitters, the light is *partially polarised*. If the polarization is consistent across the spectrum of the source, partially polarised light can be described as a superposition of a completely unpolarised component, and a completely polarised one. One may then describe the light in terms of the degree of polarization, and the parameters of the polarization ellipse.

Light reflected by shiny transparent materials is partly or fully polarised, except when the light is normal (perpendicular) to the surface. It was this effect that allowed the mathematician Étienne-Louis Malus to make the measurements that allowed for his development of the first mathematical models for polarised light. Polarization occurs when light is scattered in the atmosphere. The scattered light produces the brightness and colour in clear skies. This partial polarization of scattered light can be taken advantage of using polarizing filters to darken the sky in photographs. Optical polarization is principally of importance in chemistry due to circular dichroism and optical rotation ("*circular birefringence*") exhibited by optically active (chiral) molecules.

Modern Optics

Modern optics encompasses the areas of optical science and engineering that became popular in the 20th century. These areas of optical science typically relate to the electromagnetic or quantum properties of light but do include other topics. A major subfield of modern optics, quantum optics, deals with specifically quantum mechanical properties of light. Quantum optics is not just theoretical; some modern devices, such as lasers, have principles of operation that depend on quantum mechanics. Light detectors, such as photomultipliers and channeltrons, respond to individual photons. Electronic image sensors, such as CCDs, exhibit shot noise corresponding to the statistics of individual photon events. Light-emitting diodes and photovoltaic cells, too, cannot be understood without quantum mechanics. In the study of these devices, quantum optics often overlaps with quantum electronics.

Specialty areas of optics research include the study of how light interacts with specific materials as in crystal optics and metamaterials. Other research focuses on the phenomenology of electromagnetic waves as in singular optics, non-imaging optics, non-linear optics, statistical optics, and radiometry. Additionally, computer engineers have taken an interest in integrated optics, machine vision, and photonic computing as possible components of the "next generation" of computers.

Today, the pure science of optics is called optical science or optical physics to distinguish it from applied optical sciences, which are referred to as optical engineering. Prominent subfields of optical engineering include illumination engineering, photonics, and optoelectronics with practical applications like lens design, fabrication and testing of optical components, and image processing. Some of these fields overlap, with nebulous boundaries between the subjects terms that mean slightly different things in different parts of the world and in different areas of industry. A professional community of researchers in nonlinear optics has developed in the last several decades due to advances in laser technology.

Lasers

Experiments such as this one with high-power lasers are part of the modern optics research.

A laser is a device that emits light (electromagnetic radiation) through a process called *stimulated emission*. The term *laser* is an acronym for *Light Amplification by Stimulated Emission of Radiation*. Laser light is usually spatially coherent, which means that the light either is emitted in a narrow, low-divergence beam, or can be converted into one with the help of optical components such as lenses. Because the microwave equivalent of the laser, the *maser*, was developed first, devices that emit microwave and radio frequencies are usually called *masers*.

VLT's laser guided star.

The first working laser was demonstrated on 16 May 1960 by Theodore Maiman at Hughes Research Laboratories. When first invented, they were called "a solution looking for a problem". Since then, lasers have become a multibillion-dollar industry, finding utility in thousands of highly varied applications. The first application of lasers visible in the daily lives of the general population was the supermarket barcode scanner, introduced in 1974. The laserdisc player, introduced in 1978, was the first successful consumer product to include a laser, but the compact disc player was the first laser-equipped device to become truly common in consumers' homes, beginning in 1982. These optical storage devices use a semiconductor laser less than a millimetre wide to scan the surface of the disc for data retrieval. Fibre-optic communication relies on lasers to transmit large amounts of information at the speed of light. Other common applications of lasers include laser printers and laser pointers. Lasers are used in medicine in areas such as bloodless surgery, laser eye surgery, and laser capture microdissection and in military applications such as missile defence systems, electro-optical countermeasures (EOCM), and lidar. Lasers are also used in holograms, bubblegrams, laser light shows, and laser hair removal.

Kapitsa–Dirac Effect

The Kapitsa–Dirac effect causes beams of particles to diffract as the result of meeting a standing wave of light. Light can be used to position matter using various phenomena.

Applications

Optics is part of everyday life. The ubiquity of visual systems in biology indicates the central role optics plays as the science of one of the five senses. Many people benefit from eyeglasses or contact lenses, and optics are integral to the functioning of many consumer goods including cameras. Rainbows and mirages are examples of optical phenomena. Optical communication provides the backbone for both the Internet and modern telephony.

Human Eye

The human eye functions by focusing light onto a layer of photoreceptor cells called the retina, which forms the inner lining of the back of the eye. The focusing is accomplished by a series of transparent media. Light entering the eye passes first through the cornea, which provides much of the eye's optical power. The light then continues through the fluid just behind the cornea—the anterior chamber, then passes through the pupil. The light then passes through the lens, which focuses the light further and allows adjustment of focus. The light then passes through the main body of fluid in the eye—the vitreous humour, and reaches the retina. The cells in the retina line the back of the eye, except for where the optic nerve exits; this results in a blind spot.

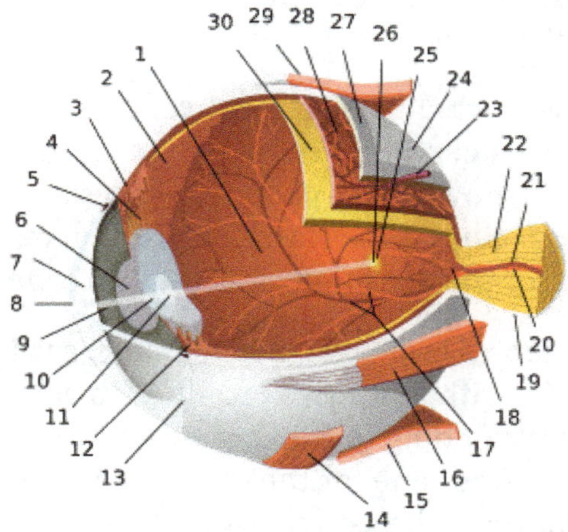

Model of a human eye. Features mentioned in this article are 3. ciliary muscle, 6. pupil, 8. cornea, 10. lens cortex, 22. optic nerve, 26. fovea, 30. retina

There are two types of photoreceptor cells, rods and cones, which are sensitive to different aspects of light. Rod cells are sensitive to the intensity of light over a wide frequency range, thus are responsible for black-and-white vision. Rod cells are not present on the fovea, the area of the retina responsible for central vision, and are not as responsive as cone cells to spatial and temporal changes in light. There are, however, twenty times more rod cells than cone cells in the retina because the rod cells are present across a wider area. Because of their wider distribution, rods are responsible for peripheral vision.

In contrast, cone cells are less sensitive to the overall intensity of light, but come in three varieties that are sensitive to different frequency-ranges and thus are used in the perception of colour and photopic vision. Cone cells are highly concentrated in the fovea and have a high visual acuity meaning that they are better at spatial resolution than rod cells. Since cone cells are not as sensitive to dim light as rod cells, most night vision is limited to rod cells. Likewise, since cone cells are in the fovea, central vision (including the vision needed to do most reading, fine detail work such as sewing, or careful examination of objects) is done by cone cells.

Ciliary muscles around the lens allow the eye's focus to be adjusted. This process is known as accommodation. The near point and far point define the nearest and farthest distances from the eye at which an object can be brought into sharp focus. For a person with normal vision, the far point is located at infinity. The near point's location depends on how much the muscles can increase the curvature of the lens, and how inflexible the lens has become with age. Optometrists, ophthalmologists, and opticians usually consider an appropriate near point to be closer than normal reading distance—approximately 25 cm.

Defects in vision can be explained using optical principles. As people age, the lens becomes less flexible and the near point recedes from the eye, a condition known as presbyopia. Similarly, people suffering from hyperopia cannot decrease the focal length of their lens enough to allow for nearby objects to be imaged on their retina. Conversely, people who cannot increase the focal length of their lens enough to allow for distant objects to be imaged on the retina suffer from myopia and have a far point that is considerably closer than infinity. A condition known as astigmatism results when the cornea is not spherical but instead is more curved in one direction. This causes horizontally extended objects to be focused on different parts of the retina than vertically extended objects, and results in distorted images.

All of these conditions can be corrected using corrective lenses. For presbyopia and hyperopia, a converging lens provides the extra curvature necessary to bring the near point closer to the eye while for myopia a diverging lens provides the curvature necessary to send the far point to infinity. Astigmatism is corrected with a cylindrical surface lens that curves more strongly in one direction than in another, compensating for the non-uniformity of the cornea.

The optical power of corrective lenses is measured in diopters, a value equal to the reciprocal of the focal length measured in metres; with a positive focal length corresponding to a converging lens and a negative focal length corresponding to a diverging lens. For lenses that correct for astigmatism as well, three numbers are given: one for the spherical power, one for the cylindrical power, and one for the angle of orientation of the astigmatism.

Visual Effects

The Ponzo Illusion relies on the fact that parallel lines appear to converge as they approach infinity.

Optical illusions (also called visual illusions) are characterized by visually perceived images that differ from objective reality. The information gathered by the eye is processed in the brain to give a percept that differs from the object being imaged. Optical illusions can be the result of a variety of phenomena including physical effects that create images that are different from the objects that make them, the physiological effects on the eyes and brain of excessive stimulation (e.g. brightness, tilt, colour, movement), and cognitive illusions where the eye and brain make unconscious inferences.

Cognitive illusions include some which result from the unconscious misapplication of certain optical principles. For example, the Ames room, Hering, Müller-Lyer, Orbison, Ponzo, Sander, and Wundt illusions all rely on the suggestion of the appearance of distance by using converging and diverging lines, in the same way that parallel light rays (or indeed any set of parallel lines) appear to converge at a vanishing point at infinity in two-dimensionally rendered images with artistic perspective. This suggestion is also responsible for the famous moon illusion where the moon, despite having essentially the same angular size, appears much larger near the horizon than it does at zenith. This illusion so confounded Ptolemy that he incorrectly attributed it to atmospheric refraction when he described it in his treatise, *Optics*.

Another type of optical illusion exploits broken patterns to trick the mind into perceiving symmetries or asymmetries that are not present. Examples include the café wall, Ehrenstein, Fraser spiral, Poggendorff, and Zöllner illusions. Related, but not strictly illusions, are patterns that occur due to the superimposition of periodic structures. For example, transparent tissues with a grid structure produce shapes known as moiré patterns, while the superimposition of periodic transparent patterns comprising parallel opaque lines or curves produces line moiré patterns.

Optical Instruments

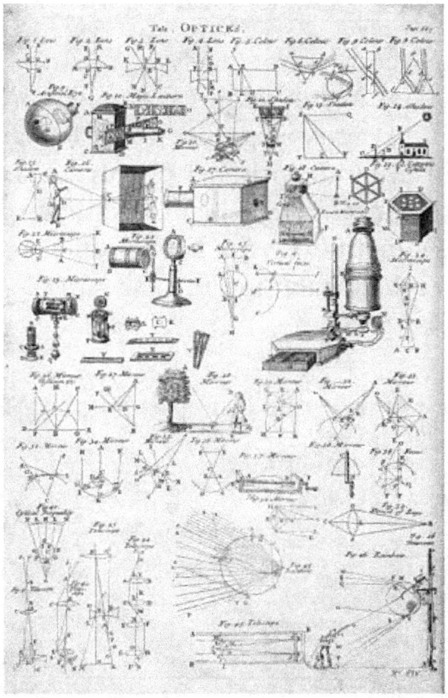

Illustrations of various optical instruments

Single lenses have a variety of applications including photographic lenses, corrective lenses, and magnifying glasses while single mirrors are used in parabolic reflectors and rear-view mirrors. Combining a number of mirrors, prisms, and lenses produces compound optical instruments which have practical uses. For example, a periscope is simply two plane mirrors aligned to allow for viewing around obstructions. The most famous compound optical instruments in science are the microscope and the telescope which were both invented by the Dutch in the late 16th century.

Microscopes were first developed with just two lenses: an objective lens and an eyepiece. The objective lens is essentially a magnifying glass and was designed with a very small focal length while the eyepiece generally has a longer focal length. This has the effect of producing magnified images of close objects. Generally, an additional source of illumination is used since magnified images are dimmer due to the conservation of energy and the spreading of light rays over a larger surface area. Modern microscopes, known as *compound microscopes* have many lenses in them (typically four) to optimize the functionality and enhance image stability. A slightly different variety of microscope, the comparison microscope, looks at side-by-side images to produce a stereoscopic binocular view that appears three dimensional when used by humans.

The first telescopes, called *refracting telescopes* were also developed with a single objective and eyepiece lens. In contrast to the microscope, the objective lens of the telescope was designed with a large focal length to avoid optical aberrations. The objective focuses an image of a distant object at its focal point which is adjusted to be at the focal point of an eyepiece of a much smaller focal length. The main goal of a telescope is not necessarily magnification, but rather collection of light which is determined by the physical size of the objective lens. Thus, telescopes are normally indicated by the diameters of their objectives rather than by the magnification which can be changed by switching eyepieces. Because the magnification of a telescope is equal to the focal length of the objective divided by the focal length of the eyepiece, smaller focal-length eyepieces cause greater magnification.

Since crafting large lenses is much more difficult than crafting large mirrors, most modern telescopes are *reflecting telescopes*, that is, telescopes that use a primary mirror rather than an objective lens. The same general optical considerations apply to reflecting telescopes that applied to refracting telescopes, namely, the larger the primary mirror, the more light collected, and the magnification is still equal to the focal length of the primary mirror divided by the focal length of the eyepiece. Professional telescopes generally do not have eyepieces and instead place an instrument (often a charge-coupled device) at the focal point instead.

Photography

Photograph taken with aperture $f/32$

Photograph taken with aperture $f/5$

The optics of photography involves both lenses and the medium in which the electromagnetic radiation is recorded, whether it be a plate, film, or charge-coupled device. Photographers must consider the reciprocity of the camera and the shot which is summarized by the relation

$$\text{Exposure} \propto \text{ApertureArea} \times \text{ExposureTime} \times \text{SceneLuminance}$$

In other words, the smaller the aperture (giving greater depth of focus), the less light coming in, so the length of time has to be increased (leading to possible blurriness if motion occurs). An example of the use of the law of reciprocity is the Sunny 16 rule which gives a rough estimate for the settings needed to estimate the proper exposure in daylight.

A camera's aperture is measured by a unitless number called the f-number or f-stop, $f/\#$, often notated as N, and given by

$$f/\# = N = \frac{f}{D}$$

where f is the focal length, and D is the diameter of the entrance pupil. By convention, "$f/\#$" is treated as a single symbol, and specific values of $f/\#$ are written by replacing the number sign with the value. The two ways to increase the f-stop are to either decrease the diameter of the entrance pupil or change to a longer focal length (in the case of a zoom lens, this can be done by simply adjusting the lens). Higher f-numbers also have a larger depth of field due to the lens approaching the limit of a pinhole camera which is able to focus all images perfectly, regardless of distance, but requires very long exposure times.

The field of view that the lens will provide changes with the focal length of the lens. There are three basic classifications based on the relationship to the diagonal size of the film or sensor size of the camera to the focal length of the lens:

- Normal lens: angle of view of about 50° (called *normal* because this angle considered roughly equivalent to human vision) and a focal length approximately equal to the diagonal of the film or sensor.
- Wide-angle lens: angle of view wider than 60° and focal length shorter than a normal lens.
- Long focus lens: angle of view narrower than a normal lens. This is any lens with a focal length longer than the diagonal measure of the film or sensor. The most common type of long focus lens is the telephoto lens, a design that uses a special *telephoto group* to be physically shorter than its focal length.

Modern zoom lenses may have some or all of these attributes.

The absolute value for the exposure time required depends on how sensitive to light the medium being used is (measured by the film speed, or, for digital media, by the quantum efficiency). Early photography used media that had very low light sensitivity, and so exposure times had to be long even for very bright shots. As technology has improved, so has the sensitivity through film cameras and digital cameras.

Other results from physical and geometrical optics apply to camera optics. For example, the maximum resolution capability of a particular camera set-up is determined by the diffraction limit associated with the pupil size and given, roughly, by the Rayleigh criterion.

Atmospheric Optics

The unique optical properties of the atmosphere cause a wide range of spectacular optical phenomena. The blue colour of the sky is a direct result of Rayleigh scattering which redirects higher frequency (blue) sunlight back into the field of view of the observer. Because blue light is scattered more easily than red light, the sun takes on a reddish hue when it is observed through a thick atmosphere, as during a sunrise or sunset. Additional particulate matter in the sky can scatter different colours at different angles creating colourful glowing skies at dusk and dawn. Scattering off of ice crystals and other particles in the atmosphere are responsible for halos, afterglows, coronas, rays of sunlight, and sun dogs. The variation in these kinds of phenomena is due to different particle sizes and geometries.

A colourful sky is often due to scattering of light off particulates and pollution

Mirages are optical phenomena in which light rays are bent due to thermal variations in the refraction index of air, producing displaced or heavily distorted images of distant objects. Other dramatic optical phenomena associated with this include the Novaya Zemlya effect where the sun appears to rise earlier than predicted with a distorted shape. A spectacular form of refraction occurs with a temperature inversion called the Fata Morgana where objects on the horizon or even beyond the horizon, such as islands, cliffs, ships or icebergs, appear elongated and elevated, like "fairy tale castles".

Rainbows are the result of a combination of internal reflection and dispersive refraction of light in raindrops. A single reflection off the backs of an array of raindrops produces a rainbow with an angular size on the sky that ranges from 40° to 42° with red on the outside. Double rainbows are produced by two internal reflections with angular size of 50.5° to 54° with violet on the outside. Because rainbows are seen with the sun 180° away from the centre of the rainbow, rainbows are more prominent the closer the sun is to the horizon.

Astrophysics

Astrophysics is the branch of physics that studies phenomena on large scales: the sun, the planets, the stars, the galaxies, the matter and energy that are between them, and the universe as a whole.

Astrophysical research requires familiarity and use of all branches of physics: high-energy, nuclear, atomic, and condensed-matter physics, thermodynamics and hydrodynamics, special and general relativity, to name only some.

The broad aim of astrophysics is to see what the universe is made of, to understand how physical processes have made it the way it is, and to explain how it works and what will be its future evolution. Some examples include: (i) the geometry of space-time and the components of the universe -- radiation, normal matter, dark matter, and dark energy; (ii) the formation of the first stars and structures after the Big Bang; (iii) black holes of all scales and their role in the formation and development of galaxies; stellar evolution, cosmic explosions, compact stellar remnants, and their contributions to element production, and to the creation and acceleration of high-energy cosmic particles; (iv) interstellar processes, including the formation of complex molecules, dust particles, and stars; and (v) the discovery and understanding of planets, their properties and habitability to life, their formation and destruction.

Observational Astrophysics

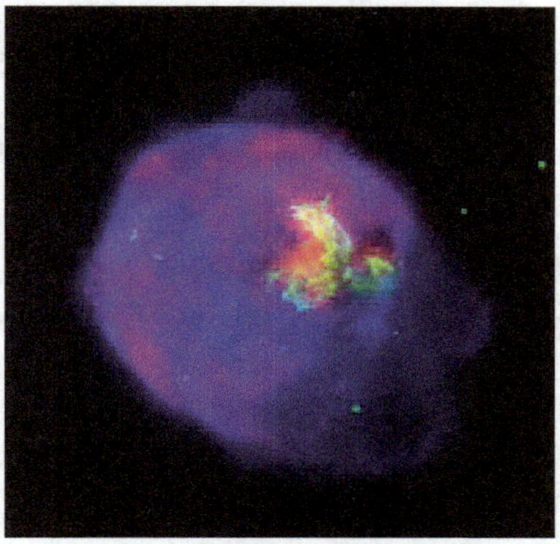

Supernova remnant LMC N 63A imaged in x-ray (blue), optical (green) and radio (red) wavelengths. The X-ray glow is from material heated to about ten million degrees Celsius by a shock wave generated by the supernova explosion.

Observational astronomy is a division of the astronomical science that is concerned with recording data, in contrast with theoretical astrophysics, which is mainly concerned with finding out the measurable implications of physical models. It is the practice of observing celestial objects by using telescopes and other astronomical apparatus.

The majority of astrophysical observations are made using the electromagnetic spectrum.

- Radio astronomy studies radiation with a wavelength greater than a few millimeters. Example areas of study are radio waves, usually emitted by cold objects such as interstellar gas and dust clouds; the cosmic microwave background radiation which is the redshifted light from the Big Bang; pulsars, which were first detected at microwave frequencies. The study of these waves requires very large radio telescopes.
- Infrared astronomy studies radiation with a wavelength that is too long to be visible to the naked eye but is shorter than radio waves. Infrared observations are usually made with telescopes similar to the familiar optical telescopes. Objects colder than stars (such as planets) are normally studied at infrared frequencies.
- Optical astronomy is the oldest kind of astronomy. Telescopes paired with a charge-coupled device or spectroscopes are the most common instruments used. The Earth's atmosphere

interferes somewhat with optical observations, so adaptive optics and space telescopes are used to obtain the highest possible image quality. In this wavelength range, stars are highly visible, and many chemical spectra can be observed to study the chemical composition of stars, galaxies and nebulae.

- Ultraviolet, X-ray and gamma ray astronomy study very energetic processes such as binary pulsars, black holes, magnetars, and many others. These kinds of radiation do not penetrate the Earth's atmosphere well. There are two methods in use to observe this part of the electromagnetic spectrum—space-based telescopes and ground-based imaging air Cherenkov telescopes (IACT). Examples of Observatories of the first type are RXTE, the Chandra X-ray Observatory and the Compton Gamma Ray Observatory. Examples of IACTs are the High Energy Stereoscopic System (H.E.S.S.) and the MAGIC telescope.

Other than electromagnetic radiation, few things may be observed from the Earth that originate from great distances. A few gravitational wave observatories have been constructed, but gravitational waves are extremely difficult to detect. Neutrino observatories have also been built, primarily to study our Sun. Cosmic rays consisting of very high energy particles can be observed hitting the Earth's atmosphere.

Observations can also vary in their time scale. Most optical observations take minutes to hours, so phenomena that change faster than this cannot readily be observed. However, historical data on some objects is available, spanning centuries or millennia. On the other hand, radio observations may look at events on a millisecond timescale (millisecond pulsars) or combine years of data (pulsar deceleration studies). The information obtained from these different timescales is very different.

The study of our very own Sun has a special place in observational astrophysics. Due to the tremendous distance of all other stars, the Sun can be observed in a kind of detail unparalleled by any other star. Our understanding of our own Sun serves as a guide to our understanding of other stars.

The topic of how stars change, or stellar evolution, is often modeled by placing the varieties of star types in their respective positions on the Hertzsprung–Russell diagram, which can be viewed as representing the state of a stellar object, from birth to destruction.

Theoretical Astrophysics

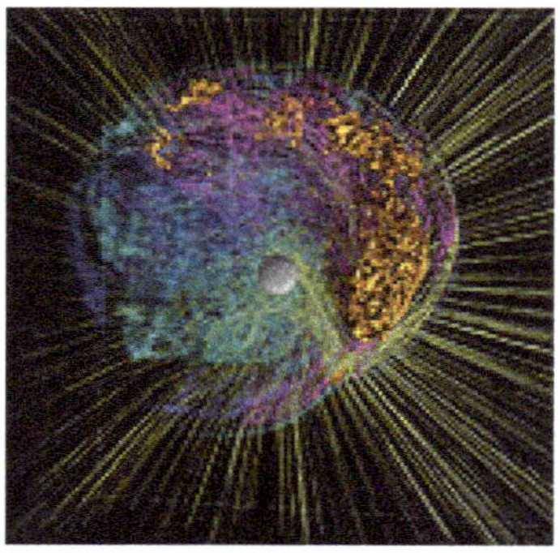

Stream lines on this simulation of a supernova show the flow of matter behind the shock wave giving clues as to the origin of pulsars

Theoretical astrophysicists use a wide variety of tools which include analytical models (for example, polytropes to approximate the behaviors of a star) and computational numerical simulations. Each has some advantages. Analytical models of a process are generally better for giving insight into the heart of what is going on. Numerical models can reveal the existence of phenomena and effects that would otherwise not be seen.

Theorists in astrophysics endeavor to create theoretical models and figure out the observational consequences of those models. This helps allow observers to look for data that can refute a model or help in choosing between several alternate or conflicting models.

Theorists also try to generate or modify models to take into account new data. In the case of an inconsistency, the general tendency is to try to make minimal modifications to the model to fit the data. In some cases, a large amount of inconsistent data over time may lead to total abandonment of a model.

Topics studied by theoretical astrophysicists include: stellar dynamics and evolution; galaxy formation and evolution; magnetohydrodynamics; large-scale structure of matter in the universe; origin of cosmic rays; general relativity and physical cosmology, including string cosmology and astroparticle physics. Astrophysical relativity serves as a tool to gauge the properties of large scale structures for which gravitation plays a significant role in physical phenomena investigated and as the basis for black hole (*astro*)physics and the study of gravitational waves.

Some widely accepted and studied theories and models in astrophysics, now included in the Lambda-CDM model, are the Big Bang, cosmic inflation, dark matter, dark energy and fundamental theories of physics. Wormholes are examples of hypotheses which are yet to be proven (or disproven).

Relativity

Special Relativity

The theory of special relativity explains how space and time are linked for objects that are moving at a consistent speed in a straight line. One of its most famous aspects concerns objects moving at the speed of light.

Simply put, as an object approaches the speed of light, its mass becomes infinite and it is unable to go any faster than light travels. This cosmic speed limit has been a subject of much discussion in physics, and even in science fiction, as people think about how to travel across vast distances.

The theory of special relativity was developed by Albert Einstein in 1905, and it forms part of the basis of modern physics. After finishing his work in special relativity, Einstein spent a decade pondering what would happen if one introduced acceleration. This formed the basis of his general relativity, published in 1915.

Before Einstein, astronomers (for the most part) understood the universe in terms of three laws of motion presented by Isaac Newton in 1686. These three laws are:

(1) Objects in motion (or at rest) remain in motion (or at rest) unless an external force imposes change.

(2) Force is equal to the change in momentum per change of time. For a constant mass, force equals mass times acceleration.

(3) For every action, there is an equal and opposite reaction.

But there were cracks in the theory for decades before Einstein's arrival on the scene. In 1865, Scottish physicist James Clerk Maxwell demonstrated that light is a wave with both electrical and magnetic components, and established the speed of light (186,000 miles per second). Scientists supposed that the light had to be transmitted through some medium, which they called the ether. (We now know that no transmission medium is required, and that light in space moves in a vacuum.)

Twenty years later, an unexpected result threw this into question. Physicist A.A. Michelson and chemist Edward Morley (both Americans at the time) calculated how Earth's motion through this "ether" affected how the speed of light is measured, and found that the speed of light is the same no matter what Earth's motion is. This led to further musings on light's behavior — and its incongruence with classical mechanics — by Austrian physicist Ernst Mach and French mathematician Henri Poincare.

Einstein began thinking of light's behavior when he was just 16 years old, in 1895. He did a thought experiment, where he rode on one light wave and looked at another light wave moving parallel to him.

Classical physics should say that the light wave Einstein was looking at would have a relative speed of zero, but this contradicted Maxwell's equations that showed light always has the same speed: 186,000 miles a second. Another problem with relative speeds is they would show that the laws of electromagnetism change depending on your vantage point, which contradicted classical physics as well (which said the laws of physics were the same for everyone.)

This led to Einstein's eventual musings on the theory of special relativity, which he broke down into the everyday example of a person standing beside a moving train, comparing observations with a person inside the train. He imagined the train being at a point in the track equally between two trees. If a bolt of lightning hit both trees at the same time, due to the motion of the train, the person on the train would see the bolt hit one tree before the other tree. But the person beside the track would see simultaneous strikes.

Einstein concluded that simultaneity is relative; events that are simultaneous for one observer may not be for another. This led him to the counterintuitive idea that time flows differently according to the state of motion, and to the conclusion that distance is also relative.

Famous Equation

Einstein's work led to some startling results, which today still seem counterintuitive at first glance even though his physics is usually introduced at the high school level.

One of the most famous equations in mathematics comes from special relativity. The equation — $E = mc^2$ — means "energy equals mass times the speed of light squared." It shows that energy (E) and mass (m) are interchangeable; they are different forms of the same thing. If mass is somehow totally converted into energy, it also shows how much energy would reside inside that mass: quite a lot. (This equation is one of the demonstrations for why an atomic bomb is so powerful, once its mass is converted to an explosion.)

This equation also shows that mass increases with speed, which effectively puts a speed limit on how fast things can move in the universe. Simply put, the speed of light (c) is the fastest velocity at which an object can travel in a vacuum. As an object moves, its mass also increases. Near the speed of light, the mass is so high that it reaches infinity, and would require infinite energy to move it, thus capping how fast an object can move. The only reason light moves at the speed it does is because photons, the quantum particles that make up light, have a mass of zero.

A special situation in the universe of the small, called "quantum entanglement," is confusing because it seems to involve quantum particles interacting with each other at speeds faster than the speed of light. Specifically, measuring the property of one particle can instantly tell you the property of another particle, no matter how far away they are.

Another strange conclusion of Einstein's work comes from the realization that time moves relative to the observer. An object in motion experiences time dilation, meaning that time moves more slowly when one is moving, than when one is standing still. Therefore, a person moving ages more slowly than a person at rest. So yes, when astronaut Scott Kelly spent nearly a year aboard the International Space Station in 2015-16, his twin astronaut brother Mark Kelly aged a little faster than Scott.

This becomes extremely apparent at speeds approaching the speed of light. Imagine a 15-year-old traveling at 99.5 percent the speed of light for five years (from the astronaut's perspective). When the 15-year-old gets back to Earth, according to NASA, he would be only 20 years old. His classmates, however, would be 65 years old.

While this time dilation sounds very theoretical, it does have practical applications as well. If you have a Global Positioning Satellite (GPS) receiver in your car, the receiver attempts to find signals from at least three satellites to coordinate your position. The GPS satellites send out timed radio signals that the receiver listens to, triangulating (or more properly speaking, trilaterating) its position based on the travel time of the signals. The challenge is, the atomic clocks on the GPS are moving and would therefore run faster than atomic clocks on Earth, creating timing issues. So, engineers need to make the clocks on a GPS tick slower, according to Richard Pogge, an astronomer at Ohio State University.

The clocks in space tick faster, according to Physics Central, because the GPS satellites are above Earth and experience weaker gravity. So even though the GPS satellites are moving and experience a seven-microsecond slowing every day because of their movement, the result of the weaker gravity causes the clocks to tick about 45 microseconds faster than a ground-based clock. Adding the two together results in the GPS satellite clock ticking faster than a ground-based clock, by about 38 microseconds daily.

Special Relativity and Quantum Mechanics

As our knowledge of physics has advanced, scientists have run into more counterintuitive situations. One is trying to reconcile general relativity — which describes well what's going on with large objects — with quantum mechanics, which is best used for very small things (such as uranium atom decay). The two fields, which excellently describe their individual fields, are incompatible with one another — which frustrated Einstein and generations of scientists after him.

"Relativity gives nonsensical answers when you try to scale it down to quantum size, eventually descending to infinite values in its description of gravity. Likewise, quantum mechanics runs into serious trouble when you blow it up to cosmic dimensions," an article in The Guardian pointed out in 2015.

"Quantum fields carry a certain amount of energy, even in seemingly empty space, and the amount of energy gets bigger as the fields get bigger. According to Einstein, energy and mass are equivalent (that's the message of $E=mc^2$), so piling up energy is exactly like piling up mass. Go big enough, and the amount of energy in the quantum fields becomes so great that it creates a black hole that causes the universe to fold in on itself."

There are several ideas to overcome this, but one approach is to imagine a quantum theory of gravity that would have a massless particle (called the graviton) to generate the force. But as physicist Dave Goldberg pointed out in io9 in 2013, there are problems with that. At the smallest scales, gravitons would have infinite energy density, creating an unimaginably powerful gravity field.

General Relativity

In 1905, Albert Einstein determined that the laws of physics are the same for all non-accelerating observers, and that the speed of light in a vacuum was independent of the motion of all observers. This was the theory of special relativity. It introduced a new framework for all of physics and proposed new concepts of space and time.

Einstein then spent 10 years trying to include acceleration in the theory and published his theory of general relativity in 1915. In it, he determined that massive objects cause a distortion in space-time, which is felt as gravity.

The Tug of Gravity

Two objects exert a force of attraction on one another known as "gravity." Sir Isaac Newton quantified the gravity between two objects when he formulated his three laws of motion. The force tugging between two bodies depends on how massive each one is and how far apart the two lie. Even as the center of the Earth is pulling you toward it (keeping you firmly lodged on the ground), your center of mass is pulling back at the Earth. But the more massive body barely feels the tug from you, while with your much smaller mass you find yourself firmly rooted thanks to that same force. Yet Newton's laws assume that gravity is an innate force of an object that can act over a distance.

Albert Einstein, in his theory of special relativity, determined that the laws of physics are the same for all non-accelerating observers, and he showed that the speed of light within a vacuum is the same no matter the speed at which an observer travels. As a result, he found that space and time were interwoven into a single continuum known as space-time. Events that occur at the same time for one observer could occur at different times for another.

As he worked out the equations for his general theory of relativity, Einstein realized that massive objects caused a distortion in space-time. Imagine setting a large body in the center of a trampoline. The body would press down into the fabric, causing it to dimple. A marble rolled around the edge would spiral inward toward the body, pulled in much the same way that the gravity of a planet pulls at rocks in space.

Condensed Matter Physics

Condensed matter physics (CMP) is the fundamental science of solids and liquids. As the largest branch of physics, it has the greatest impact on our daily lives by providing foundations for technology developments. For example, the invention of transistors and semiconductor chips have led to the widespread use of a variety of data storage, telecommunication, and multi-media devices (e.g., cellular phones, digital still and video cameras, MP3 players, GPS, flat panel HDTVs, DVDs, hard disk drives, flash drives, Blue-ray players, iPod and iPad, etc.), and personal computers. Almost every aspect of our daily life benefits from research in condensed matter physics: for example, composite materials are used in jet turbines and modern tennis rackets; superconducting magnets are used in MRI tomography for medical diagnostics; various solid state sensors and detectors for space exploration and homeland security, and lasers for entertainment, broadband communication, and various medical procedures.

Today, condensed matter physics is one of the most active and exciting research area in both basic sciences and technological applications. At the fundamental level, condensed matter physics is intellectually stimulating due to the continuing discoveries of many new phenomena and the development of new concepts and tools that are necessary to understand them. It is the field in which advances in theory can most directly be confronted with experiments. It has repeatedly served as a source or testing ground for new ideas (e.g., Josephson effect, integer and fractional quantum Hall effects, Aharanov-Bohm effect, mechanism of high-Tc superconductors, dissipative quantum physics, critical phenomena, mesoscopic physics, nonlinear dynamics, etc.). As a result, over the past 50 years, 22 Nobel Prizes in Physics were awarded to condensed matter physics and related areas, and 5 Nobel Prizes in Chemistry were awarded for subjects in condensed matter physics. In addition to its scientific value, condensed matter physics is intimately connected with industry. Condensed matter physics is one of the foundations of most modern technologies, like energy, information, defense, and manufacturing. A large number of scientists trained in condensed matter physics work in industry and found the training they received in university very rewarding.

Superconductivity

Superconductivity was discovered in 1911. It is a phenomenon of great intricacy, diversity, and elegance. It is one of the most interesting and challenging subfields of CMP. For instance, the mechanism of high-Tc superconductors remains unsolved despite two decades of persistent efforts by some of the leading scientists in the world. On the other hand, significant progress has been made in the technical application of superconductivity. Large scale applications include the world's fastest experimental magnetically levitated train, superconducting magnets in MRI (Magnetic Resonance Image) systems, the world's largest electromagnets for thermonuclear fusion experiments, and bending and focusing magnets for the world's most powerful particle accelerators. Small-scale electronic applications include the fastest operating and the least-power consuming digital logic devices and circuits - the Rapid Single Flux Quanta (RSFQ) logic family, the most sensitive and lowest-noise electronic and magnetic sensors (SQUIDs), the most accurate voltage standard (the Josephson voltage standard), and the highest resolution x-ray detectors, just to name a few. Superconducting devices are also studied in our group in the context of quantum computing.

Theoretical

Theoretical condensed matter physics involves the use of theoretical models to understand properties of states of matter. These include models to study the electronic properties of solids, such as the Drude model, the Band structure and the density functional theory. Theoretical models have also been developed to study the physics of phase transitions, such as the Ginzburg–Landau theory, critical exponents and the use of mathematical methods of quantum field theory and the renormalization group. Modern theoretical studies involve the use of numerical computation of electronic structure and mathematical tools to understand phenomena such as high-temperature superconductivity, topological phases, and gauge symmetries.

Emergence

Theoretical understanding of condensed matter physics is closely related to the notion of emergence, wherein complex assemblies of particles behave in ways dramatically different from their individual constituents. For example, a range of phenomena related to high temperature superconductivity are understood poorly, although the microscopic physics of individual electrons and lattices is well known. Similarly, models of condensed matter systems have been studied where collective excitations behave like photons and electrons, thereby describing electromagnetism as an emergent phenomenon. Emergent properties can also occur at the interface between materials: one example is the lanthanum aluminate-strontium titanate interface, where two non-magnetic insulators are joined to create conductivity, superconductivity, and ferromagnetism.

Electronic Theory of Solids

The metallic state has historically been an important building block for studying properties of solids. The first theoretical description of metals was given by Paul Drude in 1900 with the Drude model, which explained electrical and thermal properties by describing a metal as an ideal gas of then-newly discovered electrons. He was able to derive the empirical Wiedemann-Franz law and get results in close agreement with the experiments. This classical model was then improved by Arnold Sommerfeld who incorporated the Fermi–Dirac statistics of electrons and was able to explain the anomalous behavior of the specific heat of metals in the Wiedemann–Franz law. In 1912, The structure of crystalline solids was studied by Max von Laue and Paul Knipping, when they observed the X-ray diffraction pattern of crystals, and concluded that crystals get their structure from periodic lattices of atoms. In 1928, Swiss physicist Felix Bloch provided a wave function solution to the Schrödinger equation with a periodic potential, called the Bloch wave.

Calculating electronic properties of metals by solving the many-body wavefunction is often computationally hard, and hence, approximation methods are needed to obtain meaningful predictions. The Thomas–Fermi theory, developed in the 1920s, was used to estimate system energy and electronic density by treating the local electron density as a variational parameter. Later in the 1930s, Douglas Hartree, Vladimir Fock and John Slater developed the so-called Hartree–Fock wavefunction as an improvement over the Thomas–Fermi model. The Hartree–Fock method accounted for exchange statistics of single particle electron wavefunctions. In general, it's very difficult to solve the Hartree–Fock equation. Only the free electron gas case can be solved exactly. Finally in 1964–65, Walter Kohn, Pierre Hohenberg and Lu Jeu Sham proposed the density functional theory which gave realistic descriptions for bulk and surface properties of metals. The density functional theory (DFT) has been widely used since the 1970s for band structure calculations of variety of solids.

Symmetry Breaking

Some states of matter exhibit symmetry breaking, where the relevant laws of physics possess some symmetry that is broken. A common example is crystalline solids, which break continuous translational symmetry. Other examples include magnetized ferromagnets, which break rotational symmetry, and more exotic states such as the ground state of a BCS superconductor, that breaks U(1) phase rotational symmetry.

Goldstone's theorem in quantum field theory states that in a system with broken continuous symmetry, there may exist excitations with arbitrarily low energy, called the Goldstone bosons. For example, in crystalline solids, these correspond to phonons, which are quantized versions of lattice vibrations.

Phase Transition

Phase transition refers to the change of phase of a system, which is brought about by change in an external parameter such as temperature. Classical phase transition occurs at finite temperature when the order of the system was destroyed. For example, when ice melts and becomes water, the ordered crystal structure is destroyed. In quantum phase transitions, the temperature is set to absolute zero, and the non-thermal control parameter, such as pressure or magnetic field, causes the phase transitions when order is destroyed by quantum fluctuations originating from the Heisenberg uncertainty principle. Here, the different quantum phases of the system refer to distinct ground states of the Hamiltonian. Understanding the behavior of quantum phase transition is important in the difficult tasks of explaining the properties of rare-earth magnetic insulators, high-temperature superconductors, and other substances.

Two classes of phase transitions occur: first-order transitions and continuous transitions. For the later, the two phases involved do not co-exist at the transition temperature, also called critical point. Near the critical point, systems undergo critical behavior, wherein several of their properties such as correlation length, specific heat, and magnetic susceptibility diverge exponentially. These critical phenomena poses serious challenges to physicists because normal macroscopic laws are no longer valid in the region and novel ideas and methods must be invented to find the new laws that can describe the system.

The simplest theory that can describe continuous phase transitions is the Ginzburg–Landau theory, which works in the so-called mean field approximation. However, it can only roughly explain continuous phase transition for ferroelectrics and type I superconductors which involves long range microscopic interactions. For other types of systems that involves short range interactions near the critical point, a better theory is needed.

Near the critical point, the fluctuations happen over broad range of size scales while the feature of the whole system is scale invariant. Renormalization group methods successively average out the shortest wavelength fluctuations in stages while retaining their effects into the next stage. Thus, the changes of a physical system as viewed at different size scales can be investigated systematically. The methods, together with powerful computer simulation, contribute greatly to the explanation of the critical phenomena associated with continuous phase transition.

Experimental

Experimental condensed matter physics involves the use of experimental probes to try to discover new properties of materials. Such probes include effects of electric and magnetic fields, measuring response functions, transport properties and thermometry. Commonly used experimental methods include spectroscopy, with probes such as X-rays, infrared light and inelastic neutron scattering; study of thermal response, such as specific heat and measuring transport via thermal and heat conduction.

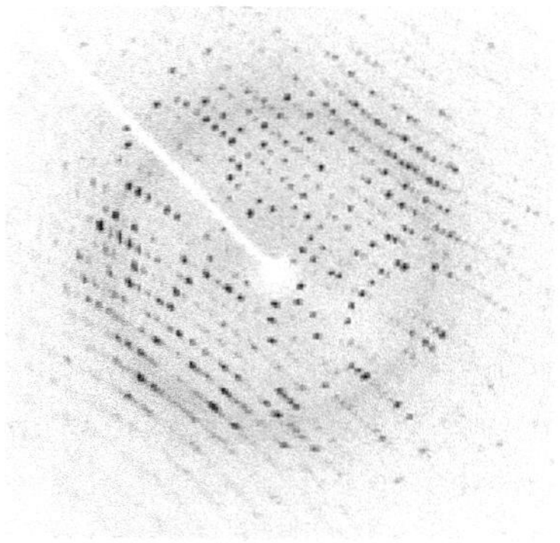

Image of X-ray diffraction pattern from a protein crystal.

Scattering

Several condensed matter experiments involve scattering of an experimental probe, such as X-ray, optical photons, neutrons, etc., on constituents of a material. The choice of scattering probe depends on the observation energy scale of interest. Visible light has energy on the scale of 1 electron volt (eV) and is used as a scattering probe to measure variations in material properties such as dielectric constant and refractive index. X-rays have energies of the order of 10 keV and hence are able to probe atomic length scales, and are used to measure variations in electron charge density.

Neutrons can also probe atomic length scales and are used to study scattering off nuclei and electron spins and magnetization (as neutrons have spin but no charge). Coulomb and Mott scattering measurements can be made by using electron beams as scattering probes. Similarly, positron annihilation can be used as an indirect measurement of local electron density. Laser spectroscopy is an excellent tool for studying the microscopic properties of a medium, for example, to study forbidden transitions in media with nonlinear optical spectroscopy.

External Magnetic Fields

In experimental condensed matter physics, external magnetic fields act as thermodynamic variables that control the state, phase transitions and properties of material systems. Nuclear magnetic resonance (NMR) is a method by which external magnetic fields are used to find resonance modes of individual electrons, thus giving information about the atomic, molecular, and bond structure of their neighborhood. NMR experiments can be made in magnetic fields with strengths up to 60 Tesla. Higher magnetic fields can improve the quality of NMR measurement data. Quantum oscillations is another experimental method where high magnetic fields are used to study material properties such as the geometry of the Fermi surface. High magnetic fields will be useful in experimentally testing of the various theoretical predictions such as the quantized magnetoelectric effect, image magnetic monopole, and the half-integer quantum Hall effect.

Cold Atomic Gases

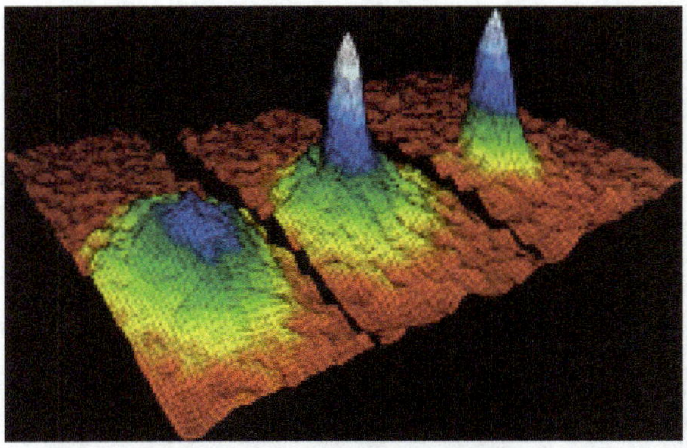

The first Bose–Einstein condensate observed in a gas of ultracold rubidium atoms.
The blue and white areas represent higher density.

Ultracold atom trapping in optical lattices is an experimental tool commonly used in condensed matter physics, and in atomic, molecular, and optical physics. The method involves using optical lasers to form an interference pattern, which acts as a *lattice*, in which ions or atoms can be placed at very low temperatures. Cold atoms in optical lattices are used as *quantum simulators*, that is, they act as controllable systems that can model behavior of more complicated systems, such as frustrated magnets. In particular, they are used to engineer one-, two- and three-dimensional lattices for a Hubbard model with pre-specified parameters, and to study phase transitions for antiferromagnetic and spin liquid ordering.

In 1995, a gas of rubidium atoms cooled down to a temperature of 170 nK was used to experimentally realize the Bose–Einstein condensate, a novel state of matter originally predicted by S. N. Bose and Albert Einstein, wherein a large number of atoms occupy one quantum state.

Applications

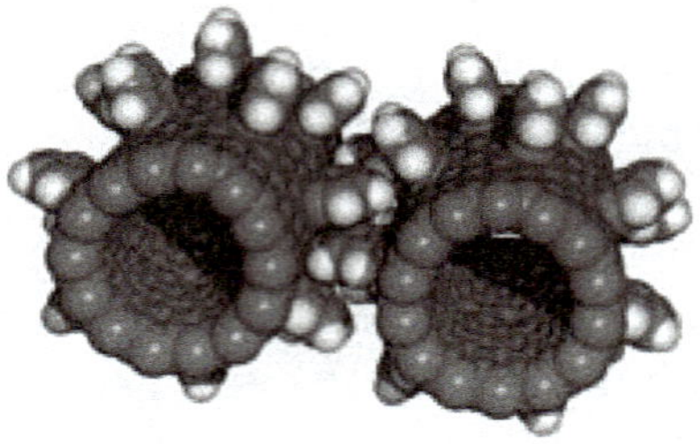

Computer simulation of *nanogears* made of fullerene molecules. It is hoped that advances in nanoscience will lead to machines working on the molecular scale.

Research in condensed matter physics has given rise to several device applications, such as the development of the semiconductor transistor, laser technology, and several phenomena studied in the context of nanotechnology. Methods such as scanning-tunneling microscopy can be used to control processes at the nanometer scale, and have given rise to the study of nanofabrication.

In quantum computation, information is represented by quantum bits, or qubits. The qubits may decohere quickly before useful computation is completed. This serious problem must be solved before quantum computing may be realized. To solve this problem, several promising approaches are proposed in condensed matter physics, including Josephson junction qubits, spintronic qubits using the spin orientation of magnetic materials, or the topological non-Abelian anyons from fractional quantum Hall effect states.

Condensed matter physics also has important uses for biophysics, for example, the experimental method of magnetic resonance imaging, which is widely used in medical diagnosis.

Particle Physics

Particle physics is the study of the fundamental particulate constituents of nature. Our knowledge of these constituents is important to understand the laws that shape our universe, how they manifest their will, and why things are the way they are. Because of the scale at which its experiments are conducted and the diverse branches of physics and mathematics its theories draw from, particle physics has many applications – ranging from solid-state physics to medical diagnostics to distributed computing.

Fundamental Particles

Fundamental, or elementary, particles are the smallest constituents of something and aren't made up of even smaller particles. Particle physics addresses both matter and force in terms of particles, so fundamental particles come in two types as well. Bosons are force-carriers: if a particle is acted upon by a force, then it can also be understood as the particle interacting with the particulate carriers of that force. For example: if the weak force is acting on a particle, then the particle is interacting with W± and Z bosons, which are carriers of the weak force. Fermions are matter-makers: they come together under the influence of others like them as well as bosons to make up matter (or antimatter).

However, a deeper classification of bosons and fermions alludes to their collective properties when they accumulate in large numbers. All particles have a spin quantum number associated with their angular momentum, expressed as integer fractions (0, 1/2, 1, 3/2, ...). All fermions have half-integer spins (1/2, 3/2, ...) and their behaviour is thus defined by the rules of Fermi-Dirac statistics. All bosons have integer spins (0, 1, 2, ...) and their behaviour is thus defined by the rules of Bose-Einstein statistics.

One commonly cited difference between the statistics is that bosons don't obey Pauli's exclusion principle while fermions do. The principle states that, given all fermions can be accorded four quantum numbers that define the particles' states, no two particles can have the same four numbers at the same time. And because bosons can possess such identical quantum numbers, they can form Bose-Einstein condensates.

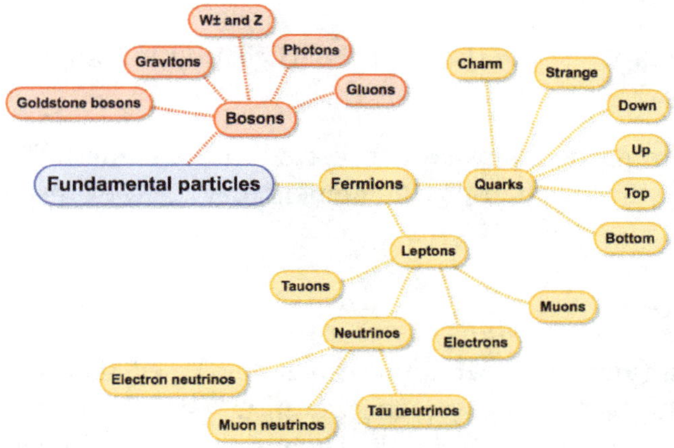

Standard Model

The current state of the classification of all elementary particles is explained by the Standard Model. It describes the strong, weak, and electromagnetic fundamental interactions, using mediating gauge bosons. The species of gauge bosons are eight gluons, W^-, W^+ and Z bosons, and the photon. The Standard Model also contains 24 fundamental fermions (12 particles and their associated anti-particles), which are the constituents of all matter. Finally, the Standard Model also predicted the existence of a type of boson known as the Higgs boson. Early in the morning on 4 July 2012, physicists with the Large Hadron Collider at CERN announced they had found a new particle that behaves similarly to what is expected from the Higgs boson.

Theory

Theoretical particle physics attempts to develop the models, theoretical framework, and mathematical tools to understand current experiments and make predictions for future experiments. There are several major interrelated efforts being made in theoretical particle physics today. One important branch attempts to better understand the Standard Model and its tests. By extracting the parameters of the Standard Model, from experiments with less uncertainty, this work probes the limits of the Standard Model and therefore expands our understanding of nature's building blocks. Those efforts are made challenging by the difficulty of calculating quantities in quantum chromodynamics. Some theorists working in this area refer to themselves as phenomenologists and they may use the tools of quantum field theory and effective field theory. Others make use of lattice field theory and call themselves *lattice theorists*.

Another major effort is in model building where model builders develop ideas for what physics may lie beyond the Standard Model (at higher energies or smaller distances). This work is often motivated by the hierarchy problem and is constrained by existing experimental data. It may involve work on supersymmetry, alternatives to the Higgs mechanism, extra spatial dimensions (such as the Randall-Sundrum models), Preon theory, combinations of these, or other ideas.

A third major effort in theoretical particle physics is string theory. *String theorists* attempt to construct a unified description of quantum mechanics and general relativity by building a theory based on small strings, and branes rather than particles. If the theory is successful, it may be considered a "Theory of Everything", or "TOE".

There are also other areas of work in theoretical particle physics ranging from particle cosmology to loop quantum gravity.

This division of efforts in particle physics is reflected in the names of categories on the arXiv, a preprint archive: hep-th (theory), hep-ph (phenomenology), hep-ex (experiments), hep-lat (lattice gauge theory).

Practical Applications

In principle, all physics (and practical applications developed therefrom) can be derived from the study of fundamental particles. In practice, even if "particle physics" is taken to mean only "high-energy atom smashers", many technologies have been developed during these pioneering investigations that later find wide uses in society. Particle accelerators are used to produce medical isotopes for research and treatment (for example, isotopes used in PET imaging), or used directly in external beam radiotherapy. The development of superconductors has been pushed forward by their use in particle physics. The World Wide Web and touchscreen technology were initially developed at CERN. Additional applications are found in medicine, national security, industry, computing, science, and workforce development, illustrating a long and growing list of beneficial practical applications with contributions from particle physics.

High Energy Physics Compared to Low Energy Physics

The term *high energy physics* requires elaboration. Intuitively, it might seem incorrect to associate "high energy" with the physics of very small, *low* mass objects, like subatomic particles. By comparison, an example of a macroscopic system, one gram of hydrogen, has ~ 6×10^{23} times the mass of a single proton. Even an entire beam of protons circulated in the LHC contains ~ 3.23×10^{14} protons, each with 6.5×10^{12} eV of energy, for a total beam energy of ~ 2.1×10^{27} eV or ~ 336.4 MJ, which is still ~ 2.7×10^{5} times lower than the mass-energy of a single gram of hydrogen. Yet, the macroscopic realm is "low energy physics", while that of quantum particles is "high energy physics".

The interactions studied in other fields of physics and science have comparatively very low energy. For example, the photon energy of visible light is about 1.8 to 3.1 eV. Similarly, the bond-dissociation energy of a carbon–carbon bond is about 3.6 eV. Other chemical reactions typically involve similar amounts of energy. Even photons with far higher energy, gamma rays of the kind produced in radioactive decay, mostly have photon energy between 10^5 eV and 10^7 eV – still two orders of magnitude lower than the mass of a single proton. Radioactive decay gamma rays are considered as part of nuclear physics, rather than high energy physics.

The proton has a mass of around 9.4×10^8 eV; some other massive quantum particles, both elementary and hadronic, have yet higher masses. Due to these very high energies *at the single particle level*, particle physics is, in fact, high-energy physics.

Nuclear Physics

The branch of science that deals with the fundamental particle namely nucleus is called Nuclear Physics. Nuclear physics basically deals with the structure and components of nucleus. The properties like radius, mass etc are dealt with and studied. Various models related to the nucleus and their development by various scientists, their experiments, theories etc. are the part of Nuclear Physics.

Father of Nuclear Physics

Ernest Rutherford is a very famous chemists. He is also known as Father of Nuclear physics as he was the first to propose the Nuclear model with a charged nucleus in the center. He found the nucleus by his very famous gold scattering experiment.

According to this model,

1. All the mass of atom is concentrated within a small positively charged region at the center called the NUCLEUS.

2. The nucleus is surrounded by the electrons which are negatively charged that move around it in circular paths. These circular paths are called the ORBITS.

3. The Rutherford model resembles our solar system. Here electrons and the nucleus are held together by Electrostatic forces of attraction.

The Radius of nucleus denoted by **R** is given by,

$R = R_o A^{\frac{1}{3}}$

Here, **A** is the mass number of the element,

(Mass number = Total no of protons + Total no of neutrons)

$\mathbf{R_o}$ = Proportionality Constant

R is directly proportional to $A^{\frac{1}{3}}$, Volume of nucleus $= \frac{4}{3}\pi R^3$,

Thus **V** α **A**

His model developed a new idea for the scientists which made Bohr to give his Bohr's model for an atom. Thus he was awarded the Nobel Prize in Chemistry in 1908.

Nuclear Physics Equations

The Mass number **(A)**, also called atomic mass number or nucleon number, is the total number of protons and neutrons in a nucleus.

$$\mathbf{A = Z + N}$$

Where **A** = Mass number,

Z = Proton number,

N = Neutron number.

When Proton and Neutron combines together in order to form nucleus then in this process some mass gets lost and this lost mass is known as Mass defect. More is the mass defect more is presumed to be stability of product nuclei.

Mathematically,

Δm = Mass of Nucleons – Mass of Nucleus.

$= A - M$

$= (Z\, m_p + N\, m_n) - M.$

$\Delta m = Z\, m_p + (A - Z)\, m_n - M$

Here M = Mass of the nucleus

Δm = Mass of the nucleons - Mass of the nucleus.

m_p = Mass of the Proton and

m_n = Mass of the Neutron.

Packing fraction is defined as Mass defect per nucleon.

Packing fraction (f) = Mass defect per nucleons

$$= \frac{\Delta m}{A}$$

$$= \frac{M - A}{A}$$

Packing Fraction (f) $= \frac{[Zm_p + (A-Z)m_n - A]}{A}$

Greater the negative value of packing fraction higher is the stability of product nuclei.

Nuclear Binding Energy

In Nucleus, Proton and Neutron are contained in small sphere. Due to such small small distance between two protons, they exert a very large force of repulsion on each other.

Therefore to prevent nucleus from bursting, certain amount of energy needed to bind the nucleus. This energy needed is known as Binding Energy. When nucleus is formed, the mass of nucleons forming it decreases. This mass defect provides the required binding energy. The energy equivalent to mass defect is called Binding Energy.

Binding energy of a nucleus may also be defined as the amount of work required to separate the nucleons at infinite distance. It is the measure of stability of nucleus.

Rutherford Gold Foil Scattering

In 1910, a physicist from New Zealand, Ernest Rutherford performed an experiment known as Rutherford's Gold Foil Experiment. This experiment was determined to find out the structure of an atom. Rutherford gave this atomic theory on the basis of following observations during the gold foil experiment:

1. Since most of the alpha particles passed through the foil without any deflection, thus it was concluded that most of the atom is empty.
2. As some particles deflected by small angles, it indicated that positively charged alpha particles were attracted by negatively charged particles means electrons.
3. Large angles of deflection experienced by the alpha particles indicated that there is a massive positively charged region present in the atom and due to repulsion alpha particles were deflected at large angles.

Compton Scattering

Compton effect was first observed by Arthur Compton in 1923. It occurs when the incident x-ray photon interacts with the electron which makes the incident photon to deflect away from its original path.

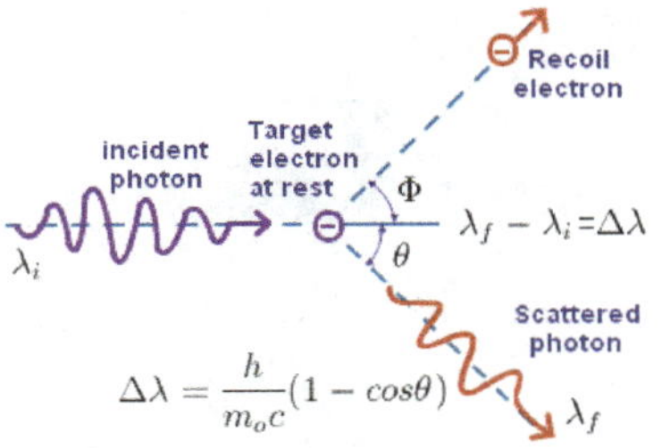

The change in wavelength is given by:

$$\lambda' - \lambda = \frac{h}{m_e c}(1 - \cos\theta)$$

Compton scattering is an inelastic scattering because the wavelength of the scattered light is different from that of the incident radiation. The change in the wavelength is called the Compton shift.

Photoelectric Effect

When light of suitable wavelength is incident on the metal plate electrons are emitted. These emitted electrons are called as photo electrons and the phenomenon is Photoelectric Effect.

Pair Production

When a photon is incident on the atomic nucleus, electron-positron pair is produced. This is called Pair-production. Pair production is a phenomenon of nature where energy is converted in to mass.

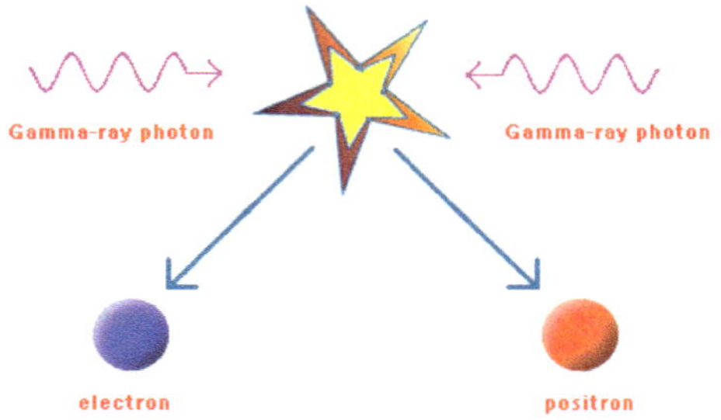

when two high energy gamma-ray photons, an electron/positron pair produced
(the energy is converted into mass, $E=mc^2$)

Cherenkov Radiation

Cherenkov radiation is electromagnetic radiation emitted when a charged particle passes through a dielectric medium at a speed greater than the phase velocity of light in that medium.

Nuclear Magnetic Resonance

Nuclear magnetic resonance (NMR) is a technique used to determine the molecular structure of the substance by applying the external magnetic field to the atomic nuclei. It has wide applications in medicine, tool analysis, analyzing quality and purity of the sample etc.

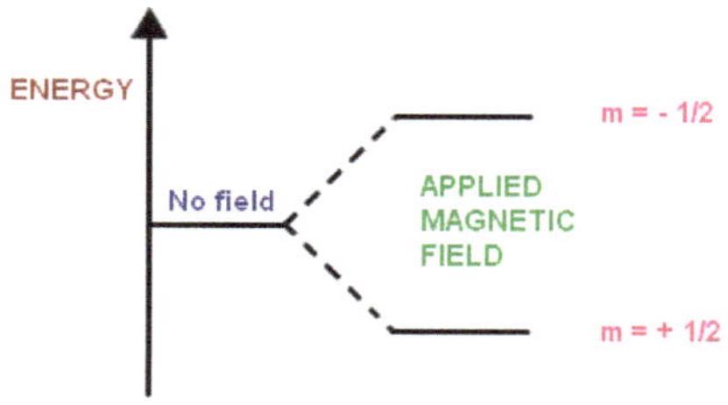

Energy level for a nucleus with spin quantum number 1/2

Radioactivity

Radioactivity is a phenomenon which was discovered by a French scientist Henri Becquerel in 1896. He verified that uranium is a powerful radiations that is capable of blackening a photographic plate. Here uranium was discovered about hundred years as these radiations are not observed. Natural radioactivity is the phenomenon of spontaneous and uncontrollable emission of radiations. This process which is considered as the disintegration are not stopped or slowed down by the use of and instrument are not observed by Henri Becquerel in discover of radioactivity.

Elementary Particles

Subatomic particles are small particles composed of nucleons or atoms. Elementary particles and composite particles are two type of sub atomicparticles. Elementary particles are those which are not measurable and are not made up of other particles and composite particles are made of some other particles also. Some of the Elementary particles are lepton, Hadron, Photon, graviton, quarks etc.

Nuclear Spin

The Nuclear spin is different from the electron spin. The nuclear spin represents the total angular momentum of the nucleus. It is represented by symbol **I**. The nucleus is although, composed of neutrons and protons but it acts as if it is a single entity which has intrinsic angular momentum. The Nuclear spin depends on the mass number, if the mass number is odd then the nucleus has half-integer spin like the electron while if the nucleus has even mass number then its spin will be integer spin.

Nuclear Stability

Nuclear stability determines how stable the nucleus is? we know that the Nucleus is made of neutrons and protons. What causes them to stick together? Why do the protons not repel each other?

Surprisingly, the mass of the nucleus is less than the sum of the masses of the individual protons and neutrons which makeup the nucleus. The lost mass (mass defect) has been changed into the energy necessary to bind the nucleus together. Suppose the mass defect is Δ m. Then the energy binding the nucleus together Δ E is given by:

$$\Delta E = \Delta m \times c^2$$

where **C** is the speed of light. If is divided by the number of nucleons in the nucleus, then the binding energy per nucleon is obtained. The higher this value is the more stable is the nucleus.

Fermi Energy

The Fermi Energy is the energy of the highest level of quantum state which is occupied by the fermions (like electrons, protons or neutrons) at the absolute zero temperature. The Fermi energy is useful in determining the thermal and electrical characteristics of the solids.

Shell Model

The shell model of the atom was first proposed by Niels Bohr in 1923. The Shell Model is the model which describes the distribution of the energy level into different shells in an atom or in the nucleus of the atom. A shell is the energy level where particles of same energy exist. The shell model describes the arrangement of the elementary particles in the different shells of the atoms. For example, the electrons are distributed in different shells based on their energy level and their direction of their rotation.

Millikan Oil Drop Model

The Millikan oil drop experiment is commonly known as Millikan experiment and was first performed by Robert Millikan and Harvey Fletcher in 1909. Robert Millikan was the professor of University of Chicago at that time. The experiment was performed by Professor Millikan with the useful inputs from Mr. Fletcher.

In this experiment a charged oil drop was suspended vertically between the two electrodes. Then the electric field is applied to such an extent that the free fall of the oil drop, under the influence of gravitational pull, was stopped. With the known electric field and other parameters Millikan was succeeded to calculate the charge on the drop. He published his findings in 1913 and was awarded Nobel prize for physics in 1923 for his work on finding the charge of the electron.

Fission

Fission is the process whereby a some matter divides and becomes multiple states. Here in Nuclear physics, Fission process is used to split up a single atomic nuclei into two or more fragments.

Nuclear Fission

The Nuclear fission is defined as the process of splitting a heavy nucleus into two nuclei of smaller mass. Eg: Consider a Uranium Nuclei.

The process of fission is represented as follows:

$$_{92}\mathbf{U}^{235} +_0 \mathbf{n}^1 \text{---}>_{92}\mathbf{U}^{236} \text{---}>_{56}\mathbf{Ba}^{141} +_{36} \mathbf{Kr}^{92} + 3_0\mathbf{n}^1 + \mathbf{Energy}$$

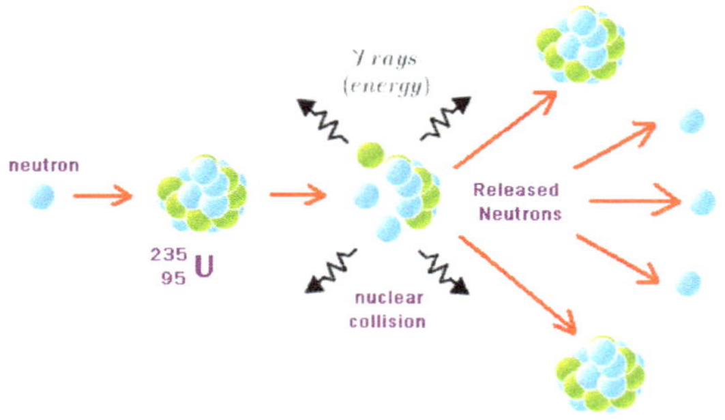

Nuclear Reactors

A Nuclear Reactor is used to extract the nuclear power from the radioactive material. The Nuclear reactor initiates and controls the nuclear chain reaction. The present day nuclear reactors are based on the Nuclear Fission Reaction.

3

Heat and Thermodynamics

Thermodynamics is a branch of physics that studies the nature of heat and temperature and their relation to work and energy. Its applications are in the fields of chemical engineering, physical chemistry, mechanical engineering, etc. This chapter includes detailed information on the core topics of heat transfer, temperature gradient, thermodynamic temperature, work and entropy.

Heat

Heat, energy that is transferred from one body to another as the result of a difference in temperature. If two bodies at different temperatures are brought together, energy is transferred—i.e., heat flows—from the hotter body to the colder. The effect of this transfer of energy usually, but not always, is an increase in the temperature of the colder body and a decrease in the temperature of the hotter body. A substance may absorb heat without an increase in temperature by changing from one physical state (or phase) to another, as from a solid to a liquid (melting), from a solid to a vapour (sublimation), from a liquid to a vapour (boiling), or from one solid form to another (usually called a crystalline transition). The important distinction between heat and temperature (heat being a form of energy and temperature a measure of the amount of that energy present in a body) was clarified during the 18th and 19th centuries.

Heat as a Form of Energy

Because all of the many forms of energy, including heat, can be converted into work, amounts of energy are expressed in units of work, such as joules, foot-pounds, kilowatt-hours, or calories. Exact relationships exist between the amounts of heat added to or removed from a body and the magnitude of the effects on the state of the body. The two units of heat most commonly used are the calorie and the British thermal unit (BTU). The calorie (or gram-calorie) is the amount of energy required to raise the temperature of one gram of water from 14.5° to 15.5° C; the BTU is the amount of energy required to raise the temperature of one pound of water from 63° to 64° F. One BTU is approximately 252 calories. Both definitions specify that the temperature changes are to be measured at a constant pressure of one atmosphere, because the amounts of energy involved depend in part on pressure. The calorie used in measuring the energy content of foods is the large calorie, or kilogram-calorie, equal to 1,000 gram-calories.

In general, the amount of energy required to raise a unit mass of a substance through a specified temperature interval is called the heat capacity, or the specific heat, of that substance. The quantity of energy necessary to raise the temperature of a body one degree varies depending upon the restraints imposed. If heat is added to a gasconfined at constant volume, the amount of heat needed to cause a one-degree temperature rise is less than if the heat is added to the same gas free to expand (as in a cylinder fitted with a movable piston) and so do work. In the first case, all the energy goes into raising the temperature of the gas, but in the second case, the energy not only contributes to the temperature increase of the gas but also provides the energy necessary for the work done by the gas on the piston. Consequently, the specific heat of a substance depends on these conditions. The most commonly determined specific heats are the specific heat at constant volume and the specific heat at constant pressure. The heat capacities of many solid elements were shown to be closely related to their atomic weights by the French scientists Pierre-Louis Dulon-gand Alexis-Thérèse Petit in 1819. The so-called law of Dulong and Petit was useful in determining the atomic weights of certain metallic elements, but there are many exceptions to it; the deviations were later found to be explainable on the basis of quantum mechanics.

It is incorrect to speak of the heat in a body, because heat is restricted to energy being transferred. Energy stored in a body is not heat (nor is it work, as work is also energy in transit). It is customary, however, to speak of sensible and latent heat. The latent heat, also called the heat of vaporization, is the amount of energy necessary to change a liquid to a vapour at constant temperature and pressure. The energy required to melt a solid to a liquid is called the heat of fusion, and the heat of sublimation is the energy necessary to change a solid directly to a vapour, these changes also taking place under conditions of constant temperature and pressure.

Air is a mixture of gases and water vapour, and it is possible for the water present in the air to change phase; i.e., it may become liquid (rain) or solid (snow). To distinguish between the energy associated with the phase change (the latent heat) and the energy required for a temperature change, the concept of sensible heat was introduced. In a mixture of water vapour and air, the sensible heat is the energy necessary to produce a particular temperature change excluding any energy required for a phase change.

Such energy transfer can occur in three ways: radiation, conduction, and convection.The standard unit of heat in the International System of Units (SI) is the calorie (cal), which is the amount of energy transfer required to raise the temperature of one gram of pure liquid water by one degree Celsius, provided the water temperature is higher than the freezing point and lower than the boiling point. Sometimes the kilocalorie (kcal) is specified as a unit of heat; 1 kcal = 1000 cal. (This is the so-called diet calorie.) Less often, the British thermal unit (Btu) is used. This is the amount of heat required to raise the temperature of one pound of pure liquid water by one degree Fahrenheit.

An example of heat by radiation is the effect of infrared (IR) energy as it strikes a surface. IR is an electromagnetic field capable of transfering energy from a source, such as a fireplace, to a destination, such as the surfaces within a room. Radiation does not require an intervening medium; it can occur through a vacuum. It is responsible for the warming of the Earth by the sun.

Heat by conduction takes place when two material media or objects are in direct contact, and the temperature of one is higher than the temperature of the other. The temperatures tend to equalize; thus the heat conduction consists of a transfer of kinetic energy from the warmer medium to the cooler one. An example is the immersion of a chilled human body in a hot bath.

Heat by convection occurs when the motion of a liquid or gas carries energy from a warmer region to a cooler region. A good example of convection is the tendency of warm air to rise and cool air to fall, equalizing the air temperature inside a room containing a hot stove. Heat convection (along with conduction) is believed to take place inside the Earth, transfering kinetic energy from the inner core through the outer core and mantle to the crust. In this situation, the outer core and the mantle behave like liquids over long periods of time.

Notation and Units

As a form of energy, heat has the unit joule (J) in the International System of Units (SI). However, in many applied fields in engineering the British thermal unit (BTU) and the calorieare often used. The standard unit for the rate of heat transferred is the watt (W), defined as one joule per second.

Use of the symbol Q for the total amount of energy transferred as heat is due to Rudolf Clausius in 1850:

> "Let the amount of heat which must be imparted during the transition of the gas in a definite manner from any given state to another, in which its volume is v and its temperature t, be called Q"

Heat released by a system into its surroundings is by convention a negative quantity ($Q < 0$); when a system absorbs heat from its surroundings, it is positive ($Q > 0$). Heat transfer rate, or heat flow per unit time, is denoted by $\dot{Q}$. This should not be confused with a time derivative of a function of state (which can also be written with the dot notation) since heat is not a function of state. Heat flux is defined as rate of heat transfer per unit cross-sectional area (units watts per square metre).

Classical Thermodynamics

Heat and Entropy

Rudolf Clausius

In 1856, Rudolf Clausius, referring to closed systems, in which transfers of matter do not occur, defined the *second fundamental theorem* (the second law of thermodynamics) in the mechanical theory of heat (thermodynamics): "if two transformations which, without necessitating any other permanent change, can mutually replace one another, be called equivalent, then the generations of the quantity of heat Q from work at the temperature T, has the *equivalence-value*:"

$$\frac{Q}{T}.$$

In 1865, he came to define the entropy symbolized by S, such that, due to the supply of the amount of heat Q at temperature T the entropy of the system is increased by

$$\Delta S = \frac{Q}{T} \quad (1)$$

In a transfer of energy as heat without work being done, there are changes of entropy in both the surroundings which lose heat and the system which gains it. The increase, ΔS, of entropy in the system may be considered to consist of two parts, an increment, $\Delta S'$ that matches, or 'compensates', the change, $-\Delta S'$, of entropy in the surroundings, and a further increment, $\Delta S''$ that may be considered to be 'generated' or 'produced' in the system, and is said therefore to be 'uncompensated'. Thus

$$\Delta S = \Delta S' + \Delta S''.$$

This may also be written

$$\Delta S_{system} = \Delta S_{compensated} + \Delta S_{uncompensated} \text{ with } \Delta S_{compensated} = -\Delta S_{surroundings}.$$

The total change of entropy in the system and surroundings is thus

$$\Delta S_{overall} = \Delta S' + \Delta S'' - \Delta S' = \Delta S''.$$

This may also be written

$$\Delta S_{overall} = \Delta S_{compensated} + \Delta S_{uncompensated} + \Delta S_{surroundings} = \Delta S_{uncompensated}.$$

It is then said that an amount of entropy $\Delta S'$ has been transferred from the surroundings to the system. Because entropy is not a conserved quantity, this is an exception to the general way of speaking, in which an amount transferred is of a conserved quantity.

From the second law of thermodynamics follows that in a spontaneous transfer of heat, in which the temperature of the system is different from that of the surroundings:

$$\Delta S_{overall} > 0.$$

For purposes of mathematical analysis of transfers, one thinks of fictive processes that are called *reversible*, with the temperature T of the system being hardly less than that of the surroundings, and the transfer taking place at an imperceptibly slow rate.

Following the definition above in formula (1), for such a fictive reversible process, a quantity of transferred heat δQ (an inexact differential) is analyzed as a quantity $T\,dS$, with dS (an exact differential):

$$T\,dS = \delta Q.$$

This equality is only valid for a fictive transfer in which there is no production of entropy, that is to say, in which there is no uncompensated entropy.

If, in contrast, the process is natural, and can really occur, with irreversibility, then there is entropy production, with $dS_{uncompensated} > 0$. The quantity $T\,dS_{uncompensated}$ was termed by Clausius the "uncompensated heat", though that does not accord with present-day terminology. Then one has

$$T\,dS = \delta Q + T\,dS_{uncompensated} > \delta Q.$$

This leads to the statement

$$T\,dS \geq \delta Q \quad (\textit{second law}).$$

which is the second law of thermodynamics for closed systems.

In non-equilibrium thermodynamics that approximates by assuming the hypothesis of local thermodynamic equilibrium, there is a special notation for this. The transfer of energy as heat is assumed to take place across an infinitesimal temperature difference, so that the system element and its surroundings have near enough the same temperature T. Then one writes

$$dS = dS_e + dS_i,$$

where by definition

$$\delta Q = T\,dS_e \;\; and \;\; dS_i \equiv dS_{uncompensated}.$$

The second law for a natural process asserts that

$$dS_i > 0.$$

Heat and Enthalpy

For a closed system (a system from which no matter can enter or exit), one version of the first law of thermodynamics states that the change in internal energy ΔU of the system is equal to the amount of heat Q supplied to the system minus the amount of work W done by system on its surroundings. The foregoing sign convention for work is used in the present article, but an alternate sign convention, followed by IUPAC, for work, is to consider the work performed on the system by its surroundings as positive. This is the convention adopted by many modern textbooks of physical chemistry, such as those by Peter Atkins and Ira Levine, but many textbooks on physics define work as work done by the system.

$$\Delta U = Q - W.$$

This formula can be re-written so as to express a definition of quantity of energy transferred as heat, based purely on the concept of adiabatic work, if it is supposed that ΔU is defined and measured solely by processes of adiabatic work:

$$Q = \Delta U + W.$$

The work done by the system includes boundary work (when the system increases its volume against an external force, such as that exerted by a piston) and other work (e.g. shaft work performed by a compressor fan), which is called isochoric work:

$$Q = \Delta U + W_{boundary} + W_{isochoric}.$$

In this Section we will neglect the "other-" or isochoric work contribution.

The internal energy, U, is a state function. In cyclical processes, such as the operation of a heat engine, state functions of the working substance return to their initial values upon completion of a cycle.

The differential, or infinitesimal increment, for the internal energy in an infinitesimal process is an exact differential dU. The symbol for exact differentials is the lowercase letter d.

In contrast, neither of the infinitesimal increments δQ nor δW in an infinitesimal process represents the state of the system. Thus, infinitesimal increments of heat and work are inexact differentials. The lowercase Greek letter delta, δ, is the symbol for inexact differentials. The integral of any inexact differential over the time it takes for a system to leave and return to the same thermodynamic state does not necessarily equal zero.

As recounted below, in the section headed Entropy, the second law of thermodynamics observes that if heat is supplied to a system in which no irreversible processes take place and which has a well-defined temperature T, the increment of heat δQ and the temperature T form the exact differential

$$dS = \frac{\delta Q}{T},$$

and that S, the entropy of the working body, is a function of state. Likewise, with a well-defined pressure, P, behind the moving boundary, the work differential, δW, and the pressure, P, combine to form the exact differential

$$dV = \frac{\delta W}{P},$$

with V the volume of the system, which is a state variable. In general, for homogeneous systems,

$$dU = TdS - PdV.$$

Associated with this differential equation is that the internal energy may be considered to be a function $U(S,V)$ of its natural variables S and V. The internal energy representation of the fundamental thermodynamic relation is written

$$U = U(S,V).$$

If V is constant

$$TdS = dU \quad (V\ constant)$$

and if P is constant

$$TdS = dH \quad (P\ constant)$$

with H the enthalpy defined by

$$H = U + PV.$$

The enthalpy may be considered to be a function $H(S,P)$ of its natural variables S and P. The enthalpy representation of the fundamental thermodynamic relation is written

$$H = H(S,P).$$

The internal energy representation and the enthalpy representation are partial Legendre transforms of one another. They contain the same physical information, written in different ways. Like the internal energy, the enthalpy stated as a function of its natural variables is a thermodynamic potential and contains all thermodynamic information about a body.

If a quantity Q of heat is added to a body while it does expansion work W on its surroundings, one has

$$\Delta H = \Delta U + \Delta(PV).$$

If this is constrained to happen at constant pressure with $\Delta P = 0$, the expansion work W done by the body is given by $W = P\,\Delta V$; recalling the first law of thermodynamics, one has

$$\Delta U = Q - W = Q - P\Delta V \quad and \quad \Delta(PV) = P\Delta V.$$

If this is constrained to happen at constant pressure with $\Delta P = 0$, the expansion work W done by the body is given by $W = P\,\Delta V$; recalling the first law of thermodynamics, one has

Consequently, by substitution one has

$$\Delta H = Q - P\Delta V + P\Delta V = Q \qquad at\ constant\ pressure.$$

In this scenario, the increase in enthalpy is equal to the quantity of heat added to the system. Since many processes do take place at constant pressure, or approximately at atmospheric pressure, the enthalpy is therefore sometimes given the misleading name of 'heat content'. It is sometimes also called the heat function.

In terms of the natural variables S and P of the state function H, this process of change of state from state 1 to state 2 can be expressed as

$$\Delta H = \int_{S_1}^{S_2} \left(\frac{\partial H}{\partial S}\right)_P dS + \int_{P_1}^{P_2} \left(\frac{\partial H}{\partial P}\right)_S dP$$

$$= \int_{S_1}^{S_2} \left(\frac{\partial H}{\partial S}\right)_P dS \qquad at\ constant\ pressure.$$

It is known that the temperature *T*(*S*, *P*) is identically stated by

$$\left(\frac{\partial H}{\partial S}\right)_P \equiv T(S,P).$$

Consequently,

$$\Delta H = \int_{S_1}^{S_2} T(S,P)dS \qquad \textit{at constant pressure.}$$

In this case, the integral specifies a quantity of heat transferred at constant pressure.

Heat Transfer

Heat Transfer between Two Bodies

Referring to conduction, Partington writes: "If a hot body is brought in conducting contact with a cold body, the temperature of the hot body falls and that of the cold body rises, and it is said that a *quantity of heat* has passed from the hot body to the cold body."

Referring to radiation, Maxwell writes: "In Radiation, the hotter body loses heat, and the colder body receives heat by means of a process occurring in some intervening medium which does not itself thereby become hot."

Maxwell writes that convection as such "is not a purely thermal phenomenon". In thermodynamics, convection in general is regarded as transport of internal energy. If, however, the convection is enclosed and circulatory, then it may be regarded as an intermediary that transfers energy as heat between source and destination bodies, because it transfers only energy and not matter from the source to the destination body.

In accordance with the first law for closed systems, energy transferred solely as heat leaves one body and enters another, changing the internal energies of each. Transfer, between bodies, of energy as work is a complementary way of changing internal energies. Though it is not logically rigorous from the viewpoint of strict physical concepts, a common form of words that expresses this is to say that heat and work are interconvertible.

Cyclically operating engines, that use only heat and work transfers, have two thermal reservoirs, a hot and a cold one. They may be classified by the range of operating temperatures of the working body, relative to those reservoirs. In a heat engine, the working body is at all times colder than the hot reservoir and hotter than the cold reservoir. In a sense, it uses heat transfer to produce work. In a heat pump, the working body, at stages of the cycle, goes both hotter than the hot reservoir, and colder than the cold reservoir. In a sense, it uses work to produce heat transfer.

Heat Engine

In classical thermodynamics, a commonly considered model is the heat engine. It consists of four bodies: the working body, the hot reservoir, the cold reservoir, and the work reservoir. A cyclic process leaves the working body in an unchanged state, and is envisaged as being repeated indefinitely often. Work transfers between the working body and the work reservoir are envisaged as reversible, and thus only one work reservoir is needed. But two thermal reservoirs are needed, because transfer of energy as heat is irreversible. A single cycle sees energy taken by the working body from the hot reservoir and sent to the two other reservoirs, the work reservoir and the cold reservoir. The hot reservoir always and only supplies energy and the cold reservoir always and only receives energy. The second law of thermodynamics requires that no cycle can occur in which no energy is received by the cold reservoir. Heat engines achieve higher efficiency when the difference between initial and final temperature is greater.

Heat Pump or Refrigerator

Another commonly considered model is the heat pump or refrigerator. Again there are four bodies: the working body, the hot reservoir, the cold reservoir, and the work reservoir. A single cycle starts with the working body colder than the cold reservoir, and then energy is taken in as heat by the working body from the cold reservoir. Then the work reservoir does work on the working body, adding more to its internal energy, making it hotter than the hot reservoir. The hot working body passes heat to the hot reservoir, but still remains hotter than the cold reservoir. Then, by allowing it to expand without doing work on another body and without passing heat to another body, the working body is made colder than the cold reservoir. It can now accept heat transfer from the cold reservoir to start another cycle.

The device has transported energy from a colder to a hotter reservoir, but this is not regarded as by an inanimate agency; rather, it is regarded as by the harnessing of work . This is because work is supplied from the work reservoir, not just by a simple thermodynamic process, but by a cycle of thermodynamic operations and processes, which may be regarded as directed by an animate or harnessing agency. Accordingly, the cycle is still in accord with the second law of thermodynamics. The efficiency of a heat pump is best when the temperature difference between the hot and cold reservoirs is least.

Functionally, such engines are used in two ways, distinguishing a target reservoir and a resource or surrounding reservoir. A heat pump transfers heat, to the hot reservoir as the target, from the resource or surrounding reservoir. A refrigerator transfers heat, from the cold reservoir as the target, to the resource or surrounding reservoir. The target reservoir may be regarded as leaking: when the target leaks hotness to the surroundings, heat pumping is used; when the target leaks coldness to the surroundings, refrigeration is used. The engines harness work to overcome the leaks.

Macroscopic View

According to Planck, there are three main conceptual approaches to heat. One is the microscopic or kinetic theory approach. The other two are macroscopic approaches. One is the approach through the law of conservation of energy taken as prior to thermodynamics, with a mechanical analysis of processes, for example in the work of Helmholtz. This mechanical view is taken in this article as currently customary for thermodynamic theory. The other macroscopic approach is the thermodynamic one, which admits heat as a primitive concept, which contributes, by scientific induction to knowledge of the law of conservation of energy. This view is widely taken as the practical one, quantity of heat being measured by calorimetry.

Bailyn also distinguishes the two macroscopic approaches as the mechanical and the thermodynamic. The thermodynamic view was taken by the founders of thermodynamics in the nineteenth century. It regards quantity of energy transferred as heat as a primitive concept coherent with a primitive concept of temperature, measured primarily by calorimetry. A calorimeter is a body in the surroundings of the system, with its own temperature and internal energy; when it is connected to the system by a path for heat transfer, changes in it measure heat transfer. The mechanical view was pioneered by Helmholtz and developed and used in the twentieth century, largely through the influence of Max Born. It regards quantity of heat transferred as heat as a derived concept, defined for closed systems as quantity of heat transferred by mechanisms other than work transfer, the latter being regarded as primitive for thermodynamics, defined by macroscopic mechanics. According to Born, the transfer of internal energy between open systems that accompanies transfer of matter "cannot be reduced to mechanics". It follows that there is no well-founded definition of quantities of energy transferred as heat or as work associated with transfer of matter.

Nevertheless, for the thermodynamical description of non-equilibrium processes, it is desired to consider the effect of a temperature gradient established by the surroundings across the system of interest when there is no physical barrier or wall between system and surroundings, that is to say, when they are open with respect to one another. The impossibility of a mechanical definition in terms of work for this circumstance does not alter the physical fact that a temperature gradient causes a diffusive flux of internal energy, a process that, in the thermodynamic view, might be proposed as a candidate concept for transfer of energy as heat.

In this circumstance, it may be expected that there may also be active other drivers of diffusive flux of internal energy, such as gradient of chemical potential which drives transfer of matter, and gradient of electric potential which drives electric current and iontophoresis; such effects usually interact with diffusive flux of internal energy driven by temperature gradient, and such interactions are known as cross-effects.

If cross-effects that result in diffusive transfer of internal energy were also labeled as heat transfers, they would sometimes violate the rule that pure heat transfer occurs only down a temperature gradient, never up one. They would also contradict the principle that all heat transfer is of one and the same kind, a principle founded on the idea of heat conduction between closed systems. One might to try to think narrowly of heat flux driven purely by temperature gradient as a conceptual component of diffusive internal energy flux, in the thermodynamic view, the concept resting specifically on careful calculations based on detailed knowledge of the processes and being indirectly assessed. In these circumstances, if perchance it happens that no transfer of matter is actualized, and there are no cross-effects, then the thermodynamic concept and the mechanical concept coincide, as if one were dealing with closed systems. But when there is transfer of matter, the exact laws by which temperature gradient drives diffusive flux of internal energy, rather than being exactly knowable, mostly need to be assumed, and in many cases are practically unverifiable. Consequently, when there is transfer of matter, the calculation of the pure 'heat flux' component of the diffusive flux of internal energy rests on practically unverifiable assumptions. This is a reason to think of heat as a specialized concept that relates primarily and precisely to closed systems, and applicable only in a very restricted way to open systems.

In many writings in this context, the term "heat flux" is used when what is meant is therefore more accurately called diffusive flux of internal energy; such usage of the term "heat flux" is a residue of older and now obsolete language usage that allowed that a body may have a "heat content".

Microscopic View

In the kinetic theory, heat is explained in terms of the microscopic motions and interactions of constituent particles, such as electrons, atoms, and molecules. The immediate meaning of the kinetic energy of the constituent particles is not as heat. It is as a component of internal energy. In microscopic terms, heat is a transfer quantity, and is described by a transport theory, not as steadily localized kinetic energy of particles. Heat transfer arises from temperature gradients or differences, through the diffuse exchange of microscopic kinetic and potential particle energy, by particle collisions and other interactions. An early and vague expression of this was made by Francis Bacon. Precise and detailed versions of it were developed in the nineteenth century.

In statistical mechanics, for a closed system (no transfer of matter), heat is the energy transfer associated with a disordered, microscopic action on the system, associated with jumps in occupation numbers of the energy levels of the system, without change in the values of the energy levels themselves. It is possible for macroscopic thermodynamic work to alter the occupation numbers without change in the values of the system energy levels themselves, but what distinguishes transfer as heat is that the transfer is entirely due to disordered, microscopic action, including radiative transfer. A mathematical definition can be formulated for small increments of quasi-static adiabatic work in terms of the statistical distribution of an ensemble of microstates.

Calorimetry

Quantity of heat transferred can be measured by calorimetry, or determined through calculations based on other quantities.

Calorimetry is the empirical basis of the idea of quantity of heat transferred in a process. The transferred heat is measured by changes in a body of known properties, for example, temperature rise, change in volume or length, or phase change, such as melting of ice.

A calculation of quantity of heat transferred can rely on a hypothetical quantity of energy transferred as adiabatic work and on the first law of thermodynamics. Such calculation is the primary approach of many theoretical studies of quantity of heat transferred.

Engineering

A red-hot iron rod from which heat transfer to the surrounding environment will be primarily through radiation

The discipline of heat transfer, typically considered an aspect of mechanical engineering and chemical engineering, deals with specific applied methods by which thermal energy in a system is generated, or converted, or transferred to another system. Although the definition of heat implicitly means the transfer of energy, the term *heat transfer*encompasses this traditional usage in many engineering disciplines and laymen language.

Heat transfer is generally described as including the mechanisms of heat conduction, heat convection, thermal radiation, but may include mass transfer and heat in processes of phase changes.

Convection may be described as the combined effects of conduction and fluid flow. From the thermodynamic point of view, heat flows into a fluid by diffusion to increase its energy, the fluid then transfers (advects) this increased internal energy (not heat) from one location to another, and this is then followed by a second thermal interaction which transfers heat to a second body or system, again by diffusion. This entire process is often regarded as an additional mechanism of heat transfer, although technically, "heat transfer" and thus heating and cooling occurs only on either end of such a conductive flow, but not as a result of flow. Thus, conduction can be said to "transfer" heat only as a net result of the process, but may not do so at every time within the complicated convective process.

Latent and Sensible Heat

Joseph Black

In an 1847 lecture entitled *On Matter, Living Force, and Heat*, James Prescott Joule characterized the terms latent heat and sensible heatas components of heat each affecting distinct physical phenomena, namely the potential and kinetic energy of particles, respectively. He described latent energy as the energy possessed via a distancing of particles where attraction was over a greater distance, i.e. a form of potential energy, and the sensible heat as an energy involving the motion of particles, i.e. kinetic energy.

Latent heat is the heat released or absorbed by a chemical substance or a thermodynamic system during a change of state that occurs without a change in temperature. Such a process may be a phase transition, such as the melting of ice or the boiling of water.

Heat Capacity

Heat capacity is a measurable physical quantity equal to the ratio of the heat added to an object to the resulting temperature change. The *molar heat capacity* is the heat capacity per unit amount

(SI unit: mole) of a pure substance, and the *specific heat capacity*, often called simply *specific heat*, is the heat capacity per unit mass of a material. Heat capacity is a physical property of a substance, which means that it depends on the state and properties of the substance under consideration.

The specific heats of monatomic gases, such as helium, are nearly constant with temperature. Diatomic gases such as hydrogen display some temperature dependence, and triatomic gases (e.g., carbon dioxide) still more.

Before the development of the laws of thermodynamics, heat was measured by changes in the states of the participating bodies.

Some general rules, with important exceptions, can be stated as follows.

In general, most bodies expand on heating. In this circumstance, heating a body at a constant volume increases the pressure it exerts on its constraining walls, while heating at a constant pressure increases its volume.

Beyond this, most substances have three ordinarily recognized states of matter, solid, liquid, and gas. Some can also exist in a plasma. Many have further, more finely differentiated, states of matter, such as for example, glass, and liquid crystal. In many cases, at fixed temperature and pressure, a substance can exist in several distinct states of matter in what might be viewed as the same 'body'. For example, ice may float in a glass of water. Then the ice and the water are said to constitute two phases within the 'body'. Definite rules are known, telling how distinct phases may coexist in a body. Mostly, at a fixed pressure, there is a definite temperature at which heating causes a solid to melt or evaporate, and a definite temperature at which heating causes a liquid to evaporate. In such cases, cooling has the reverse effects.

All of these, the commonest cases, fit with a rule that heating can be measured by changes of state of a body. Such cases supply what are called *thermometric bodies*, that allow the definition of empirical temperatures. Before 1848, all temperatures were defined in this way. There was thus a tight link, apparently logically determined, between heat and temperature, though they were recognized as conceptually thoroughly distinct, especially by Joseph Blackin the later eighteenth century.

There are important exceptions. They break the obviously apparent link between heat and temperature. They make it clear that empirical definitions of temperature are contingent on the peculiar properties of particular thermometric substances, and are thus precluded from the title 'absolute'. For example, water contracts on being heated near 277 K. It cannot be used as a thermometric substance near that temperature. Also, over a certain temperature range, ice contracts on heating. Moreover, many substances can exist in metastable states, such as with negative pressure, that survive only transiently and in very special conditions. Such facts, sometimes called 'anomalous', are some of the reasons for the thermodynamic definition of absolute temperature.

In the early days of measurement of high temperatures, another factor was important, and used by Josiah Wedgwood in his pyrometer. The temperature reached in a process was estimated by the shrinkage of a sample of clay. The higher the temperature, the more the shrinkage. This was the only available more or less reliable method of measurement of temperatures above 1000 °C. But such shrinkage is irreversible. The clay does not expand again on cooling. That is why it could be used for the measurement. But only once. It is not a thermometric material in the usual sense of the word.

Nevertheless, the thermodynamic definition of absolute temperature does make essential use of the concept of heat, with proper circumspection.

Thermodynamics

Thermodynamics is the branch of physics that deals with the relationships between heat and other forms of energy. In particular, it describes how thermal energy is converted to and from other forms of energy and how it affects matter.

Thermal energy is the energy a substance or system has due to its temperature, i.e., the energy of moving or vibrating molecules, according to the Energy Education website of the Texas Education Agency. Thermodynamics involves measuring this energy, which can be "exceedingly complicated," according to David McKee, a professor of physics at Missouri Southern State University. "The systems that we study in thermodynamics consist of very large numbers of atoms or molecules interacting in complicated ways. But, if these systems meet the right criteria, which we call equilibrium, they can be described with a very small number of measurements or numbers. Often this is idealized as the mass of the system, the pressure of the system, and the volume of the system, or some other equivalent set of numbers. Three numbers describe 10^{26} or 10^{30} nominal independent variables."

Thermodynamics has many sections under it and is considered as a broad subject because it deals with topics that exist all around us and thus classification becomes necessary.

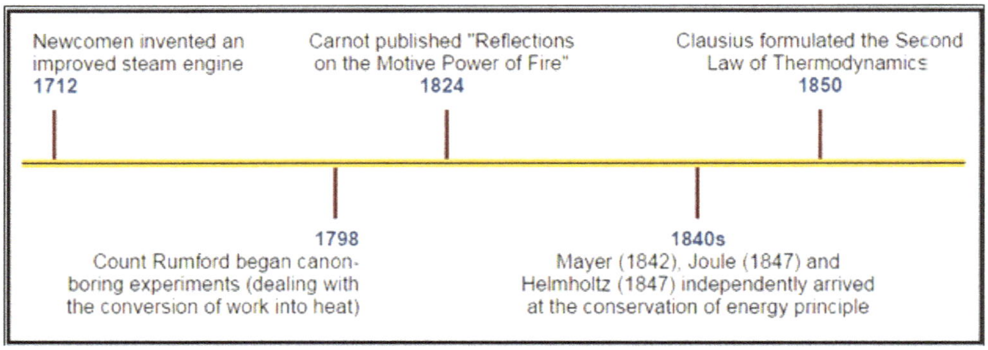

- Classical Thermodynamics – In this section, the behavior of matter is analyzed with a macroscopic approach. Units such as temperature and pressure are taken into consideration which helps the individuals to calculate other properties and to predict the characteristics of the matter that is undergoing the process.
- Statistical Thermodynamics – In this section, every molecule is under the spotlight i.e. the properties of each and every molecule and ways in which they interact are taken into consideration to characterize the behavior of a group of molecules.
- Pure Component Thermodynamics – As the name itself states, this section tries to describe the behavior of a system that has an unadulterated or pure constituent.

- Solution Thermodynamics – This section attempts to describe the behavior of a system that contains more than one chemical in the mixture.

Here, we will focus on the Classical part of the thermodynamics in our curriculum.

Laws of Thermodynamics

Thermodynamics is principally based on a set of four laws which are universally valid when applied to systems that fall within the constraints implied by each. In the various theoretical descriptions of thermodynamics these laws may be expressed in seemingly differing forms, but the most prominent formulations are the following:

- Zeroth law of thermodynamics: *If two systems are each in thermal equilibrium with a third, they are also in thermal equilibrium with each other.*

This statement implies that thermal equilibrium is an equivalence relation on the set of thermodynamic systems under consideration. Systems are said to be in equilibrium if the small, random exchanges between them (e.g. Brownian motion) do not lead to a net change in energy. This law is tacitly assumed in every measurement of temperature. Thus, if one seeks to decide if two bodies are at the same temperature, it is not necessary to bring them into contact and measure any changes of their observable properties in time. The law provides an empirical definition of temperature and justification for the construction of practical thermometers.

The zeroth law was not initially recognized as a law, as its basis in thermodynamical equilibrium was implied in the other laws. The first, second, and third laws had been explicitly stated prior and found common acceptance in the physics community. Once the importance of the zeroth law for the definition of temperature was realized, it was impracticable to renumber the other laws, hence it was numbered the *zeroth law*.

- First law of thermodynamics: *The internal energy of an isolated system is constant.*

The first law of thermodynamics is an expression of the principle of conservation of energy. It states that energy can be transformed (changed from one form to another), but cannot be created or destroyed.

The first law is usually formulated by saying that the change in the internal energy of a closed thermodynamic system is equal to the difference between the heat supplied to the system and the amount of work done by the system on its surroundings. It is important to note that internal energy is a state of the system, whereas heat and work modify the state of the system. In other words, a change of internal energy of a system may be achieved by any combination of heat and work added or removed from the system as long as those total to the change of internal energy. The manner by which a system achieves its internal energy is path independent.

- Second law of thermodynamics: *Heat cannot spontaneously flow from a colder location to a hotter location.*

The second law of thermodynamics is an expression of the universal principle of decay observable in nature. The second law is an observation of the fact that over time, differences in temperature, pressure, and chemical potential tend to even out in a physical system that is isolated from the outside world. Entropy is a measure of how much this process has progressed. The entropy of an isolated system which is not in equilibrium will tend to increase over time, approaching a maximum value at equilibrium. However, principles guiding systems that are far from equilibrium are still debatable. One of such principles is the maximum entropy production principle. It states that non-equilibrium systems behave such a way as to maximize its entropy production.

In classical thermodynamics, the second law is a basic postulate applicable to any system involving heat energy transfer; in statistical thermodynamics, the second law is a consequence of the assumed randomness of molecular chaos. There are many versions of the second law, but they all have the same effect, which is to explain the phenomenon of irreversibility in nature.

- Third law of thermodynamics: *As a system approaches absolute zero, all processes cease and the entropy of the system approaches a minimum value.*

The third law of thermodynamics is a statistical law of nature regarding entropy and the impossibility of reaching absolute zero of temperature. This law provides an absolute reference point for the determination of entropy. The entropy determined relative to this point is the absolute entropy. Alternate definitions are, "the entropy of all systems and of all states of a system is smallest at absolute zero," or equivalently "it is impossible to reach the absolute zero of temperature by any finite number of processes".

Absolute zero, at which all activity would stop if it were possible to happen, is −273.15 °C (degrees Celsius), or −459.67 °F (degrees Fahrenheit), or 0 K (kelvin), or 0° R (degrees Rankine).

System Models

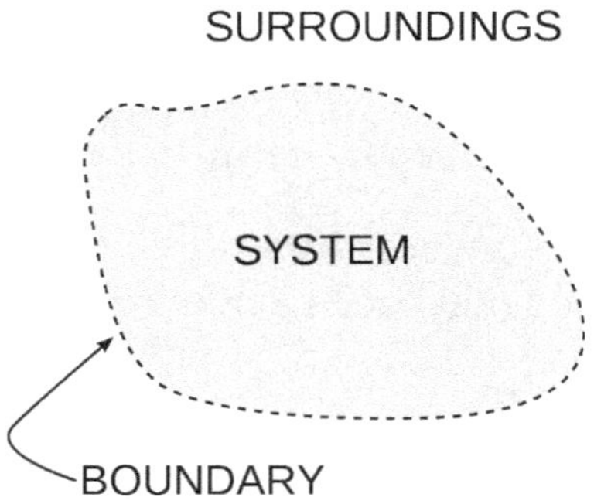

A diagram of a generic thermodynamic system

An important concept in thermodynamics is the thermodynamic system, which is a precisely defined region of the universe under study. Everything in the universe except the system is called the *surroundings*. A system is separated from the remainder of the universe by a *boundary* which may be a physical boundary or notional, but which by convention defines a finite volume. Exchanges of work, heat, or matter between the system and the surroundings take place across this boundary.

In practice, the boundary of a system is simply an imaginary dotted line drawn around a volume within which is going to be a change in the internal energy of that volume. Anything that passes across the boundary that effects a change in the internal energy of the system needs to be accounted for in the energy balance equation. The volume can be the region surrounding a single atom resonating energy, such as Max Planck defined in 1900; it can be a body of steam or air in a steam engine, such as Sadi Carnot defined in 1824; it can be the body of a tropical cyclone, such as Kerry Emanuel theorized in 1986 in the field of atmospheric thermodynamics; it could also be just one nuclide (i.e. a system of quarks) as hypothesized in quantum thermodynamics, or the event horizon of a black hole.

Boundaries are of four types: fixed, movable, real, and imaginary. For example, in an engine, a fixed boundary means the piston is locked at its position, within which a constant volume process might occur. If the piston is allowed to move that boundary is movable while the cylinder and cylinder head boundaries are fixed. For closed systems, boundaries are real while for open systems boundaries are often imaginary. In the case of a jet engine, a fixed imaginary boundary might be assumed at the intake of the engine, fixed boundaries along the surface of the case and a second fixed imaginary boundary across the exhaust nozzle.

Generally, thermodynamics distinguishes three classes of systems, defined in terms of what is allowed to cross their boundaries:

Interactions of thermodynamic systems			
Type of system	**Mass flow**	**Work**	**Heat**
Open	✓	✓	✓
Closed	✗	✓	✓
Thermally isolated	✗	✓	✗
Mechanically isolated	✗	✗	✓
Isolated	✗	✗	✗

As time passes in an isolated system, internal differences of pressures, densities, and temperatures tend to even out. A system in which all equalizing processes have gone to completion is said to be in a state of thermodynamic equilibrium.

Once in thermodynamic equilibrium, a system's properties are, by definition, unchanging in time. Systems in equilibrium are much simpler and easier to understand than are systems which are not in equilibrium. Often, when analysing a dynamic thermodynamic process, the simplifying assumption is made that each intermediate state in the process is at equilibrium, producing thermodynamic processes which develop so slowly as to allow each intermediate step to be an equilibrium state and are said to be reversible processes.

States and Processes

When a system is at equilibrium under a given set of conditions, it is said to be in a definite thermodynamic state. The state of the system can be described by a number of state quantities that do not depend on the process by which the system arrived at its state. They are called intensive variables or extensive variables according to how they change when the size of the system changes. The properties of the system can be described by an equation of state which specifies the relationship between these variables. State may be thought of as the instantaneous quantitative description of a system with a set number of variables held constant.

A thermodynamic process may be defined as the energetic evolution of a thermodynamic system proceeding from an initial state to a final state. It can be described by process quantities. Typically, each thermodynamic process is distinguished from other processes in energetic character according to what parameters, such as temperature, pressure, or volume, etc., are held fixed; Furthermore, it is useful to group these processes into pairs, in which each variable held constant is one member of a conjugate pair.

Several commonly studied thermodynamic processes are:

- Adiabatic process: occurs without loss or gain of energy by heat
- Isenthalpic process: occurs at a constant enthalpy
- Isentropic process: a reversible adiabatic process, occurs at a constant entropy
- Isobaric process: occurs at constant pressure
- Isochoric process: occurs at constant volume (also called isometric/isovolumetric)
- Isothermal process: occurs at a constant temperature
- Steady state process: occurs without a change in the internal energy

Instrumentation

There are two types of thermodynamic instruments, the meter and the reservoir. A thermodynamic meter is any device which measures any parameter of a thermodynamic system. In some cases, the thermodynamic parameter is actually defined in terms of an idealized measuring instrument. For example, the zeroth law states that if two bodies are in thermal equilibrium with a third body, they are also in thermal equilibrium with each other. This principle, as noted by James Maxwell in 1872, asserts that it is possible to measure temperature. An idealized thermometer is a sample of an ideal gas at constant pressure. From the ideal gas law $pV=nRT$, the volume of such a sample can be used as an indicator of temperature; in this manner it defines temperature. Although pressure is defined mechanically, a pressure-measuring device, called a barometer may also be constructed from a sample of an ideal gas held at a constant temperature. A calorimeter is a device which is used to measure and define the internal energy of a system.

A thermodynamic reservoir is a system which is so large that its state parameters are not appreciably altered when it is brought into contact with the system of interest. When the reservoir is brought into contact with the system, the system is brought into equilibrium with the reservoir. For example, a pressure reservoir is a system at a particular pressure, which imposes that pressure upon the system to which it is mechanically connected. The Earth's atmosphere is often used as a pressure reservoir. If ocean water is used to cool a power plant, the ocean is often a temperature reservoir in the analysis of the power plant cycle.

Conjugate Variables

The central concept of thermodynamics is that of energy, the ability to do work. By the First Law, the total energy of a system and its surroundings is conserved. Energy may be transferred into a system by heating, compression, or addition of matter, and extracted from a system by cooling, expansion, or extraction of matter. In mechanics, for example, energy transfer equals the product of the force applied to a body and the resulting displacement.

Conjugate variables are pairs of thermodynamic concepts, with the first being akin to a "force" applied to some thermodynamic system, the second being akin to the resulting "displacement," and the product of the two equalling the amount of energy transferred. The common conjugate variables are:

- Pressure-volume (the mechanical parameters);
- Temperature-entropy (thermal parameters);
- Chemical potential-particle number (material parameters).

Potentials

Thermodynamic potentials are different quantitative measures of the stored energy in a system. Potentials are used to measure the energy changes in systems as they evolve from an initial state to a final state. The potential used depends on the constraints of the system, such as constant temperature or pressure. For example, the Helmholtz and Gibbs energies are the energies available in a system to do useful work when the temperature and volume or the pressure and temperature are fixed, respectively.

The five most well known potentials are:

Name	Symbol	Formula	Natural variables
Internal energy	U	$\int (TdS - pdV + \sum_i \mu_i dN_i)$	$S, V, \{N_i\}$
Helmholtz free energy	F	$U - TS$	$T, V, \{N_i\}$
Enthalpy	H	$U + pV$	$S, p, \{N_i\}$
Gibbs free energy	G	$U + pV - TS$	$T, p, \{N_i\}$
Landau Potential (Grand potential)	Ω, Φ_G,	$U - TS - \sum_i \mu_i N_i$	$T, V, \{\mu_i\}$

where T is the temperature, S the entropy, p the pressure, V the volume, μ the chemical potential, N the number of particles in the system, and is the count of particles types in the system.

Thermodynamic potentials can be derived from the energy balance equation applied to a thermodynamic system. Other thermodynamic potentials can also be obtained through Legendre transformation.

Applied Fields

- Atmospheric thermodynamics
- Biological thermodynamics

- Black hole thermodynamics
- Chemical thermodynamics
- Classical thermodynamics
- Equilibrium thermodynamics
- Industrial ecology (re: Exergy)
- Maximum entropy thermodynamics
- Non-equilibrium thermodynamics
- Philosophy of thermal and statistical physics
- Psychrometrics
- Quantum thermodynamics
- Statistical thermodynamics
- Thermoeconomics

Enthalpy

Enthalpy is the measurement of energy in a thermodynamic system. The quantity of enthalpy equals to the total content of heat of a system, equivalent to the system's internal energy plus the product of volume and pressure.

Mathematically, the enthalpy, H, equals to the sum of the internal energy, E, and the product of the pressure, P, and volume, V, of the system.

$H = E + PV$

Entropy

Entropy is a thermodynamic quantity whose value depends on the physical state or condition of a system. In other words, it is a thermodynamic function used to measure the randomness or disorder of a system. For example, the entropy of a solid, where the particles are not free to move, is less than the entropy of a gas, where the particles will fill the container.

Heat Transfer

Heat transfer is the process of transfer of heat from high temperature reservoir to low temperature reservoir. In terms of the thermodynamic system, heat transfer is the movement of heat across the boundary of the system due to temperature difference between the system and the surroundings. The heat transfer can also take place within the system due to temperature difference at various points inside the system. The difference in temperature is considered to be 'potential' that causes the flow of heat and the heat itself is called as flux.

Modes of Heat Transfer

- Conduction: An energy transfer across a system boundary due to a temperature difference by the mechanism of intermolecular interactions. Conduction needs matter and does not require any bulk motion of matter.

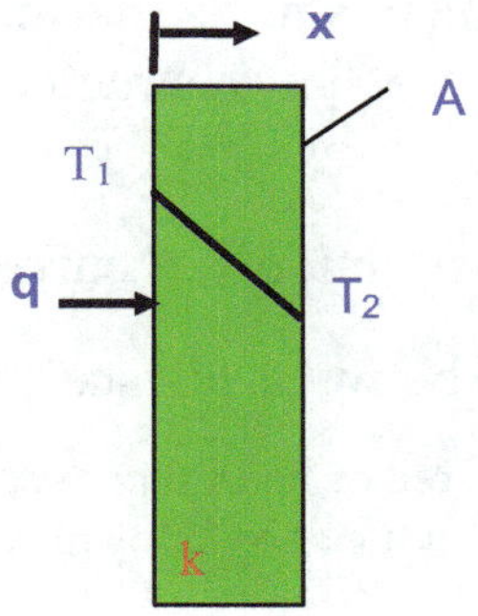

Conduction rate equation is described by the Fourier Law:

$$q'' = -kA\nabla T$$

where: q = heat flow vector, (W)

k = thermal conductivity, a thermodynamic property of the material.

(W/m K)

A = Cross sectional area in direction of heat flow. (m^2)

∇T = Gradient of temperature (K/m)

$= \partial T / \partial x\, i + \partial T / \partial y\, j + \partial T / \partial z\, k$

Note: Since this is a vector equation, it is often convenient to work with one component of the vector. For example, in the x direction:

$$q_x = -k A_x dT/dx$$

In circular coordinates it may convenient to work in the radial direction

$$qr = -k A_r dT/dr$$

- Convection: An energy transfer across a system boundary due to a temperature difference by the combined mechanisms of intermolecular interactions and bulk transport. Convection needs fluid matter.

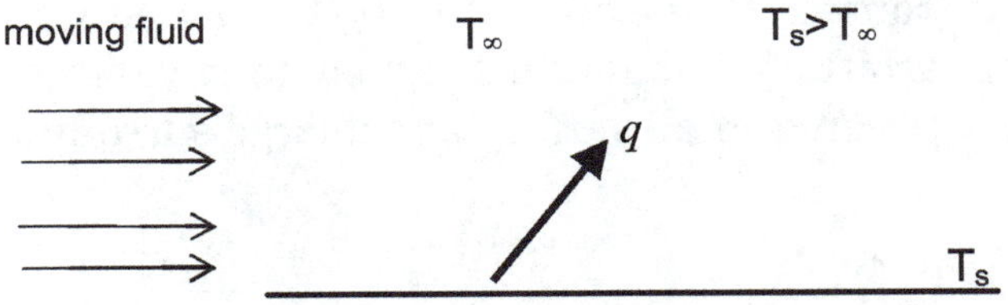

Newton's Law of Cooling:

$$q = h\,As\;\Delta T$$

where: q = heat flow from surface, a scalar, (W)

h = heat transfer coefficient (which is not a thermodynamic property of the material, but may depend on geometry of surface, flow characteristics, thermodynamic properties of the fluid, etc. ($W/m^2\ K$)

As = Surface area from which convection is occurring. (m^2)

$\Delta T = T_S - T_\varpi$ = Temperature Difference between surface and coolant. (K)

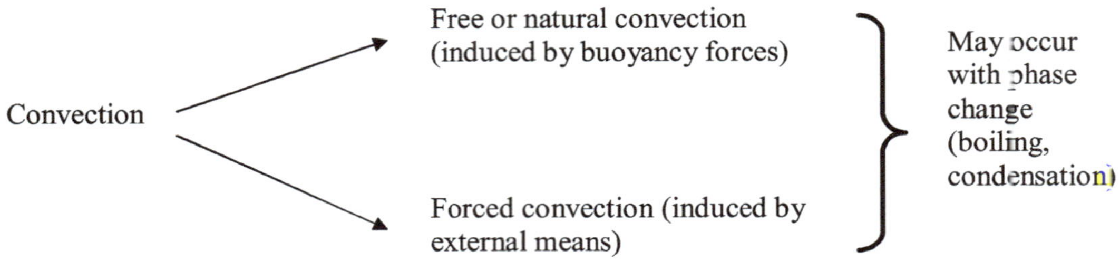

Table- Typical values of h ($W/m^2\ K$)

Free convection	gases: 2 - 25 liquid: 50 – 100
Forced convection	gases: 25 - 250 liquid: 50 - 20,000
Boiling/Condensation	2500 -100,000

- Radiation: Radiation heat transfer involves the transfer of heat by electromagnetic radiation that arises due to the temperature of the body. Radiation does not need matter. Emissive power of a surface:

$$E = \sigma \varepsilon T^4_S\ (W/m^2)$$

where: ε= emissivity, which is a surface property (ε= 1 is black body).

σ = Steffan Boltzman constant = $5.67 \times 10^{-8}\ W/m^2\ K^2$.

Ts = Absolute temperature of the surface (K).

The above equation is derived from Stefan Boltzman law, which describes a gross heat emission rather than heat transfer. The expression for the actual radiation heat transfer rate between surfaces having arbitrary orientations can be quite complex. However, the rate of radiation heat exchange between a small surface and a large surrounding is given by the following expression:

$$q = \varepsilon \cdot \sigma \cdot A \cdot (T_s^4 - T_{sur}^4)$$

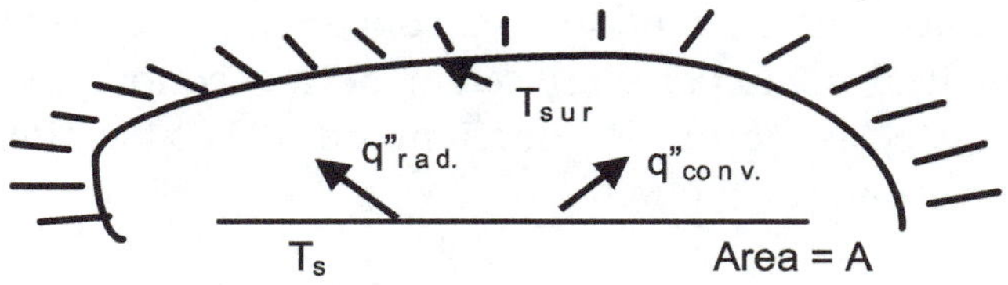

where:

ε = Surface Emissivity

A= Surface Area

T_s = Absolute temperature of surface. (K)

T_{sur} = Absolute temperature of surroundings.(K)

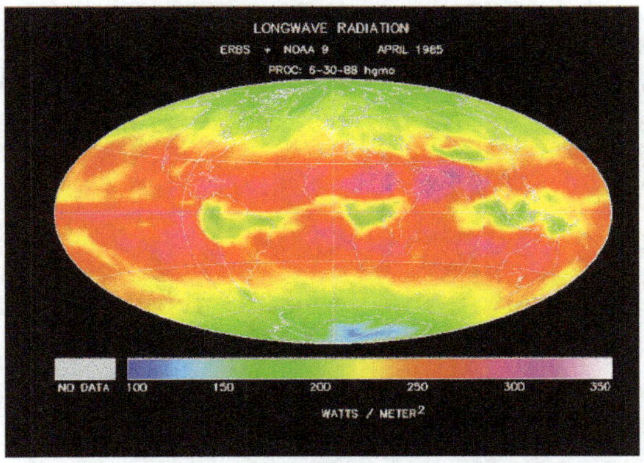

Earth's long wave thermal radiation intensity, from clouds, atmosphere and surface

Heat is defined in physics as the transfer of thermal energy across a well-defined boundary around a thermodynamic system. The thermodynamic free energy is the amount of work that a thermodynamic system can perform. Enthalpy is a thermodynamic potential, designated by the letter "H", that is the sum of the internal energy of the system (U) plus the product of pressure (P) and volume(V). Joule is a unit to quantify energy, work, or the amount of heat.

Heat transfer is a process function (or path function), as opposed to functions of state; therefore, the amount of heat transferred in a thermodynamic process that changes the state of a system depends on how that process occurs, not only the net difference between the initial and final states of the process.

Thermodynamic and mechanical heat transfer is calculated with the heat transfer coefficient, the proportionality between the heat flux and the thermodynamic driving force for the flow of heat. Heat flux is a quantitative, vectorial representation of heat-flow through a surface.

In engineering contexts, the term *heat* is taken as synonymous to thermal energy. This usage has its origin in the historical interpretation of heat as a fluid (*Caloric*) that can be transferred by various causes, and that is also common in the language of laymen and everyday life.

The transport equations for thermal energy (Fourier's law), mechanical momentum (Newton's law for fluids), and mass transfer (Fick's laws of diffusion) are similar, and analogies among these three transport processes have been developed to facilitate prediction of conversion from any one to the others.

Thermal engineering concerns the generation, use, conversion, and exchange of heat transfer. As such, heat transfer is involved in almost every sector of the economy. Heat transfer is classified into various mechanisms, such as thermal conduction, thermal convection, thermal radiation, and transfer of energy by phase changes.

Mechanisms

The fundamental modes of heat transfer are:

Advection

Advection is the transport mechanism of a fluid from one location to another, and is dependent on motion and momentum of that fluid.

Conduction or Diffusion

The transfer of energy between objects that are in physical contact. Thermal conductivityis the property of a material to conduct heat and evaluated primarily in terms of Fourier's Law for heat conduction.

Convection

The transfer of energy between an object and its environment, due to fluid motion. The average temperature is a reference for evaluating properties related to convective heat transfer.

Radiation

The transfer of energy by the emission of electromagnetic radiation.

Advection

By transferring matter, energy—including thermal energy—is moved by the physical transfer of a hot or cold object from one place to another. This can be as simple as placing hot water in a bottle and heating a bed, or the movement of an iceberg in changing ocean currents. A practical example is thermal hydraulics. This can be described by the formula:

$$Q = v\rho c_p \Delta T$$

where

- Q is heat flux (W/m^2),
- ρ is density (kg/m^3),
- c_p is heat capacity at constant pressure (J/kg·K),
- ΔT is the change in temperature (K),
- v is velocity (m/s).

Conduction

On a microscopic scale, heat conduction occurs as hot, rapidly moving or vibrating atoms and molecules interact with neighboring atoms and molecules, transferring some of their energy (heat) to these neighboring particles. In other words, heat is transferred by conduction when adjacent atoms vibrate against one another, or as electrons move from one atom to another. Conduction is the most significant means of heat transfer within a solid or between solid objects in thermal contact. Fluids—especially gases—are less conductive. Thermal contact conductance is the study of heat conduction between solid bodies in contact. The process of heat transfer from one place to another place without the movement of particles is called conduction. Example: Heat transfer through Metal rods. *Steady state conduction* is a form of conduction that happens when the temperature difference driving the conduction is constant, so that after an equilibration time, the spatial distribution of temperatures in the conducting object does not change any further. In steady state conduction, the amount of heat entering a section is equal to amount of heat coming out.

Transient conduction occurs when the temperature within an object changes as a function of time. Analysis of transient systems is more complex and often calls for the application of approximation theories or numerical analysis by computer.

Convection

The flow of fluid may be forced by external processes, or sometimes (in gravitational fields) by buoyancy forces caused when thermal energy expands the fluid (for example in a fire plume), thus influencing its own transfer. The latter process is often called "natural convection". All convective processes also move heat partly by diffusion, as well. Another form of convection is forced convection. In this case the fluid is forced to flow by using a pump, fan or other mechanical means.

Convective heat transfer, or convection, is the transfer of heat from one place to another by the movement of fluids, a process that is essentially the transfer of heat via mass transfer. Bulk motion of fluid enhances heat transfer in many physical situations, such as (for example) between a solid surface and the fluid. Convection is usually the dominant form of heat transfer in liquids and gases. Although sometimes discussed as a third method of heat transfer, convection is usually used to describe the combined effects of heat conduction within the fluid (diffusion) and heat transference by bulk fluid flow streaming. The process of transport by fluid streaming is known as advection, but pure advection is a term that is generally associated only with mass transport in fluids, such as advection of pebbles in a river. In the case of heat transfer in fluids, where transport by advection in a fluid is always also accompanied by transport via heat diffusion (also known as heat conduction) the process of heat convection is understood to refer to the sum of heat transport by advection and diffusion/conduction.

Free, or natural, convection occurs when bulk fluid motions (streams and currents) are caused by buoyancy forces that result from density variations due to variations of temperature in the fluid. *Forced* convection is a term used when the streams and currents in the fluid are induced by external means—such as fans, stirrers, and pumps—creating an artificially induced convection current.

Convection-cooling

Convective cooling is sometimes described as Newton's law of cooling:

> *The rate of heat loss of a body is proportional to the temperature difference between the body and its surroundings.*

However, by definition, the validity of Newton's law of Cooling requires that the rate of heat loss from convection be a linear function of ("proportional to") the temperature difference that drives heat transfer, and in convective cooling this is sometimes not the case. In general, convection is not linearly dependent on temperature gradients, and in some cases is strongly nonlinear. In these cases, Newton's law does not apply.

Convection vs. Conduction

In a body of fluid that is heated from underneath its container, conduction and convection can be considered to compete for dominance. If heat conduction is too great, fluid moving down by convection is heated by conduction so fast that its downward movement will be stopped due to its buoyancy, while fluid moving up by convection is cooled by conduction so fast that its driving buoyancy will diminish. On the other hand, if heat conduction is very low, a large temperature gradient may be formed and convection might be very strong.

The Rayleigh number (Ra) is the product of the Grashof (*Gr*) and Prandtl (*Pr*) numbers. It is a measure which determines the relative strength of conduction and convection.

$$Ra = GrPr = \frac{g\Delta\rho L^3}{\mu\alpha} = \frac{g\beta\Delta T L^3}{\nu\alpha}$$

where

- g is acceleration due to gravity,
- ρ is the density with $\Delta\rho$ being the density difference between the lower and upper ends,
- μ is the dynamic viscosity,
- α is the Thermal diffusivity,
- β is the volume thermal expansivity (sometimes denoted α elsewhere),
- T is the temperature,
- ν is the kinematic viscosity, and
- L is characteristic length.

The Rayleigh number can be understood as the ratio between the rate of heat transfer by convection to the rate of heat transfer by conduction; or, equivalently, the ratio between the corresponding timescales (i.e. conduction timescale divided by convection timescale), up to a numerical factor. This can be seen as follows, where all calculations are up to numerical factors depending on the geometry of the system.

The buoyancy force driving the convection is roughly $g\Delta\rho L^3$, so the corresponding pressure is roughly $g\Delta\rho L$. In steady state, this is canceled by the shear stress due to viscosity, and therefore roughly equals $\mu V / L = \mu / T_{conv}$, where V is the typical fluid velocity due to convection and T_{conv} the order of its timescale. The conduction timescale, on the other hand, is of the order of $T_{cond} = L^2 / \alpha$.

Convection occurs when the Rayleigh number is above 1,000–2,000.

Radiation

Thermal radiation occurs through a vacuum or any transparent medium (solid or fluid). It is the transfer of energy by means of photons in electromagnetic waves governed by the same laws.

Thermal radiation is energy emitted by matter as electromagnetic waves, due to the pool of thermal energy in all matter with a temperature above absolute zero. Thermal radiation propagates without the presence of matter through the vacuum of space.

Thermal radiation is a direct result of the random movements of atoms and molecules in matter. Since these atoms and molecules are composed of charged particles (protons and electrons), their movement results in the emission of electromagnetic radiation, which carries energy away from the surface.

The Stefan-Boltzmann equation, which describes the rate of transfer of radiant energy, is as follows for an object in a vacuum :

$$Q = \epsilon \sigma T^4.$$

For radiative transfer between two objects, the equation is as follows:

$$Q = \epsilon \sigma (T_a^4 - T_b^4),$$

where Q is the heat flux, ε is the emissivity (unity for a black body), σ is the Stefan–Boltzmann constant, and T is the absolute temperature (in kelvins or degrees Rankine).

Radiation is typically only important for very hot objects, or for objects with a large temperature difference.

Radiation from the sun, or solar radiation, can be harvested for heat and power. Unlike conductive and convective forms of heat transfer, thermal radiation can be concentrated in a small spot by using reflecting mirrors, which is exploited in concentrating solar powergeneration. For example, the sunlight reflected from mirrors heats the PS10 solar power tower and during the day it can heat water to 285 °C (545 °F).

Phase Transition

Lightning is a highly visible form of energy transfer and is an example of plasma present at Earth's surface. Plasma temperatures in lightning can approach 28,000 kelvins (27,726.85 °C) (49,940.33 °F) and electron densities may exceed 10^{24} m^{-3}

Phase transition or phase change, takes place in a thermodynamic system from one phase or state of matter to another one by heat transfer. Phase change examples are the melting of ice or the boiling of water. The Mason equationexplains the growth of a water droplet based on the effects of heat transport on evaporation and condensation.

Phase transitions involve the four fundamental states of matter:

- Solid – Deposition, freezing and solid to solid transformation.
- Gas – Boiling/evaporation, recombination/deionization, and sublimation.
- Liquid – Condensation and melting/fusion.
- Plasma – Ionization.

Boiling

Nucleate boiling of water

The boiling point of a substance is the temperature at which the vapor pressure of the liquid equals the pressure surrounding the liquid and the liquid evaporates resulting in an abrupt change in vapor volume.

Saturation temperature means boiling point. The saturation temperature is the temperature for a corresponding saturation pressure at which a liquid boils into its vapor phase. The liquid can be said to be saturated with thermal energy. Any addition of thermal energy results in a phase transition.

At standard atmospheric pressure and low temperatures, no boiling occurs and the heat transfer rate is controlled by the usual single-phase mechanisms. As the surface temperature is increased, local boiling occurs and vapor bubbles nucleate, grow into the surrounding cooler fluid, and collapse. This is sub-cooled nucleate boiling, and is a very efficient heat transfer mechanism. At high bubble generation rates, the bubbles begin to interfere and the heat flux no longer increases rapidly with surface temperature (this is the departure from nucleate boiling, or DNB).

At similar standard atmospheric pressure and high temperatures, the hydrodynamically-quieter regime of film boiling is reached. Heat fluxes across the stable vapor layers are low, but rise slowly with temperature. Any contact between fluid and the surface that may be seen probably leads to the extremely rapid nucleation of a fresh vapor layer ("spontaneous nucleation"). At higher temperatures still, a maximum in the heat flux is reached (the critical heat flux, or CHF).

The Leidenfrost Effect demonstrates how nucleate boiling slows heat transfer due to gas bubbles on the heater's surface. As mentioned, gas-phase thermal conductivity is much lower than liquid-phase thermal conductivity, so the outcome is a kind of "gas thermal barrier".

Condensation

Condensation occurs when a vapor is cooled and changes its phase to a liquid. During condensation, the latent heat of vaporization must be released. The amount of the heat is the same as that absorbed during vaporization at the same fluid pressure.

There are several types of condensation:

- Homogeneous condensation, as during a formation of fog.
- Condensation in direct contact with subcooled liquid.
- Condensation on direct contact with a cooling wall of a heat exchanger: This is the most common mode used in industry:
    - Filmwise condensation is when a liquid film is formed on the subcooled surface, and usually occurs when the liquid wets the surface.
    - Dropwise condensation is when liquid drops are formed on the subcooled surface, and usually occurs when the liquid does not wet the surface.

Dropwise condensation is difficult to sustain reliably; therefore, industrial equipment is normally designed to operate in filmwise condensation mode.

Melting

Ice melting

Melting is a thermal process that results in the phase transition of a substance from a solid to a liquid. The internal energy of a substance is increased, typically with heat or pressure, resulting in a rise of its temperature to the melting point, at which the ordering of ionic or molecular entities in the solid breaks down to a less ordered state and the solid liquefies. Molten substances generally have reduced viscosity with elevated temperature; an exception to this maxim is the element sulfur, whose viscosity increases to a point due to polymerization and then decreases with higher temperatures in its molten state.

Modeling Approaches

Heat transfer can be modeled in the following ways.

Climate Models

Climate models study the radiant heat transfer by using quantitative methods to simulate the interactions of the atmosphere, oceans, land surface, and ice.

Heat Equation

The heat equation is an important partial differential equation that describes the distribution of heat (or variation in temperature) in a given region over time. In some cases, exact solutions of the equation are available; in other cases the equation must be solved numerically using computational methods.

Lumped System Analysis

Lumped system analysis often reduces the complexity of the equations to one first-order linear differential equation, in which case heating and cooling are described by a simple exponential solution, often referred to as Newton's law of cooling.

System analysis by the lumped capacitance model is a common approximation in transient conduction that may be used whenever heat conduction within an object is much faster than heat conduction across the boundary of the object. This is a method of approximation that reduces one aspect of the transient conduction system—that within the object—to an equivalent steady state system. That is, the method assumes that the temperature within the object is completely uniform, although its value may be changing in time.

In this method, the ratio of the conductive heat resistance within the object to the convective heat transfer resistance across the object's boundary, known as the *Biot number*, is calculated. For small Biot numbers, the approximation of *spatially uniform temperature within the object* can be used: it can be presumed that heat transferred into the object has time to uniformly distribute itself, due to the lower resistance to doing so, as compared with the resistance to heat entering the object.

Engineering

Heat transfer has broad application to the functioning of numerous devices and systems. Heat-transfer principles may be used to preserve, increase, or decrease temperature in a wide variety of circumstances. Heat transfer methods are used in numerous disciplines, such as automotive engineering, thermal management of electronic devices and systems, climate control, insulation, materials processing, and power station engineering.

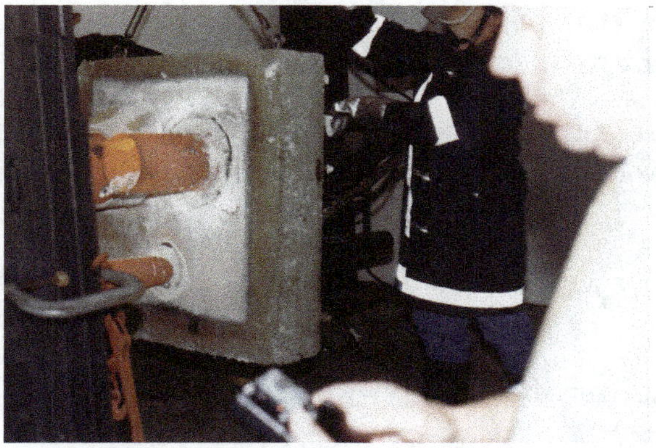

Heat exposure as part of a fire test for firestop products

Insulation, Radiance and Resistance

Thermal insulators are materials specifically designed to reduce the flow of heat by limiting conduction, convection, or both. Thermal resistance is a heat property and the measurement by which an object or material resists to heat flow (heat per time unit or thermal resistance) to temperature difference.

Radiance or spectral radiance are measures of the quantity of radiation that passes through or is emitted. Radiant barriers are materials that reflect radiation, and therefore reduce the flow of heat from radiation sources. Good insulators are not necessarily good radiant barriers, and vice versa. Metal, for instance, is an excellent reflector and a poor insulator.

The effectiveness of a radiant barrier is indicated by its reflectivity, which is the fraction of radiation reflected. A material with a high reflectivity (at a given wavelength) has a low emissivity (at that same wavelength), and vice versa. At any specific wavelength, reflectivity=1 - emissivity. An ideal radiant barrier would have a reflectivity of 1, and would therefore reflect 100 percent of incoming radiation. Vacuum flasks, or Dewars, are silvered to approach this ideal. In the vacuum of space, satellites use multi-layer insulation, which consists of many layers of aluminized (shiny) Mylar to greatly reduce radiation heat transfer and control satellite temperature.

Devices

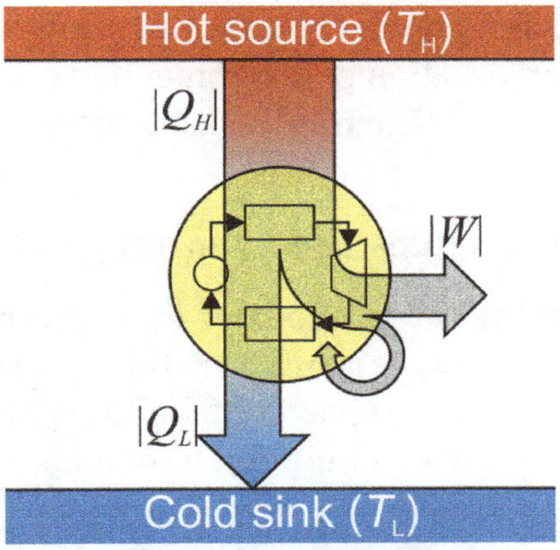

Schematic flow of energy in a heat engine

A heat engine is a system that performs the conversion of a flow of thermal energy (heat) to mechanical energy to perform mechanical work.

A thermocouple is a temperature-measuring device and widely used type of temperature sensor for measurement and control, and can also be used to convert heat into electric power.

A thermoelectric cooler is a solid state electronic device that pumps (transfers) heat from one side of the device to the other when electric current is passed through it. It is based on the Peltier effect.

A thermal diode or thermal rectifier is a device that causes heat to flow preferentially in one direction.

Heat Exchangers

A heat exchanger is used for more efficient heat transfer or to dissipate heat. Heat exchangers are widely used in refrigeration, air conditioning, space heating, power generation, and chemical processing. One common example of a heat exchanger is a car's radiator, in which the hot coolant fluid is cooled by the flow of air over the radiator's surface.

Common types of heat exchanger flows include parallel flow, counter flow, and cross flow. In parallel flow, both fluids move in the same direction while transferring heat; in counter flow, the fluids move in opposite directions; and in cross flow, the fluids move at right angles to each other. Common constructions for heat exchanger include shell and tube, double pipe, extruded finned pipe, spiral fin pipe, u-tube, and stacked plate.

A heat sink is a component that transfers heat generated within a solid material to a fluid medium, such as air or a liquid. Examples of heat sinks are the heat exchangers used in refrigeration and air conditioning systems or the radiator in a car. A heat pipe is another heat-transfer device that combines thermal conductivity and phase transition to efficiently transfer heat between two solid interfaces.

Applications

Architecture

Efficient energy use is the goal to reduce the amount of energy required in heating or cooling. In architecture, condensation and air currents can cause cosmetic or structural damage. An energy audit can help to assess the implementation of recommended corrective procedures. For instance, insulation improvements, air sealing of structural leaks or the addition of energy-efficient windows and doors.

- Smart meter is a device that records electric energy consumption in intervals.
- Thermal transmittance is the rate of transfer of heat through a structure divided by the difference in temperature across the structure. It is expressed in watts per square meter per kelvin, or $W/(m^2K)$. Well-insulated parts of a building have a low thermal transmittance, whereas poorly-insulated parts of a building have a high thermal transmittance.
- Thermostat is a device to monitor and control temperature.

Climate Engineering

Climate engineering consists of carbon dioxide removal and solar radiation management. Since the amount of carbon dioxide determines the radiative balance of Earth atmosphere, carbon dioxide removal techniques can be applied to reduce the radiative forcing. Solar radiation management is the attempt to absorb less solar radiation to offset the effects of greenhouse gases.

An example application in climate engineering includes the creation of Biochar through the pyrolysisprocess. Thus, storing greenhouse gases in carbon reduces the radiative forcing capacity in the atmosphere, causing more long-wave (infrared) radiation out to Space

Greenhouse Effect

The greenhouse effect is a process by which thermal radiation from a planetary surface is absorbed by atmospheric greenhouse gases, and is re-radiated in all directions. Since part of this re-radiation is back towards the surface and the lower atmosphere, it results in an elevation of the average surface temperature above what it would be in the absence of the gases.

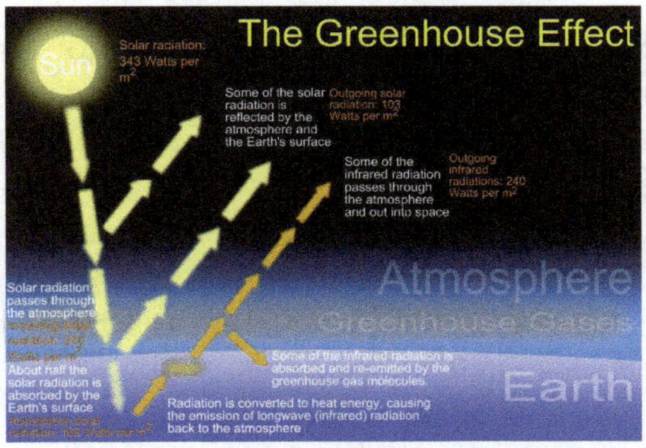

A representation of the exchanges of energy between the source (the Sun), the Earth's surface, the Earth's atmosphere, and the ultimate sink outer space. The ability of the atmosphere to capture and recycle energy emitted by the Earth surface is the defining characteristic of the greenhouse effect

Heat Transfer in the Human Body

The principles of heat transfer in engineering systems can be applied to the human body in order to determine how the body transfers heat. Heat is produced in the body by the continuous metabolism of nutrients which provides energy for the systems of the body. The human body must maintain a consistent internal temperature in order to maintain healthy bodily functions. Therefore, excess heat must be dissipated from the body to keep it from overheating. When a person engages in elevated levels of physical activity, the body requires additional fuel which increases the metabolic rate and the rate of heat production. The body must then use additional methods to remove the additional heat produced in order to keep the internal temperature at a healthy level.

Heat transfer by convection is driven by the movement of fluids over the surface of the body. This convective fluid can be either a liquid or a gas. For heat transfer from the outer surface of the body, the convection mechanism is dependent on the surface area of the body, the velocity of the air, and the temperature gradient between the surface of the skin and the ambient air. The normal temperature of the body is approximately 37 °C. Heat transfer occurs more readily when the temperature of the surroundings is significantly less than the normal body temperature. This concept explains why a person feels "cold" when not enough covering is worn when exposed to a cold environment. Clothing can be considered an insulator which provides thermal resistance to heat flow over the covered portion of the body. This thermal resistance causes the temperature on the surface of the clothing to be less than the temperature on the surface of the skin. This smaller temperature gradient between the surface temperature and the ambient temperature will cause a lower rate of heat transfer than if the skin were not covered.

In order to ensure that one portion of the body is not significantly hotter than another portion, heat must be distributed evenly through the bodily tissues. Blood flowing through blood vessels acts as a convective fluid and helps to prevent any buildup of excess heat inside the tissues of the body. This flow of blood through the vessels can be modeled as pipe flow in an engineering system. The heat carried by the blood is determined by the temperature of the surrounding tissue, the diameter of the blood vessel, the thickness of the fluid, velocity of the flow, and the heat transfer coefficient of the blood. The velocity, blood vessel diameter, and the fluid thickness can all be related with the Reynolds Number, a dimensionless number used in fluid mechanics to characterize the flow of fluids.

Latent heat loss, also known as evaporative heat loss, accounts for a large fraction of heat loss from the body. When the core temperature of the body increases, the body triggers sweat glands in the skin to bring additional moisture to the surface of the skin. The liquid is then transformed into vapor which removes heat from the surface of the body. The rate of evaporation heat loss is directly related to the vapor pressure at the skin surface and the amount of moisture present on the skin. Therefore, the maximum of heat transfer will occur when the skin is completely wet. The body continuously loses water by evaporation but the most significant amount of heat loss occurs during periods of increased physical activity.

Cooling Techniques

Evaporative Cooling

Evaporative cooling happens when water vapor is added to the surrounding air. The energy needed to evaporate the water is taken from the air in the form of sensible heat and converted into latent heat, while the air remains at a constant enthalpy. Latent heat describes the amount of heat that is needed to evaporate the liquid; this heat comes from the liquid itself and the surrounding gas and surfaces. The greater the difference between the two temperatures, the greater the evaporative cooling effect. When the temperatures are the same, no net evaporation of water in air occurs; thus, there is no cooling effect.

A traditional air cooler in Mirzapur, Uttar Pradesh, India

Laser Cooling

In quantum physics, laser cooling is used to achieve temperatures of near absolute zero (−273.15 °C, −459.67 °F) of atomic and molecular samples to observe unique quantum effects that can only occur at this heat level.

- Doppler cooling is the most common method of laser cooling.
- Sympathetic cooling is a process in which particles of one type cool particles of another type. Typically, atomic ions that can be directly laser-cooled are used to cool nearby ions or atoms. This technique allows cooling of ions and atoms that cannot be laser cooled directly.

Magnetic Cooling

Magnetic evaporative cooling is a process for lowering the temperature of a group of atoms, after pre-cooled by methods such as laser cooling. Magnetic refrigeration cools below 0.3K, by making use of the magnetocaloric effect.

Radiative Cooling

Radiative cooling is the process by which a body loses heat by radiation. Outgoing energy is an important effect in the Earth's energy budget. In the case of the Earth-atmosphere system, it refers to the process by which long-wave (infrared) radiation is emitted to balance the absorption of short-wave (visible) energy from the Sun. Convective transport of heat and evaporative transport of latent heat both remove heat from the surface and redistribute it in the atmosphere.

Thermal Energy Storage

Thermal energy storage includes technologies for collecting and storing energy for later use. It may be employed to balance energy demand between day and nighttime. The thermal reservoir may be maintained at a temperature above or below that of the ambient environment. Applications include space heating, domestic or process hot water systems, or generating electricity.

Temperature Gradient

This radiosonde was found in the Isar Valley. The instruments are attached to the lower end and transmit the soundings back down to earth via radio

The temperature gradient (called temp hereafter) tells us how much colder the air gets as we travel vertically up through the atmosphere. It is normally drawn as a graph in a coordinate system, where Y is the altitude (or pressure) and X is the temperature. Vertical resolution is often set to 100m.

The faster the temperature decreases with altitude, the higher the instability of the atmosphere – and that is what makes the temp so interesting to us, as we generally want to have an unstable atmosphere so that thermals may form.

If the temperature decreases slowly with altitude, or even briefly increases, the atmosphere is stable. The more stable it becomes, the unlikelier it is to have good thermal development.

The temp is a great tool for predicting a number of things about the flying conditions of any given day.

If we have the temp, the humidity at ground level and the predicted maximum temperature of the day we may already deduct the following values:

- Cloud base altitude,
- Trigger temperature (the temperature that air pockets must reach in order to be released as thermals),
- The likelihood of thunderstorms, the amount of cloud cover and the expected day quality in terms of thermal strength.

The temperature as function of altitude is physically acquired by releasing balloons (called radiosondes) sending their measurements (soundings) back to earth via radio.

The actual coordinate system, with its complex combination of isotherms, saturated and dry adiabats and the mixing ratio, will not be described in detail – my aim is to point to the general usefulness of the temperature gradient for free flyers.

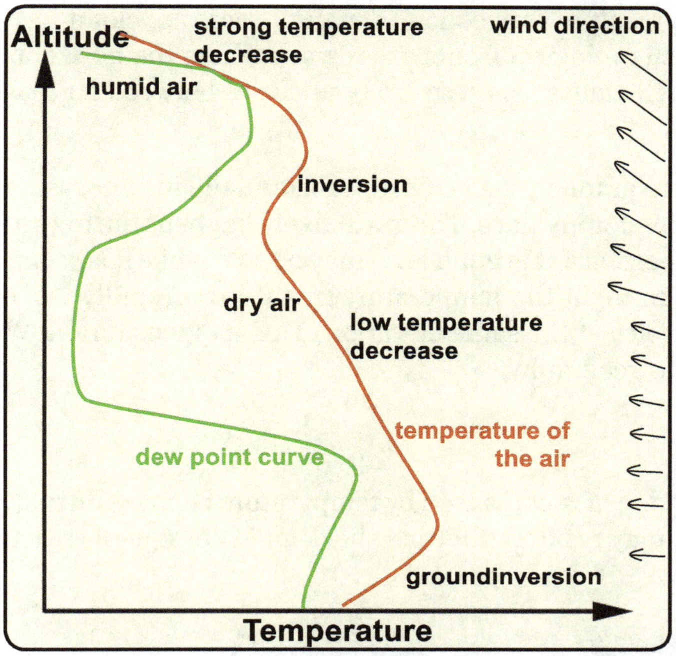

Illustration- The red graph shows the temperature as function of altitude/pressure, the green graph is the dew point. The harder the red curve leans left, the greater the instability. If the red graph takes a turn to the right it means that the temperature is increasing at that level, so there's an inversion there

Mathematical Description

Assuming that the temperature T is an intensive quantity, i.e., a single-valued, continuous and differentiable function of three-dimensional space (often called a scalar field), i.e., that

$$T = T(x, y, z)$$

where x, y and z are the coordinates of the location of interest, then the temperature gradient is the vector quantity defined as

$$\nabla T = \left(\frac{\partial T}{\partial x}, \frac{\partial T}{\partial y}, \frac{\partial T}{\partial z} \right).$$

Physical Processes

Climatology

On a global and annual basis, the dynamics of the atmosphere (and the oceans) can be understood as attempting to reduce the large difference of temperature between the polesand the equator by redistributing warm and cold air and water, known as Earth's heat engine.

Meteorology

Differences in air temperature between different locations are critical in weather forecasting and climate. The absorption of solar light at or near the planetary surface increases the temperature gradient and may result in convection (a major process of cloud formation, often associated with precipitation). Meteorological fronts are regions where the horizontal temperature gradient may reach relatively high values, as these are boundaries between air masses with rather distinct properties.

Clearly, the temperature gradient may change substantially in time, as a result of diurnal or seasonal heating and cooling for instance. This most likely happens during an inversion. For instance, during the day the temperature at ground level may be cold while it's warmer up in the atmosphere. As the day shifts over to night the temperature might drop rapidly while at other places on the land stay warmer or cooler at the same elevation. This happens on the West Coast of the United States sometimes due to geography.

Weathering

Expansion and contraction of rock, caused by temperature changes during a wildfire, through thermal stress weathering, may result in thermal shock and subsequent structure failure.

Thermodynamic Temperature

In everyday life, we usually think of temperature in terms of comparisons: How hot or cold something is relative to some physical property or reference, as in "Bundle up, it's below freezing out there!" or "My daughter is running a 103-degree fever."

Thermodynamic temperature, by contrast, is an absolute measure of the average total internal energy of an object or objects—namely its kinetic energy (energy of motion) plus contributions from other factors.

It applies to the average energy of a collection of atoms or subatomic particles—for example, the atoms in a block of iron, or the air molecules in a room. It is expressed in number of kelvins above absolute zero, the theoretical point at which nothing can get colder.

Usually, most of an object's kinetic energy is embodied in what is called "translational motion," which occurs when it moves around in space. In gases, atoms or molecules fly around in all directions, colliding with barriers and with each other. (Their individual speeds are different, but for any given set of objects at a given temperature, one range of speeds is more probable than others.) The average speed of the atoms and molecules depends on their energy content.

In solids, where atoms are constrained by bonds and cannot move independently, kinetic energy takes the form of collective motions called phonons. Thermal energy can also flow through solids in the motion of unbound, mobile electrons. In addition, objects or parts of complex objects can have kinetic energy in the form of *vibration* and *rotation*, and these, too, add to the total energy content.

Thermodynamic temperature is proportional to the average of all the energies in all the ways in which it is possible for an object to move. These are known as "degrees of freedom." So, for a single helium atom, there are only three degrees of freedom: motion in the up-down, left-right, back-forth directions. A two-atom molecule of nitrogen, however, there are two additional degrees of freedom: one of rotation and one of vibration. In general, the more components there are in a complex object such as a large molecule, the greater the number of possible motions and degrees of freedom.

It is extremely difficult to measure this internal energy directly. Customarily, scientists measure its effect when it moves as heat (thermal energy in transit) to or from a system of objects and a thermometer until heat no longer flows between them—in other words, when they are in thermal equilibrium. That is temperature.

Thus, internal energy and temperature are different, though directly related. The SI unit of energy is the joule. A "derived" SI unit, the joule itself is defined in terms of three SI base units — the kilogram, the meter and the second. But thermodynamic temperature is expressed in kelvins. There needs to be a way to connect the two.

The bridge between those two realms is the Boltzmann constant (k_B, or often just k), which relates the kinetic energy content (E) of matter to its temperature (T): $E = k_B T$. For the simplest collection of particles such as atoms, the average kinetic energy is $\frac{1}{2}mv^2$ distributed over the three degrees of freedom, where m is the mass and v is the velocity, so the total translational energy is $3/2\ k_B T$.

Temperature is a measure of the random submicroscopic motions and vibrations of the particle constituents of matter. These motions comprise the internal energy of a substance. More specifically, the thermodynamic temperature of any bulk quantity of matter is the measure of the average kinetic energy per classical (i.e., non-quantum) degree of freedom of its constituent particles. "Translational motions" are almost always in the classical regime. Translational motions are ordinary, whole-body movements in three-dimensional space in which particles move about and exchange energy in collisions. Thermodynamic temperature's null point, absolute zero, is the temperature at which the particle constituents of matter are as close as possible to complete rest; that is, they have minimal motion, retaining only quantum mechanical motion. Zero kinetic energy remains in a substance at absolute zero.

Throughout the scientific world where measurements are made in SI units, thermodynamic temperature is measured in kelvins (symbol: K). Many engineering fields in the U.S. however, measure thermodynamic temperature using the Rankine scale.

By international agreement, the unit *kelvin* and its scale are defined by two points: absolute zero, and the triple point of Vienna Standard Mean Ocean Water (water with a specified blend of hydrogen and oxygen isotopes). Absolute zero, the lowest possible temperature, is defined as being precisely 0 K *and* −273.15 °C. The triple point of water is defined as being precisely 273.16 K *and* 0.01 °C. This definition does three things:

1. It fixes the magnitude of the kelvin unit as being precisely 1 part in 273.16 parts the difference between absolute zero and the triple point of water;

2. It establishes that one kelvin has precisely the same magnitude as a one-degree increment on the Celsius scale; and
3. It establishes the difference between the two scales' null points as being precisely 273.15 kelvins (0 K = −273.15 °C and 273.16 K = 0.01 °C).

Temperatures expressed in kelvins (T_K) are converted to degrees Rankine ($T_{°R}$) simply by multiplying by 1.8 ($T_{°R} = 1.8 \times T_K$). Temperatures expressed in degrees Rankine are converted to kelvins by *dividing* by 1.8 ($T_K = T_{°R} \div 1.8$).

Practical Realization

Although the kelvin and Celsius scales are defined using absolute zero (0 K) and the triple point of water (273.16 K and 0.01 °C), it is impractical to use this definition at temperatures that are very different from the triple point of water. ITS-90 is then designed to represent the thermodynamic temperature as closely as possible throughout its range. Many different thermometer designs are required to cover the entire range. These include helium vapor pressure thermometers, helium gas thermometers, standard platinum resistance thermometers (known as SPRTs, PRTs or Platinum RTDs) and monochromatic radiation thermometers.

For some types of thermometer the relationship between the property observed (e.g., length of a mercury column) and temperature, is close to linear, so for most purposes a linear scale is sufficient, without point-by-point calibration. For others a calibration curve or equation is required. The mercury thermometer, invented before the thermodynamic temperature was understood, originally *defined* the temperature scale; its linearity made readings correlate well with true temperature, i.e. the "mercury" temperature scale was a close fit to the true scale.

The Relationship of Temperature, Motions, Conduction and Thermal Energy

The Nature of Kinetic Energy, Translational Motion and Temperature

The thermodynamic temperature is a measure of the average energy of the translational, vibrational, and rotational motions of matter's particle constituents (molecules, atoms, and subatomic particles). The full variety of these kinetic motions, along with potential energies of particles, and also occasionally certain other types of particle energy in equilibrium with these, contribute the total internal energy (loosely, the thermal energy) of a substance. Thus, internal energy may be stored in a number of ways (degrees of freedom) within a substance. When the degrees of freedom are in the classical regime ("unfrozen") the temperature is very simply related to the average energy of those degrees of freedom at equilibrium. The three translational degrees of freedom are unfrozen except for the very lowest temperatures, and their kinetic energy is simply related to the thermodynamic temperature over the widest range.

The relationship of kinetic energy, mass, and velocity is given by the formula $E_k = \frac{1}{2}mv^2$. Accordingly, particles with one unit of mass moving at one unit of velocity have precisely the same kinetic energy, and precisely the same temperature, as those with four times the mass but half the velocity.

Except in the quantum regime at extremely low temperatures, the thermodynamic temperature of any *bulk quantity* of a substance (a statistically significant quantity of particles) is directly proportional to the mean average kinetic energy of a specific kind of particle motion known as *translational motion*. These simple movements in the three *x*, *y*, and *z*–axis dimensions of space means the particles move in the three spatial *degrees of freedom*. The temperature derived from this translational kinetic energy is sometimes referred to as *kinetic temperature* and is equal to the thermodynamic temperature over a very wide range of temperatures. Since there are three translational degrees of freedom (e.g., motion along the x, y, and z axes), the translational kinetic energy is related to the kinetic temperature by:

$$\bar{E} = \frac{3}{2} k_B T_k$$

where:

- $\bar{E}$ is the mean kinetic energy in joules (J) and is pronounced "E bar"
- $k_B = 1.3806504(24) \times 10^{-23}$ J/K is the Boltzmann constant and is pronounced "Kay sub bee"
- T_k is the kinetic temperature in kelvins (K) and is pronounced "Tee sub kay"

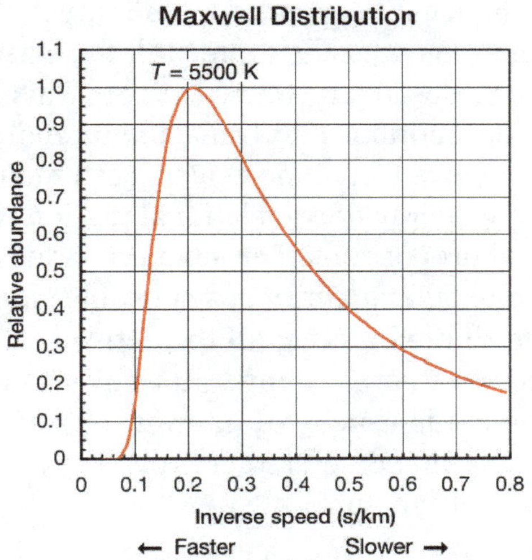

Figure- The translational motions of helium atoms occur across a range of speeds

While the Boltzmann constant is useful for finding the mean kinetic energy of a particle, it's important to note that even when a substance is isolated and in thermodynamic equilibrium (all parts are at a uniform temperature and no heat is going into or out of it), the translational motions of individual atoms and molecules occur across a wide range of speeds. At any one instant, the proportion of particles moving at a given speed within this range is determined by probability as described by the Maxwell–Boltzmann distribution. The graph shown here in *Figure* shows the speed distribution of 5500 K helium atoms. They have a *most probable* speed of 4.780 km/s. However, a certain proportion of atoms at any given instant are moving faster while others are moving relatively slowly; some are momentarily at a virtual standstill (off the *x*–axis to the right).

The High Speeds of Translational Motion

Although very specialized laboratory equipment is required to directly detect translational motions, the resultant collisions by atoms or molecules with small particles suspended in a fluid produces Brownian motion that can be seen with an ordinary microscope. The translational motions of elementary particles are *very* fast and temperatures close to absolute zero are required to directly observe them. For instance, when scientists at the NISTachieved a record-setting cold temperature of 700 nK (billionths of a kelvin) in 1994, they used optical lattice laser equipment to adiabatically cool caesium atoms. They then turned off the entrapment lasers and directly measured atom velocities of 7 mm per second in order to calculate their temperature.

The Internal Motions of Molecules and Specific Heat

There are other forms of internal energy besides the kinetic energy of translational motion. As can be seen in the animation at right, molecules are complex objects; they are a population of atoms and thermal agitation can strain their internal chemical bonds in three different ways: via rotation, bond length, and bond angle movements. These are all types of *internal degrees of freedom*. This makes molecules distinct from *monatomic*substances (consisting of individual atoms) like the noble gases helium and argon, which have only the three translational degrees of freedom. Kinetic energy is stored in molecules' internal degrees of freedom, which gives them an *internal temperature*. Even though these motions are called *internal*, the external portions of molecules still move—rather like the jiggling of a stationary water balloon. This permits the two-way exchange of kinetic energy between internal motions and translational motions with each molecular collision. Accordingly, as energy is removed from molecules, both their kinetic temperature (the temperature derived from the kinetic energy of translational motion) and their internal temperature simultaneously diminish in equal proportions. This phenomenon is described by the equipartition theorem, which states that for any bulk quantity of a substance in equilibrium, the kinetic energy of particle motion is evenly distributed among all the active (i.e. unfrozen) degrees of freedom available to the particles. Since the internal temperature of molecules are usually equal to their kinetic temperature, the distinction is usually of interest only in the detailed study of non-local thermodynamic equilibrium (LTE) phenomena such as combustion, the sublimation of solids, and the diffusionof hot gases in a partial vacuum.

The kinetic energy stored internally in molecules causes substances to contain more internal energy at any given temperature and to absorb additional internal energy for a given temperature increase. This is because any kinetic energy that is, at a given instant, bound in internal motions is not at that same instant contributing to the molecules' translational motions. This extra thermal energy simply increases the amount of energy a substance absorbs for a given temperature rise. This property is known as a substance's specific heat capacity.

Different molecules absorb different amounts of thermal energy for each incremental increase in temperature; that is, they have different specific heat capacities. High specific heat capacity arises, in part, because certain substances' molecules possess more internal degrees of freedom than others do. For instance, nitrogen, which is a diatomic molecule, has *five* active degrees of freedom at room temperature: the three comprising translational motion plus two rotational degrees of freedom internally. Since the two internal degrees of freedom are essentially unfrozen, in accordance with the equipartition theorem, nitrogen has five-thirds the specific heat capacity per mole (a specific number of molecules) as do the monatomic gases. Another example is gasoline. Gasoline can absorb a large amount of thermal energy per mole with only a modest temperature change because each molecule comprises an average of 21 atoms and therefore has many internal degrees of freedom. Even larger, more complex molecules can have dozens of internal degrees of freedom.

The Diffusion of Thermal Energy: Entropy, Phonons, and Mobile Conduction Electrons

Heat conduction is the diffusion of thermal energy from hot parts of a system to cold. A system can be either a single bulk entity or a plurality of discrete bulk entities. The term *bulk* in this context means a statistically significant quantity of particles (which can be a microscopic amount). Whenever thermal energy diffuses within an isolated system, temperature differences within the system decrease (and entropy increases).

One particular heat conduction mechanism occurs when translational motion, the particle motion underlying temperature, transfers momentum from particle to particle in collisions. In gases, these translational motions are of the nature. Not only does momentum (heat) diffuse throughout the volume of the gas through serial collisions, but entire molecules or atoms can move forward into new territory, bringing their kinetic energy with them. Consequently, temperature differences equalize throughout gases very quickly—especially for light atoms or molecules; convection speeds this process even more.

Translational motion in *solids,* however, takes the form of *phonons.* Phonons are constrained, quantized wave packets that travel at a given substance's speed of sound. The manner in which phonons interact within a solid determines a variety of its properties, including its thermal conductivity. In electrically insulating solids, phonon-based heat conduction is *usually* inefficient and such solids are considered *thermal insulators* (such as glass, plastic, rubber, ceramic, and rock). This is because in solids, atoms and molecules are locked into place relative to their neighbors and are not free to roam.

Metals however, are not restricted to only phonon-based heat conduction. Thermal energy conducts through metals extraordinarily quickly because instead of direct molecule-to-molecule collisions, the vast majority of thermal energy is mediated via very light, mobile *conduction electrons.* This is why there is a near-perfect correlation between metals' thermal conductivity and their electrical conductivity. Conduction electrons imbue metals with their extraordinary conductivity because they are *delocalized* (i.e., not tied to a specific atom) and behave rather like a sort of quantum gas due to the effects of *zero-point energy.* Furthermore, electrons are relatively light with a rest mass only $1/1836^{th}$ that of a proton. This is about the same ratio as a .22 Short bullet (29 grains or 1.88 g) compared to the rifle that shoots it. As Isaac Newton wrote with his third law of motion,

All forces occur in pairs, and these two forces are equal in magnitude and opposite in direction.

However, a bullet accelerates faster than a rifle given an equal force. Since kinetic energy increases as the square of velocity, nearly all the kinetic energy goes into the bullet, not the rifle, even though both experience the same force from the expanding propellant gases. In the same manner, because they are much less massive, thermal energy is readily borne by mobile conduction electrons. Additionally, because they're delocalized and *very* fast, kinetic thermal energy conducts extremely quickly through metals with abundant conduction electrons.

The Diffusion of Thermal Energy: Black-body Radiation

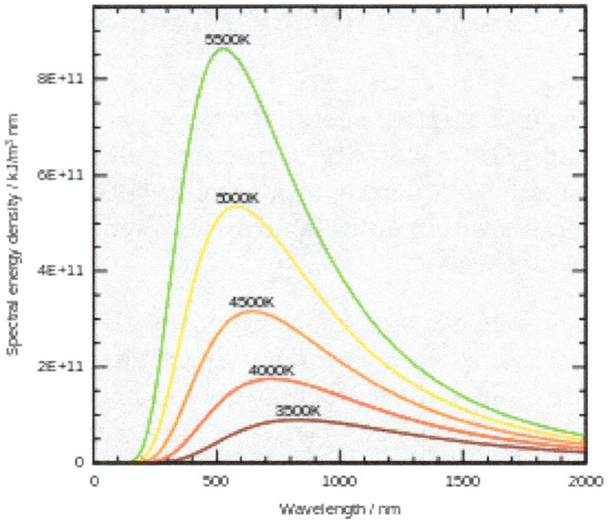

Figure- The spectrum of black-body radiation has the form of a Planck curve.
A 5500 K black-body has a peak emittance wavelength of 527 nm

Thermal radiation is a byproduct of the collisions arising from various vibrational motions of atoms. These collisions cause the electrons of the atoms to emit thermal photons (known as black-bodyradiation). Photons are emitted anytime an electric charge is accelerated (as happens when electron clouds of two atoms collide). Even *individual molecules* with internal temperatures greater than absolute zero also emit black-body radiation from their atoms. In any bulk quantity of a substance at equilibrium, black-body photons are emitted across a range of wavelengths in a spectrum that has a bell curve-like shape called a Planck curve. The top of a Planck curve (the peak emittance wavelength) is located in a particular part of the electromagnetic spectrumdepending on the temperature of the black-body. Substances at extreme cryogenic temperatures emit at long radio wavelengths whereas extremely hot temperatures produce short gamma rays.

Black-body radiation diffuses thermal energy throughout a substance as the photons are absorbed by neighboring atoms, transferring momentum in the process. Black-body photons also easily escape from a substance and can be absorbed by the ambient environment; kinetic energy is lost in the process.

As established by the Stefan–Boltzmann law, the intensity of black-body radiation increases as the fourth power of absolute temperature. Thus, a black-body at 824 K (just short of glowing dull red) emits *60 times* the radiant power as it does at 296 K (room temperature). This is why one can so easily feel the radiant heat from hot objects at a distance. At higher temperatures, such as those found in an incandescent lamp, black-body radiation can be the principal mechanism by which thermal energy escapes a system.

Table of Thermodynamic Temperatures

The full range of the thermodynamic temperature scale, from absolute zero to absolute hot, and some notable points between them are shown in the table below.

	Kelvin	Peak emittance wave length of black-body photons
Absolute zero (precisely by definition)	0 K	∞
Coldest measured temperature	450 pK	6,400 kilometers
One millikelvin (precisely by definition)	0.001 K	2.897 77 meters (Radio, FM band)
Cosmic Microwave Background Radiation	2.725 48(57) K	1.063 mm(peak wavelength)
Water's triple point (precisely by definition)	273.16 K	10,608.3 nm (Long wavelength I.R.)
Incandescent lamp[B]	2500 K	1160 nm (Near infrared)[C]
Sun's visible surface[C]	5778 K	501.5 nm (Green light)
Lightning bolt's channel	28,000 K	100 nm (Far Ultravioletlight)
Sun's core	16 MK	0.18 nm (X-rays)
Thermonuclear explosion (peak temperature)	350 MK	8.3×10^{-3} nm (Gamma rays)
Sandia National Labs' Z machine [D]	2 GK	1.4×10^{-3} nm (Gamma rays)
Core of a high–mass star on its last day	3 GK	1×10^{-3} nm (Gamma rays)
Merging binary neutron star system	350 GK	8×10^{-6} nm (Gamma rays)
Gamma-ray burst progenitors	1 TK	3×10^{-6} nm (Gamma rays)
Relativistic Heavy Ion Collider	1 TK	3×10^{-6} nm (Gamma rays)
CERN's proton vs. nucleus collisions	10 TK	3×10^{-7} nm (Gamma rays)
Universe 5.391×10^{-44} s after the Big Bang	1.417×10^{32} K	1.616×10^{-26}nm (Planck frequency)

[A] The 2500 K value is approximate.
[B] For a true blackbody (which tungsten filaments are not). Tungsten filaments' emissivity is greater at shorter wavelengths, which makes them appear whiter.
[C] Effective photosphere temperature.
[D] For a true blackbody (which the plasma was not). The Z machine's dominant emission originated from 40 MK electrons (soft x–ray emissions) within the plasma.

The Heat of Phase Changes

Figure- Ice and water: two phases of the same substance

The kinetic energy of particle motion is just one contributor to the total thermal energy in a substance; another is *phase transitions*, which are the potential energy of molecular bonds that can form in a substance as it cools (such as during condensing and freezing). The thermal energy required for a phase transition is called *latent heat*. This phenomenon may more easily be grasped by considering it in the reverse direction: latent heat is the energy required to *break* chemical bonds (such as during evaporation and melting). Almost everyone is familiar with the effects of phase transitions; for instance, steam at 100 °C can cause severe burns much faster than the 100 °C air from a hair dryer. This occurs because a large amount of latent heat is liberated as steam condenses into liquid water on the skin.

Even though thermal energy is liberated or absorbed during phase transitions, pure chemical elements, compounds, and eutectic alloys *exhibit no temperature change whatsoever* while they undergo them. Consider one particular type of phase transition: melting. When a solid is melting, crystal lattice chemical bonds are being broken apart; the substance is transitioning from what is known as a *more ordered state* to a *less ordered state*. In *figure,* the melting of ice is shown within the lower left box heading from blue to green.

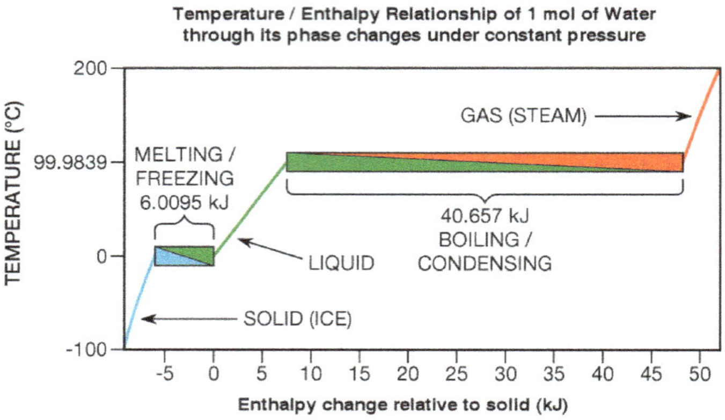

Figure- Water's temperature does not change during phase transitions as heat flows into or out of it. The total heat capacity of a mole of water in its liquid phase (the green line) is 7.5507 kJ

At one specific thermodynamic point, the melting point (which is 0 °C across a wide pressure range in the case of water), all the atoms or molecules are, on average, at the maximum energy threshold their chemical bonds can withstand without breaking away from the lattice. Chemical bonds are all-or-nothing forces: they either hold fast, or break; there is no in-between state. Consequently, when a substance is at its melting point, every joule of added thermal energy only breaks the bonds of a specific quantity of its atoms or molecules, converting them into a liquid of precisely the same temperature; no kinetic energy is added to translational motion (which is what gives substances their temperature). The effect is rather like popcorn: at a certain temperature, additional thermal energy can't make the kernels any hotter until the transition (popping) is complete. If the process is reversed (as in the freezing of a liquid), thermal energy must be removed from a substance.

As stated above, the thermal energy required for a phase transition is called *latent heat*. In the specific cases of melting and freezing, it's called *enthalpy of fusion* or *heat of fusion*. If the molecular bonds in a crystal lattice are strong, the heat of fusion can be relatively great, typically in the range of 6 to 30 kJ per mole for water and most of the metallic elements. If the substance is one of the monatomic gases, (which have little tendency to form molecular bonds) the heat of fusion is more modest, ranging from 0.021 to 2.3 kJ per mole. Relatively speaking, phase transitions can be truly energetic events. To completely melt ice at 0 °C into water at 0 °C, one must add roughly 80 times the thermal energy as is required to increase the temperature of the same mass of liquid water by one degree Celsius. The metals' ratios are even greater, typically in the range of 400 to 1200 times. And the phase transition of boiling is much more energetic than freezing. For instance, the energy required to completely boil or vaporize water (what is known as *enthalpy of vaporization*) is roughly *540 times* that required for a one-degree increase.

Water's sizable enthalpy of vaporization is why one's skin can be burned so quickly as steam condenses on it. In the opposite direction, this is why one's skin feels cool as liquid water on it evaporates (a process that occurs at a sub-ambient wet-bulb temperature that is dependent on relative humidity). Water's highly energetic enthalpy of vaporization is also an important factor underlying why *solar pool covers* (floating, insulated blankets that cover swimming pools when not in use) are so effective at reducing heating costs: they prevent evaporation. For instance, the evaporation of just 20 mm of water from a 1.29-meter-deep pool chills its water 8.4 degrees Celsius (15.1 °F).

Internal Energy

The total energy of all particle motion translational and internal, including that of conduction electrons, plus the potential energy of phase changes, plus zero-point energy comprise the *internal energy* of a substance.

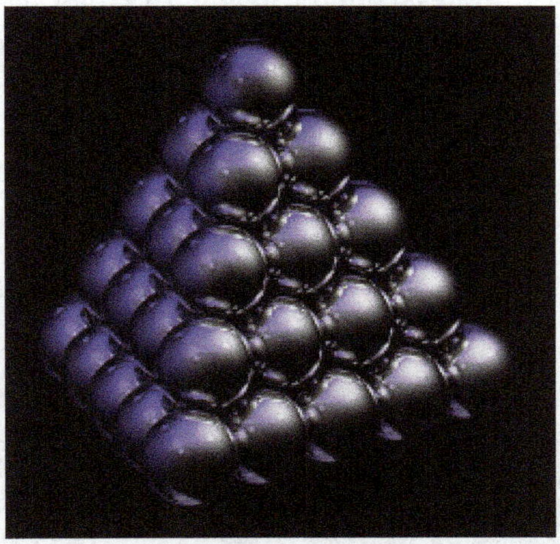

Figure- When many of the chemical elements, such as the noble gases and platinum-group metals, freeze to a solid — the most ordered state of matter — their crystal structures have a *closest-packed arrangement*. This yields the greatest possible packing density and the lowest energy state

Internal Energy at Absolute Zero

As a substance cools, different forms of internal energy and their related effects simultaneously decrease in magnitude: the latent heat of available phase transitions is liberated as a substance changes from a less ordered state to a more ordered state; the translational motions of atoms and molecules diminish (their kinetic temperature decreases); the internal motions of molecules diminish (their internal temperature decreases); conduction electrons (if the substance is an electrical conductor) travel *somewhat* slower; and black-body radiation's peak emittance wavelength increases (the photons' energy decreases). When the particles of a substance are as close as possible to complete rest and retain only ZPE-induced quantum mechanical motion, the substance is at the temperature of absolute zero (T=0).

Note that whereas absolute zero is the point of zero thermodynamic temperature and is also the point at which the particle constituents of matter have minimal motion, absolute zero is not necessarily the point at which a substance contains zero thermal energy; one must be very precise with what one means by *internal energy*. Often, all the phase changes that *can* occur in a substance, *will* have occurred by the time it reaches absolute zero. However, this is not always the case. Notably, T=0 helium remains liquid at room pressure and must be under a pressure of at least 25 bar (2.5 MPa) to crystallize. This is because helium's heat of fusion (the energy required to melt helium ice) is so low (only 21 joules per mole) that the motion-inducing effect of zero-point energy is sufficient to prevent it from freezing at lower pressures. Only if under at least 25 bar (2.5 MPa) of pressure will this latent thermal energy be liberated as helium freezes while approaching absolute zero. A further complication is that many solids change their crystal structure to more compact arrangements at extremely high pressures (up to millions of bars, or hundreds of gigapascals). These are known as *solid-solid phase transitions* wherein latent heat is liberated as a crystal lattice changes to a more thermodynamically favorable, compact one.

The above complexities make for rather cumbersome blanket statements regarding the internal energy in T=0 substances. Regardless of pressure though, what *can* be said is that at absolute zero, all solids with a lowest-energy crystal lattice such those with a *closest-packed arrangement* contain minimal internal energy, retaining only that due to the ever-present background of zero-point energy. One can also say that for a given substance at constant pressure, absolute zero is the point of lowest *enthalpy* (a measure of work potential that takes internal energy, pressure, and volume into consideration). Lastly, it is always true to say that all T=0 substances contain zero kinetic thermal energy.

Practical Applications for Thermodynamic Temperature

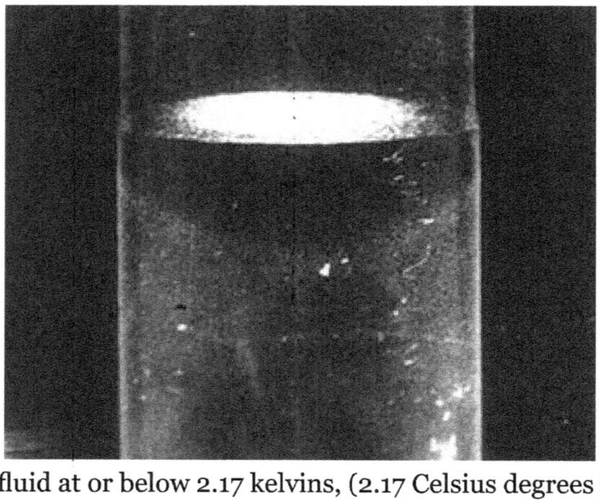

Helium-4, is a superfluid at or below 2.17 kelvins, (2.17 Celsius degrees above absolute zero)

Thermodynamic temperature is useful not only for scientists, it can also be useful for lay-people in many disciplines involving gases. By expressing variables in absolute terms and applying Gay–Lussac's law of temperature/pressure proportionality, solutions to everyday problems are straightforward; for instance, calculating how a temperature change affects the pressure inside an automobile tire. If the tire has a relatively cold pressure of 200 kPa-gage , then in absolute terms (relative to a vacuum), its pressure is 300 kPa-absolute. Room temperature ("cold" in tire terms) is 296 K. If the tire pressure is 20 °C hotter (20 kelvins), the solution is calculated as $^{316}\text{K}/_{296\text{ K}}$ = 6.8% greater thermodynamic temperature *and* absolute pressure; that is, a pressure of 320 kPa-absolute, which is 220 kPa-gage.

Definition of Thermodynamic Temperature

The thermodynamic temperature is defined by the second law of thermodynamicsand its consequences. The thermodynamic temperature can be shown to have special properties, and in particular can be seen to be uniquely defined (up to some constant multiplicative factor) by considering the efficiency of idealized heat engines. Thus the *ratio* T_2/T_1 of two temperaturesT_1 andT_2 is the same in all absolute scales.

Strictly speaking, the temperature of a system is well-defined only if it is at thermal equilibrium. From a microscopic viewpoint, a material is at thermal equilibrium if the quantity of heat between its individual particles cancel out. There are many possible scales of temperature, derived from a variety of observations of physical phenomena.

Loosely stated, temperature differences dictate the direction of heat between two systems such that their combined energy is maximally distributed among their lowest possible states. We call this distribution "entropy". To better understand the relationship between temperature and entropy, consider the relationship between heat, work and temperature illustrated in the Carnot heat engine. The engine converts heat into work by directing a temperature gradient between a higher temperature heat source, T_H, and a lower temperature heat sync, T_C, through a gas filled piston. The work done per cycle is equal to the difference between the heat supplied to the engine by T_H, q_H, and the heat supplied to T_C by the engine, q_C. The efficiency of the engine is the work divided by the heat put into the system or

$$Efficiency = \frac{w_{cy}}{q_H} = \frac{q_H - q_C}{q_H} = 1 - \frac{q_C}{q_H} \qquad (1)$$

where w_{cy} is the work done per cycle. Thus the efficiency depends only on q_C/q_H.

Carnot's theorem states that all reversible engines operating between the same heat reservoirs are equally efficient. Thus, any reversible heat engine operating between temperatures T_1 and T_2 must have the same efficiency, that is to say, the efficiency is the function of only temperatures

$$\frac{q_C}{q_H} = f(T_H, T_C) \qquad (2).$$

In addition, a reversible heat engine operating between temperatures T_1 and T_3must have the same efficiency as one consisting of two cycles, one between $T_1$and another (intermediate) temperature T_2, and the second between T_2 andT_3. If this were not the case, then energy (in the form of Q) will be wasted or gained, resulting in different overall efficiencies every time a cycle is split into component cycles; clearly a cycle can be composed of any number of smaller cycles.

With this understanding of Q_1, Q_2 and Q_3, we note also that mathematically,

$$f(T_1,T_3)=\frac{q_3}{q_1}=\frac{q_2q_3}{q_1q_2}=f(T_1,T_2)f(T_2,T_3).$$

But the first function is *NOT* a function of T_2, therefore the product of the final two functions *MUST* result in the removal of T_2 as a variable. The only way is therefore to define the function f as follows:

$$f(T_1,T_2)=\frac{g(T_2)}{g(T_1)}.$$

and

$$f(T_2,T_3)=\frac{g(T_3)}{g(T_2)}.$$

so that

$$f(T_1,T_3)=\frac{g(T_3)}{g(T_1)}=\frac{q_3}{q_1}.$$

i.e. The ratio of heat exchanged is a function of the respective temperatures at which they occur. We can choose any monotonic function for our $g(T)$; it is a matter of convenience and convention that we choose $g(T)=T$. Choosing then *one* fixed reference temperature (i.e. triple point of water), we establish the thermodynamic temperature scale.

It is to be noted that such a definition coincides with that of the ideal gas derivation; also it is this *definition* of the thermodynamic temperature that enables us to represent the Carnot efficiency in terms of T_H and T_C, and hence derive that the (complete) Carnot cycle is isentropic:

$$\frac{q_C}{q_H}=f(T_H,T_C)=\frac{T_C}{T_H}. \qquad (3).$$

Substituting this back into our first formula for efficiency yields a relationship in terms of temperature:

$$Efficiency=1-\frac{q_C}{q_H}=1-\frac{T_C}{T_H} \qquad (4).$$

Notice that for T_C=0 the efficiency is 100% and that efficiency becomes greater than 100% for T_C<0, which cases are unrealistic. Subtracting the right hand side of Equation 4 from the middle portion and rearranging gives

$$\frac{q_H}{T_H}-\frac{q_C}{T_C}=0,$$

where the negative sign indicates heat ejected from the system. The generalization of this equation is Clausius theorem, which suggests the existence of a state function S (i.e., a function which depends only on the state of the system, not on how it reached that state) defined (up to an additive constant) by

$$dS = \frac{dq_{rev}}{T} \qquad (5),$$

where the subscript indicates heat transfer in a reversible process. The function Scorresponds to the entropy of the system, mentioned previously, and the change of S around any cycle is zero (as is necessary for any state function). Equation 5 can be rearranged to get an alternative definition for temperature in terms of entropy and heat (to avoid logic loop, we should first define entropy through statistical mechanics):

$$T = \frac{dq_{rev}}{dS}.$$

For a system in which the entropy S is a function $S(E)$ of its energy E, the thermodynamic temperature T is therefore given by

$$\frac{1}{T} = \frac{dS}{dE},$$

so that the reciprocal of the thermodynamic temperature is the rate of increase of entropy with energy.

Work

In thermodynamics, work performed by a system is the energy transferred by the system to its surroundings. Kinetic energy, potential energy and internal energy are forms of energy that are properties of a system. Work is a form of energy, but it is energy in transit. A system contains no work, work is a process done by or on a system. In general, work is defined for mechanical systems as the action of a force on an object through a distance.

$$W = F \,.\, d$$

where:

W = work (J)

F = force (N)

d = displacement (m)

The most important takeaway from this equation is that work is proportional to the displacement as well as the magnitude of the force used. Different versions of the work equation can be used depending on the type of force involved. Some examples of doing work include:

- A person lifting books from the ground to a shelf does work against gravity.
- A battery pushing electrical current through a circuit does work against resistance.
- A child pushing a box along the floor does work against friction.

In thermodynamics, we are mainly interested in work done by expanding or compressing gases.

Pressure-volume Work: Work Done by a Gas

Gases can do work through expansion or compression against a constant external pressure. Work done by gases is also sometimes called pressure-volume or PV work for reasons that will hopefully become more clear in this section.

Let's consider gas contained in a piston.

If the gas is heated, energy is added to the gas molecules. We can observe the increase in average kinetic energy of the molecules by measuring how the temperature of the gas increases. As the gas molecules move faster, they also collide with the piston more often. These increasingly frequent collisions transfer energy to the piston and move it against an external pressure, increasing the overall volume of the gas. In this example, the gas has done work on the surroundings, which includes the piston and the rest of the universe.

To calculate how much work a gas has done (or has done to it) against a constant external pressure, we use a variation on the previous equation:

$$work = w = -Pexternal \times \Delta V$$

where P external is the external pressure (as opposed to the pressure of the gas in the system) and ΔV is the change in the volume of the gas, which can be calculated from the initial and final volume of the gas:

$$\Delta V = Vfinal - Vinitia$$

Since work is energy, it has units of Joules (where 1 $J = 1\frac{kg.m^2}{S^2}$). You may also see other units used, such as atmospheres for pressure and liters for volume, resulting in atm as the unit for work. We can convert L · atm to convert to Joules using the conversion factor of $\frac{101.325J}{1L \cdot atm}$.

The Sign of Work

As a matter of convention, negative work occurs when a system does work on the surroundings.

- When the gas does work the volume of a gas increases (ΔV>0) and the work done is negative.
- When work is done on the gas, the volume of the gas decreases (ΔV<0) and work is positive.

One way to remember the sign convention is to always think about the change in energy from the point of view of the gas. When the gas expands against an external pressure, the gas has to transfer some energy to the surroundings. Thus, the negative work decreases the overall energy of the gas. When the gas is compressed, energy is transferred to the gas so the energy of the gas increases due to positive work.

Example: Calculating Work Done on a Gas

To illustrate how to use the equation for PV work, let's imagine a bicycle pump. We will assume that the air in the bicycle pump can be approximated as an ideal gas in a piston. We can do work on the air in the pump by compressing it. Initially, the gas has a volume of 3.00 L. We apply a constant external pressure of 1.10 atm to push down the handle of the bike pump until the gas is compressed to a volume of 2.50 L. How much work did we do on the gas?

We can use the equation from the previous section to calculate how much work was done to compress the gas:

$$w = -P_{external} \times \Delta V$$

$$= -P_{external} \times (V_{final} - V_{initial})$$

If we plug in the values for $P_{external}, V_{final},\ and\ V_{initial}$ for our example, we get:

$$w = -1.10 atm \times (2.50L - 3.00L)$$

$$= -1.10 atm \times -0.50L$$

$$= 0.55 L \cdot atm$$

Let's check the sign for the work to make sure it makes sense. We know the gas had work done on it, since the volume of the gas decreased. That means the value of work we calculated should be positive, which matches our result. We can also convert our calculated work to Joules using the conversion factor:

$$W = 0.55\ \cancel{L\text{-}atm} \times \frac{101.325\,\text{J}}{1\ \cancel{L\text{-}atm}} = 56\,\text{J}$$

Thus, we did 56 J of work to compress the gas in the bicycle pump from 3.00 L to 2.50 L.

Work when Volume or Pressure is Constant

There are a few common scenarios where we might want to calculate work in chemistry class, and it helps to be able to recognize them when they come up. We will discuss how work is calculated for these cases.

Constant Volume Processes

Sometimes reactions or processes occur in a rigid, sealed container such as a bomb calorimeter.

When there is no change in volume possible, it is also not possible for gases to do work because $\Delta V=0$. In these = o and change in energy for the system must occur in other ways such as heat.

Formal Definition

In thermodynamics, the quantity of work done by a closed system on its surroundings is defined by factors strictly confined to the interface of the surroundings with the system and to the surroundings of the system, for example, an extended gravitational field in which the system sits, that is to say, to things external to the system. There are a few especially important kinds of thermodynamic work.

A simple example of one of those important kinds is pressure–volume work. The pressure of concern is that exerted by the surroundings on the surface of the system, and the volume of interest is the negative of the increment of volume gained by the system from the surroundings. It is usually arranged that the pressure exerted by the surroundings on the surface of the system is well defined and equal to the pressure exerted by the system on the surroundings. This arrangement for transfer of energy as work can be varied in a particular way that depends on the strictly mechanical nature of pressure–volume work. The variation consists in letting the coupling between the system and surroundings be through a rigid rod that links pistons of different areas for the system and surroundings. Then for a given amount of work transferred, the exchange of volumes involves different pressures, inversely with the piston areas, for mechanical equilibrium. This cannot be done for the transfer of energy as heat because of its non-mechanical nature.

Another important kind of work is isochoric work, that is to say work that involves no eventual overall change of volume of the system between the initial and the final states of the process. Examples are friction on the surface of the system as in Rumford's experiment; shaft work such as in Joule's experiments; and slow vibrational action on the system that leaves its eventual volume unchanged, but involves friction within the system. Isochoric work for a body in its own state of internal thermodynamic equilibrium is done only by the surroundings on the body, not by the body on the surroundings, so that the sign of isochoric work with the present sign convention is always negative.

When work is done by a closed system that cannot pass heat in or out because it is adiabatically isolated, the work is referred to as being adiabatic in character. Adiabatic work can be of the pressure–volume kind or of the isochoric kind, or both.

Sign Convention

Classically, a negative value of work indicates that a positive amount of work done *by* the system leads to energy being lost from the system. This sign convention has historically been used in many physics textbooks and will be used in the present article.

According to the first law of thermodynamics for a closed system, any net increase in the internal energy U must be fully accounted for, in terms of heat δQ entering the system and the work δW done by the system:

$$dU = \delta Q - \delta W.$$

An alternate sign convention is to consider the work performed *on* the system by its surroundings as positive. This leads to a change in sign of the work, so that $dU = \delta Q + \delta W$. This convention has historically been used in chemistry, but has been adopted in several modern physics textbooks.

In the above, the letter *d* indicates an exact differential, expressing that internal energy *U* is a property of the state of the system; they depend only on the original state and the final state, and not upon the path taken. In contrast, the Greek deltas(δ's) in this equation reflect the fact that the heat transfer and the work transfer are *not* properties of the final state of the system. Given only the initial state and the final state of the system, one can only say what the total change in internal energy was, not how much of the energy went out as heat, and how much as work. This can be summarized by saying that heat and work are not state functions of the system. This is in contrast to classical mechanics, where net work exerted by a particle is a state function.

Other Mechanical Types of Work

There are several ways of doing mechanical work, each in some way related to a force acting through a distance. In basic mechanics, the work done by a constant force F on a body displaced a distance s in the direction of the force is given by

$$W = Fs$$

If the force is not constant, the work done is obtained by integrating the differential amount of work,

$$W = \int_1^2 F\,ds.$$

Shaft Work

Energy transmission with a rotating shaft is very common in engineering practice. Often the torque T applied to the shaft is constant which means that the force F applied is constant. For a specified constant torque, the work done during n revolutions is determined as follows: A force F acting through a moment arm r generates a torque T

$$T = Fr \rightarrow F = \frac{T}{r}$$

This force acts through a distance s, which is related to the radius r by

$$s = 2(r\pi n)$$

The shaft work is then determined from:

$$W_s = Fs = 2\pi nT$$

The power transmitted through the shaft is the shaft work done per unit time, which is expressed as

$$\dot{W}_s = 2\pi T\dot{n}$$

Spring Work

When a force is applied on a spring, and the length of the spring changes by a differential amount dx, the work done is

$$\partial w_s = F dx$$

For linear elastic springs, the displacement x is proportional to the force applied

$$F = Kx,$$

where K is the spring constant and has the unit of N/m. The displacement x is measured from the undisturbed position of the spring (that is, X=0 when F=0). Substituting the two equations

$$W_s = \frac{1}{2}k\left(x_1^2 - x_2^2\right),$$

where x_1 and x_2 are the initial and the final displacement of the spring respectively, measured from the undisturbed position of the spring

Work Done on Elastic Solid Bars

Solids are often modeled as linear springs because under the action of a force they contract or elongate, and when the force is lifted, they return to their original lengths, like a spring. This is true as long as the force is in the elastic range, that is, not large enough to cause permanent or plastic deformation. Therefore, the equations given for a linear spring can also be used for elastic solid bars. Alternately, we can determine the work associated with the expansion or contraction of an elastic solid bar by replacing the pressure P by its counterpart in solids, normal stress σ=F/A in the work expansion.

$$W = \int_1^2 F\,dx.$$

$$W = \int A\sigma dx$$

where A is the cross sectional area of the bar.

Work Associated with the Stretching of Liquid Film

Consider a liquid film such as a soap film suspended on a wire frame. Some force is required to stretch this film by the movable portion of the wire frame. This force is used to overcome the microscopic forces between molecules at the liquid-air interface. These microscopic forces are perpendicular to any line in the surface and the force generated by these forces per unit length is called the surface tension σ whose unit is N/m. Therefore, the work associated with the stretching of a film is called surface tension work, and is determined from

$$W_s = \int_1^2 \sigma_s \, dA.$$

where dA=2b dx is the change in the surface area of the film. The factor 2 is due to the fact that the film has two surfaces in contact with air. The force acting on the moveable wire as a result of surface tension effects is F=2b σ, where σ is the surface tension force per unit length.

Non-mechanical Forms of Work

Non-mechanical work in thermodynamics is work determined by long-range forces penetrating into the system as force fields. The action of such forces can be initiated by events in the surroundings of the system, or by thermodynamic operations on the shielding walls of the system. The long-range forces are forces in the ordinary physical sense of the word, not the so-called 'thermodynamic forces' of non-equilibrium thermodynamic terminology.

The non-mechanical work of long-range forces can have either positive or negative sign, work being done by the system on the surroundings, or *vice versa*. Work done by long-range forces can be done indefinitely slowly, so as to approach the fictive reversible quasi-static ideal, in which entropy is not created in the system by the process.

In thermodynamics, non-mechanical work is to be contrasted with mechanical work that is done by forces in immediate contact between the system and its surroundings. If the putative 'work' of a process cannot be defined as either long-range work or else as contact work, then sometimes it cannot be described by the thermodynamic formalism as work at all. Nevertheless, the thermodynamic formalism allows that energy can be transferred between an open system and its surroundings by processes for which work is not defined. An example is when the wall between the system and its surrounds is not considered as idealized and vanishingly thin, so that processes can occur within the wall, such as friction affecting the transfer of matter across the wall; in this case, the forces of transfer are neither strictly long-range nor strictly due to contact between the system and its surrounds; the transfer of energy can then be considered as by convection, and assessed in sum just as transfer of internal energy. This is conceptually different from transfer of energy as heat through a thick fluid-filled wall in the presence of a gravitational field, between a closed system and its surroundings; in this case there may convective circulation within the wall but the process may still be considered as transfer of energy as heat between the system and its surroundings; if the whole wall is moved by the application of force from the surroundings, without change of volume of the wall, so as to change the volume of the system, then it is also at the same time transferring energy as work. A chemical reaction within a system can lead to electrical long-range forces and to electric current flow, which transfer energy as work between system and surroundings, though the system's chemical reactions themselves (except for the special limiting case in which in they are driven through devices in the surroundings so as to occur along a line of thermodynamic equilibrium) are always irreversible and do not directly interact with the surroundings of the system.

Non-mechanical work contrasts with pressure–volume work. Pressure–volume work is one of the two mainly considered kinds of mechanical contact work. A force acts on the interfacing wall between system and surroundings. The force is that due to the pressure exerted on the interfacing wall by the material inside the system; that pressure is an internal state variable of the system, but is properly measured by external devices at the wall. The work is due to change of system volume by expansion or contraction of the system. If the system expands, in the present article it is said to do positive work on the surroundings. If the system contracts, in the present article it is said to do negative work on the surroundings. Pressure–volume work is a kind of contact work, because it occurs through direct material contact with the surrounding wall or matter at the boundary of the system. It is accurately described by changes in state variables of the system, such as the time courses of changes in the pressure and volume of the system. The volume of the system is classified as a "deformation variable", and is properly measured externally to the system, in the surroundings. Pressure–volume work can have either positive or negative sign. Pressure–volume work, performed slowly enough, can be made to approach the fictive reversible quasi-static ideal.

Non-mechanical work also contrasts with shaft work. Shaft work is the other of the two mainly considered kinds of mechanical contact work. It transfers energy by rotation, but it does not eventually change the shape or volume of the system. Because it does not change the volume of the system it is not measured as pressure–volume work, and it is called isochoric work. Considered solely in terms of the eventual difference between initial and final shapes and volumes of the system, shaft work does not make a change. During the process of shaft work, for example the rotation of a paddle, the shape of the system changes cyclically, but this does not make an eventual change in the shape or volume of the system. Shaft work is a kind of contact work, because it occurs through direct material contact with the surrounding matter at the boundary of the system. A system that is initially in a state of thermodynamic equilibrium cannot initiate any change in its internal energy. In particular, it cannot initiate shaft work. This explains the curious use of the phrase "inanimate material agency" by Kelvin in one of his statements of the second law of thermodynamics. Thermodynamic operations or changes in the surroundings are considered to be able to create elaborate changes such as indefinitely prolonged, varied, or ceased rotation of a driving shaft, while a system that starts in a state of thermodynamic equilibrium is inanimate and cannot spontaneously do that. Thus the sign of shaft work is always negative, work being done on the system by the surroundings. Shaft work can hardly be done indefinitely slowly; consequently it always produces entropy within the system, because it relies on friction or viscosity within the system for its transfer. The foregoing comments about shaft work apply only when one ignores that the system can store angular momentum and its related energy.

Examples of non-mechanical work modes include

- Electrical work – where the force is defined by the surroundings' *voltage* (the electrical potential) and the generalized displacement is change of spatial distribution of *electrical charge*
- Magnetic work – where the force is defined by the surroundings' *magnetic field strength* and the generalized displacement is change of total *magnetic dipole moment*

- Electrical polarization work – where the force is defined by the surroundings' *electric field strength* and the generalized displacement is change of the *polarization of the medium* (the sum of the electric dipole moments of the molecules)
- Gravitational work – where the force is defined by the surroundings' gravitational field and the generalized displacement is change of the spatial distribution of the matter within the system.

Free Energy and Exergy

The amount of useful work which *can* be extracted from a thermodynamic system is discussed in the article *Second Law of Thermodynamics.* Under many practical situations this can be represented by the thermodynamic Availability or Exergy function. Two important cases are thermodynamic systems where the temperature and volume are held constant in which the measure of "useful" work attainable reduces to the Helmholtz free energy function; and systems where the temperature and pressure are held constant in which the measure of "useful" work attainable reduces to the Gibbs free energy.

Entropy

The concept of entropy emerged from the mid-19th century discussion of the efficiency of heat engines. Generations of students struggled with Carnot's cycle and various types of expansion of ideal and real gases, and never really understood why they were doing so.

Many earlier textbooks took the approach of defining a change in entropy, ΔS, via the equation:

$$\Delta S = Q_{\text{reversible}/T} \quad (i)$$

where Q is the quantity of heat and T the thermodynamic temperature. However, it is more common today to find entropy explained in terms of the degree of disorder in the system and to define the entropy change, ΔS, as:

$$\Delta S = -\Delta H/T \, (ii)$$

where ΔH is the enthalpy change. This more modern approach has two disadvantages. First the units of entropy are Joules per Kelvin but the degree of disorder has no units. Secondly, the equation (*ii*) defining entropy change does not recognise that the system has to be at equilibrium for it to be valid.

We prefer to consider that the entropy of a system corresponds to the molecular distribution of its molecular energy among the available energy levels and that systems tends to adopt the broadest possible distribution. Alongside this it is important to bear in mind the three laws of thermodynamics. The first law deals with the conservation of energy, the second law is concerned with the direction in which spontaneous reactions go, and the third law establishes that at absolute zero, all pure substances have the same entropy, which is taken, by convention, to be zero.

In classical terms, systems at absolute zero have no energy and the atoms or molecules would be close packed together. As energy, in any form is supplied to a system, its molecules begin to rotate, vibrate and translate, which is observable as a rise in temperature. The rise in temperature caused by a given quantity of heat will be different for different substances and depends on the heat capacity of the substance.

Since different substances have different heat capacities and because some compounds will have melted or vaporised, by the time they have reached their standard states at 298 K, their standard entropies will be different. For example, the standard entropy of graphite is 6 J K^{-1} mol^{-1}, whereas that for water it is 70 J K^{-1} mol^{-1}, and for nitrogen it is 192 J K^{-1} mol^{-1}.

The Distribution of Energy

To obtain some idea of what entropy is, it is helpful to imagine what happens when a small quantity of energy is supplied to a very small system. Suppose there are 20 units of energy to be supplied to a system of just 10 identical particles. The average energy is easily calculated to be two units per particle. Although this is the average energy, this state of the system can only arise if every particle has two units of energy. If all the energy were concentrated in one particle, any one of the 10 particles could take it up and this state of the system is 10 times more probable than every particle having two units each.

If the energy is shared between two molecules, half of the energy could be given to any one of the 10 particles and the other half to any one of the remaining nine particles, giving 90 ways in which this state of the system could be achieved. However, since the particles are indistinguishable, there would be only 45 ways in which this state can be achieved. Sharing the energy equally between four particles increases the number of possible ways this energy can be distributed to 210. This kind of calculation quickly becomes very tedious, even for a very small system of only 10 particles, and completely impracticable for systems where the number of particles approaches the number of atoms/molecules in even a very small fraction of one mole.

The very large numbers involved allow a number of simplifying approximations to be made and Ludwig Boltzmann was able to show that as the energy is spread more widely in a system, the number of possible distributions increases to a peak - corresponding to the most probable distribution - and then declines.

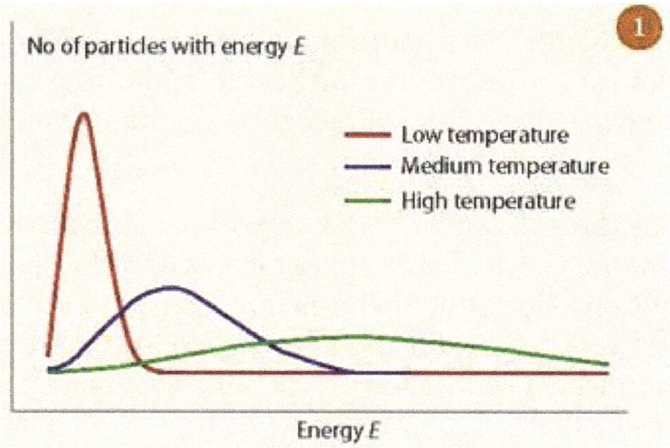

Figure- Boltzmann distribution of the number of particles with energy E at low, medium and high temperatures

At low temperatures, when the system has only a small amount of energy, comparatively few energy levels can be explored by the molecules because there is simply not enough energy in the system. Most of the molecules will have close to the average energy but there will be some with more than the average energy which occurs because there are some other molecules which have less than the average amount of energy.

As energy is supplied to the system, higher energy levels can be explored and the peak shifts towards higher energies. The temperature rises but, because the amount of substance in the system is constant, the height of the peak decreases as the particles with the most probable energy take up more energy and move into these higher energy levels. This spreading out of energy into a variety of energy levels is an essential characteristic of entropy.

Another important feature of entropy is that it is dynamic - although the amount of energy and amount of substance in an isolated system is constant - the energy is constantly being exchanged between different energy levels as a result of the collisions between molecules. This is in direct contrast to the common analogy used to describe entropy - *ie* the chaos of a teenager's room is often quoted as a system with high entropy, but the socks under the bed never exchange energy with the shirt hanging on the bed-post.

Standard Entropy

When considering calculations, it is important to remember that the defining equation (*i*) requires the quantity of heat to be taken up reversibly. (This is often overlooked by most textbooks written for A-level specifications.) When a system, initially at 0 K, absorbs this quantity of energy, its temperature rises by δT to $0 + \delta T$ or T_1. A further quantity of energy, q, brings about an entropy change, as defined by equation (*i*), and the temperature rises to $T_1 + \delta T$ or T_2. This process of adding energy to the system by a series of reversible absorptions eventually brings the system to 298 K and the sum of these operations gives the standard entropy of the system (which is the difference between the entropy of the substance in its standard state and its entropy at 0 K).

Rather than tediously adding these minute increments, the use of calculus allows us to achieve the same result. Entropy changes accompanying phase changes - *eg* melting, vaporisation *etc* - can be calculated without the use of calculus if the two phases are in equilibrium (*ie* at the melting point or boiling point). Under these conditions, the entropy change can be calculated from equation (*ii*).

Tables of standard entropies list the entropy of 1 mole of an element or compound in its standard state at 298 K - the standard state of a solid element or compound is the pure solid, that of a liquid is the pure liquid in equilibrium with its vapour and that of a gas is when it is at a pressure of 105 N m^{-2}or 1 standard atmosphere.

The use of an increase in total entropy as a criterion of spontaneity of reaction has to be treated with some caution. Its chief disadvantage is that it requires knowledge of the entropy changes in both the system and its surroundings.

$$\Delta S_{total} = \Delta S_{surrounding} + \Delta S_{system}$$

Moreover, since the standard entropies relate to standard states at 298 K, it is not appropriate, in most cases, to use these values at different temperatures. Whereas data can generally be obtained directly from the system itself, this is not the case with the surroundings which are generally assumed to be large enough not to be significantly affected by the changes occurring in the system.

Thermodynamics is an elegant and powerful tool but it has to be used with caution. Chemical reactions involve reactants and products in reactions which may not necessarily go to completion. Why this should be so involves both the enthalpy and entropy changes as reactants turn into products.

Exothermic reactions result in less energy in the system but are usually (not always) associated with a reduction in the entropy of the system, endothermic reactions are generally the reverse of this. When there are two opposing changes, a balance point has to be attained and the result is a mixture of reactants and products.

The key to determining which way and how far a chemical reaction will go is provided by another thermodynamic property - the Gibbs free energy (*G*) and it should come as no surprise to find that it is related to both the enthalpy and the entropy of the system:

$$\Delta G = \Delta H - T\Delta S$$

When the concept of entropy is being introduced, it is important that misconceptions should be avoided, particularly the idea that entropy represents the degree of disorder in the system. Entropy is dynamic - the energy of the system is constantly being redistributed among the possible distributions as a result of molecular collisions - and this is implicit in the dimensions of entropy being energy and reciprocal temperature, with units of J K^{-1}, whereas the degree of disorder is a dimensionless number.

Definitions and Descriptions

Any method involving the notion of entropy, the very existence of which depends on the second law of thermodynamics, will doubtless seem to many far-fetched, and may repel beginners as obscure and difficult of comprehension.

There are two related definitions of entropy: the thermodynamicdefinition and the statistical mechanicsdefinition. Historically, the classical thermodynamics definition developed first. In the classical thermodynamics viewpoint, the system is composed of very large numbers of constituents (atoms, molecules) and the state of the system is described by the average thermodynamic properties of those constituents; the details of the system's constituents are not directly considered, but their behavior is described by macroscopically averaged properties, e.g. temperature, pressure, entropy, heat capacity. The early classical definition of the properties of the system assumed equilibrium. The classical thermodynamic definition of entropy has more recently been extended into the area of non-equilibrium thermodynamics. Later, the thermodynamic properties, including entropy, were given an alternative definition in terms of the statistics of the motions of the microscopic constituents of a system — modeled at first classically, e.g. Newtonian particles constituting a gas, and later quantum-mechanically (photons, phonons, spins, etc.). The statistical mechanics description of the behavior of a system is necessary as the definition of the properties of a system using classical thermodynamics become an increasingly unreliable method of predicting the final state of a system that is subject to some process.

Function of State

There are many thermodynamic properties that are functions of state. This means that at a particular thermodynamic state (which should not be confused with the microscopic state of a system), these properties have a certain value. Often, if two properties of the system are determined, then the state is determined and the other properties' values can also be determined. For instance, a quantity of gas at a particular temperature and pressure has its state fixed by those values and thus has a specific volume that is determined by those values. As another instance, a system composed of a pure substance of a single phase at a particular uniform temperature and pressure is determined (and is thus a particular state) and is at not only a particular volume but also at a particular entropy. The fact that entropy is a function of state is one reason it is useful. In the Carnot cycle, the working fluid returns to the same state it had at the start of the cycle, hence the line integral of any state function, such as entropy, over this reversible cycle is zero.

Reversible Process

Entropy is conserved for a reversible process. A reversible process is one that does not deviate from thermodynamic equilibrium, while producing the maximum work. Any process which happens quickly enough to deviate from thermal equilibrium cannot be reversible. In these cases energy is lost to heat, total entropy increases, and the potential for maximum work to be done in the transition is also lost. More specifically, total entropy is conserved in a reversible process and not conserved in an irreversible process. For example, in the Carnot cycle, while the heat flow from the hot reservoir to the cold reservoir represents an increase in entropy, the work output, if reversibly and perfectly stored in some energy storage mechanism, represents a decrease in entropy that could be used to operate the heat engine in reverse and return to the previous state, thus the *total* entropy change is still zero at all times if the entire process is reversible. An irreversible process increases entropy.

Carnot Cycle

The concept of entropy arose from Rudolf Clausius's study of the Carnot cycle.In a Carnot cycle, heat Q_H is absorbed isothermally at temperature T_H from a 'hot' reservoir and given up isothermally as heat Q_C to a 'cold' reservoir at T_C. According to Carnot's principle, work can only be produced by the system when there is a temperature difference, and the work should be some function of the difference in temperature and the heat absorbed (Q_H). Carnot did not distinguish between Q_H and Q_C, since he was using the incorrect hypothesis that caloric theory was valid, and hence heat was conserved (the incorrect assumption that Q_H and Q_C were equal) when, in fact, Q_H is greater than Q_C. Through the efforts of Clausius and Kelvin, it is now known that the maximum work that a heat engine can produce is the product of the Carnot efficiency and the heat absorbed from the hot reservoir:

$$W = \left(\frac{T_H - T_C}{T_H}\right) Q_H = \left(1 - \frac{T_C}{T_H}\right) Q_H \qquad (1)$$

To derive the Carnot efficiency, which is 1-(T_C/T_H) (a number less than one), Kelvin had to evaluate the ratio of the work output to the heat absorbed during the isothermal expansion with the help of the Carnot-Clapeyron equation which contained an unknown function, known as the Carnot function. The possibility that the Carnot function could be the temperature as measured from a zero temperature, was suggested by Joule in a letter to Kelvin. This allowed Kelvin to establish his absolute temperature scale. It is also known that the work produced by the system is the difference between the heat absorbed from the hot reservoir and the heat given up to the cold reservoir:

$$W = Q_H - Q_C \qquad (2)$$

Since the latter is valid over the entire cycle, this gave Clausius the hint that at each stage of the cycle, work and heat would not be equal, but rather their difference would be a state function that would vanish upon completion of the cycle. The state function was called the internal energy and it became the first law of thermodynamics.

Now equating (1) and (2) gives

$$\frac{Q_H}{T_H} - \frac{Q_C}{T_C} = 0$$

or

$$\frac{Q_H}{T_H} = \frac{Q_C}{T_C}$$

This implies that there is a function of state which is conserved over a complete cycle of the Carnot cycle. Clausius called this state function *entropy*. One can see that entropy was discovered through mathematics rather than through laboratory results. It is a mathematical construct and has no easy physical analogy. This makes the concept somewhat obscure or abstract, akin to how the concept of energy arose.

Clausius then asked what would happen if there should be less work produced by the system than that predicted by Carnot's principle. The right-hand side of the first equation would be the upper bound of the work output by the system, which would now be converted into an inequality

$$W < \left(1 - \frac{T_C}{T_H}\right) Q_H$$

When the second equation is used to express the work as a difference in heats, we get

$$Q_H - Q_C < \left(1 - \frac{T_C}{T_H}\right) Q_H$$

or

$$Q_C > \frac{T_C}{T_H} Q_H$$

So more heat is given up to the cold reservoir than in the Carnot cycle. If we denote the entropies by $S_{i=Qi/Ti}$ for the two states, then the above inequality can be written as a decrease in the entropy

$$S_H - S_C < 0$$

or

$$S_H < S_C$$

The entropy that leaves the system is greater than the entropy that enters the system, implying that some irreversible process prevents the cycle from producing the maximum amount of work predicted by the Carnot equation.

The Carnot cycle and efficiency are useful because they define the upper bound of the possible work output and the efficiency of any classical thermodynamic system. Other cycles, such as the Otto cycle, Diesel cycle and Brayton cycle, can be analyzed from the standpoint of the Carnot cycle. Any machine or process that converts heat to work and is claimed to produce an efficiency greater than the Carnot efficiency is not viable because it violates the second law of thermodynamics. For very small numbers of particles in the system, statistical thermodynamics must be used. The efficiency of devices such as photovoltaic cells require an analysis from the standpoint of quantum mechanics.

Classical Thermodynamics

Conjugate variables of thermodynamics	
Pressure	Volume
(Stress)	(Strain)
Temperature	Entropy
Chemical potential	Particle number

The thermodynamic definition of entropy was developed in the early 1850s by Rudolf Clausiusand essentially describes how to measure the entropy of an isolated system in thermodynamic equilibrium with its parts. Clausius created the term entropy as an extensive thermodynamic variable that was shown to be useful in characterizing the Carnot cycle. Heat transfer along the isotherm steps of the Carnot cycle was found to be proportional to the temperature of a system (known as its absolute temperature). This relationship was expressed in increments of entropy equal to the ratio of incremental heat transfer divided by temperature, which was found to vary in the thermodynamic cycle but eventually return to the same value at the end of every cycle. Thus it was found to be a function of state, specifically a thermodynamic state of the system.

While Clausius based his definition on a reversible process, there are also irreversible processes that change entropy. Following the second law of thermodynamics, entropy of an isolated system always increases for irreversible processes. The difference between an isolated system and closed system is that heat may *not* flow to and from an isolated system, but heat flow to and from a closed system is possible. Nevertheless, for both closed and isolated systems, and indeed, also in open systems, irreversible thermodynamics processes may occur.

According to the Clausius equality, for a reversible cyclic process: $\oint \frac{\delta Q_{rev}}{T} = 0$. This means the line integral $\int_L \frac{\delta Q_{rev}}{T}$ is path-independent.

So we can define a state function S called entropy, which satisfies

$$dS = \frac{\delta Q_{rev}}{T}.$$

Clausius coined the name *entropy* (*Entropie*) for S in 1865. He gives "transformational content" (*Verwandlungsinhalt*) as a synonym, paralleling his "thermal and ergonal content" (*Wärme- und Werkinhalt*) as the name of U, but preferring the term *entropy* as a close parallel of *energy*, formed by replacing the root of of ργον "work" by that of τροπή "transformation".

To find the entropy difference between any two states of a system, the integral must be evaluated for some reversible path between the initial and final states.Since entropy is a state function, the entropy change of the system for an irreversible path is the same as for a reversible path between the same two states. However, the entropy change of the surroundings will be different.

We can only obtain the change of entropy by integrating the above formula. To obtain the absolute value of the entropy, we need the third law of thermodynamics, which states that $S = 0$ at absolute zero for perfect crystals.

From a macroscopic perspective, in classical thermodynamics the entropy is interpreted as a state function of a thermodynamic system: that is, a property depending only on the current state of the system, independent of how that state came to be achieved. In any process where the system gives up energy ΔE, and its entropy falls by ΔS, a quantity at least $T_R \Delta S$ of that energy must be given up to the system's surroundings as unusable heat (T_R is the temperature of the system's external surroundings). Otherwise the process cannot go forward. In classical thermodynamics, the entropy of a system is defined only if it is in thermodynamic equilibrium.

Statistical Mechanics

The statistical definition was developed by Ludwig Boltzmann in the 1870s by analyzing the statistical behavior of the microscopic components of the system. Boltzmann showed that this definition of entropy was equivalent to the thermodynamic entropy to within a constant number which has since been known as Boltzmann's constant. In summary, the thermodynamic definition of entropy provides the experimental definition of entropy, while the statistical definition of entropy extends the concept, providing an explanation and a deeper understanding of its nature.

The interpretation of entropy in statistical mechanics is the measure of uncertainty, or *mixedupness* in the phrase of Gibbs, which remains about a system after its observable macroscopic properties, such as temperature, pressure and volume, have been taken into account. For a given set of macroscopic variables, the entropy measures the degree to which the probability of the system is spread out over different possible microstates. In contrast to the macrostate, which characterizes plainly observable average quantities, a microstate specifies all molecular details about the system including the position and velocity of every molecule. The more such states available to the system with appreciable probability, the greater the entropy. In statistical mechanics, entropy is a measure of the number of ways in which a system may be arranged, often taken to be a measure of "disorder" (the higher the entropy, the higher the disorder).This definition describes the entropy as being proportional to the natural logarithm of the number of possible microscopic configurations of the individual atoms and molecules of the system (microstates) which could give rise to the observed macroscopic state (macrostate) of the system. The constant of proportionality is the Boltzmann constant.

Specifically, entropy is a logarithmic measure of the number of states with significant probability of being occupied:

$$S = -k_B \sum_i p_i \log p_i,$$

or, equivalently, the expected value of the logarithm of the probability that a microstate will be occupied

$$S = -k_B \, \mathrm{E}_i(\log p_i)$$

where k_B is the Boltzmann constant, equal to 1.380 65×10^{-23} J/K. The summation is over all the possible microstates of the system, and p_i is the probability that the system is in the i-th microstate. This definition assumes that the basis set of states has been picked so that there is no information on their relative phases. In a different basis set, the more general expression is

$$S = -k_B \, \mathrm{Tr}(\hat{\rho} \log(\hat{\rho})),$$

where $\hat{\rho}$ is the density matrix, Tr is trace and $\log$ is the matrix logarithm. This density matrix formulation is not needed in cases of thermal equilibrium so long as the basis states are chosen to be energy eigenstates. For most practical purposes, this can be taken as the fundamental definition of entropy since all other formulas for S can be mathematically derived from it, but not vice versa.

In what has been called *the fundamental assumption of statistical thermodynamics* or *the fundamental postulate in statistical mechanics*, the occupation of any microstate is assumed to be equally probable (i.e. $p_i = 1/\Omega$, where Ω is the number of microstates); this assumption is usually justified for an isolated system in equilibrium. Then the previous equation reduces to

$$S = k_B \log \Omega.$$

In thermodynamics, such a system is one in which the volume, number of molecules, and internal energy are fixed (the microcanonical ensemble).

The most general interpretation of entropy is as a measure of our uncertainty about a system. The equilibrium state of a system maximizes the entropy because we have lost all information about the initial conditions except for the conserved variables; maximizing the entropy maximizes our ignorance about the details of the system. This uncertainty is not of the everyday subjective kind, but rather the uncertainty inherent to the experimental method and interpretative model.

The interpretative model has a central role in determining entropy. The qualifier "for a given set of macroscopic variables" above has deep implications: if two observers use different sets of macroscopic variables, they see different entropies. For example, if observer A uses the variables *U*, *V* and *W*, and observer B uses *U*, *V*, *W*, *X*, then, by changing *X*, observer B can cause an effect that looks like a violation of the second law of thermodynamics to observer A. In other words: the set of macroscopic variables one chooses must include everything that may change in the experiment, otherwise one might see decreasing entropy.

Entropy can be defined for any Markov processes with reversible dynamics and the detailed balance property.

In Boltzmann's 1896 *Lectures on Gas Theory*, he showed that this expression gives a measure of entropy for systems of atoms and molecules in the gas phase, thus providing a measure for the entropy of classical thermodynamics.

Entropy of a System

Entropy arises directly from the Carnot cycle. It can also be described as the reversible heat divided by temperature. Entropy is a fundamental function of state.

In a thermodynamic system, pressure, density, and temperature tend to become uniform over time because the equilibrium state has higher probability (more possible combinations of microstates) than any other state.

As an example, for a glass of ice water in air at room temperature, the difference in temperature between a warm room (the surroundings) and cold glass of ice and water (the system and not part of the room), begins to equalize as portions of the thermal energy from the warm surroundings spread to the cooler system of ice and water. Over time the temperature of the glass and its contents and the temperature of the room become equal. In other words, the entropy of the room has decreased as some of its energy has been dispersed to the ice and water.

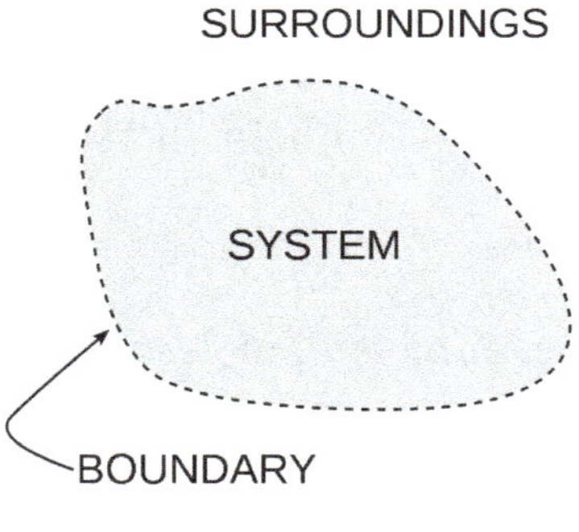

A thermodynamic system

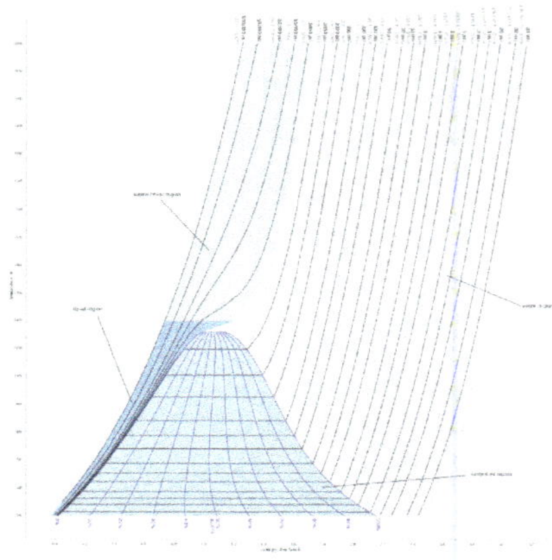

A temperature–entropy diagram for steam. The vertical axis represents uniform temperature, and the horizontal axis represents specific entropy. (Dark-blue is liquid water, light-blue is liquid-steam mixture, and faint-blue is steam. Grey-blue represents supercritical liquid water.)

However, as calculated in the example, the entropy of the system of ice and water has increased more than the entropy of the surrounding room has decreased. In an isolated system such as the room and ice water taken together, the dispersal of energy from warmer to cooler always results in a net increase in entropy. Thus, when the "universe" of the room and ice water system has reached a temperature equilibrium, the entropy change from the initial state is at a maximum. The entropy of the thermodynamic system is a measure of how far the equalization has progressed.

Thermodynamic entropy is a non-conserved state function that is of great importance in the sciences of physics and chemistry. Historically, the concept of entropy evolved to explain why some processes (permitted by conservation laws) occur spontaneously while their time reversals (also permitted by conservation laws) do not; systems tend to progress in the direction of increasing entropy. For isolated systems, entropy never decreases. This fact has several important consequences in science: first, it prohibits "perpetual motion" machines; and second, it implies the arrow of entropy has the same direction as the arrow of time. Increases in entropy correspond to irreversible changes in a system, because some energy is expended as waste heat, limiting the amount of work a system can do.

Unlike many other functions of state, entropy cannot be directly observed but must be calculated. Entropy can be calculated for a substance as the standard molar entropy from absolute zero (also known as absolute entropy) or as a difference in entropy from some other reference state which is defined as zero entropy. Entropy has the dimension of energy divided by temperature, which has a unit of joulesper kelvin (J/K) in the International System of Units. While these are the same units as heat capacity, the two concepts are distinct. Entropy is not a conserved quantity: for example, in an isolated system with non-uniform temperature, heat might irreversibly flow and the temperature become more uniform such that entropy increases. The second law of thermodynamics states that a closed system has entropy which may increase or otherwise remain constant. Chemical reactions cause changes in entropy and entropy plays an important role in determining in which direction a chemical reaction spontaneously proceeds.

One dictionary definition of entropy is that it is "a measure of thermal energy per unit temperature that is not available for useful work". For instance, a substance at uniform temperature is at maximum entropy and cannot drive a heat engine. A substance at non-uniform temperature is at a lower entropy (than if the heat distribution is allowed to even out) and some of the thermal energy can drive a heat engine.

A special case of entropy increase, the entropy of mixing, occurs when two or more different substances are mixed. If the substances are at the same temperature and pressure, there is no net exchange of heat or work – the entropy change is entirely due to the mixing of the different substances. At a statistical mechanical level, this results due to the change in available volume per particle with mixing.

Second Law of Thermodynamics

The second law of thermodynamics requires that, in general, the total entropy of any system can't decrease other than by increasing the entropy of some other system. Hence, in a system isolated from its environment, the entropy of that system tends not to decrease. It follows that heat can't flow from a colder body to a hotter body without the application of work (the imposition of order) to the colder body. Secondly, it is impossible for any device operating on a cycle to produce net work from a single temperature reservoir; the production of net work requires flow of heat from a hotter reservoir to a colder reservoir, or a single expanding reservoir undergoing adiabatic cooling, which performs adiabatic work. As a result, there is no possibility of a perpetual motion system. It follows that a reduction in the increase of entropy in a specified process, such as a chemical reaction, means that it is energetically more efficient.

It follows from the second law of thermodynamics that the entropy of a system that is not isolated may decrease. An air conditioner, for example, may cool the air in a room, thus reducing the entropy of the air of that system. The heat expelled from the room (the system), which the air conditioner transports and discharges to the outside air, always makes a bigger contribution to the entropy of the environment than the decrease of the entropy of the air of that system. Thus, the total of entropy of the room plus the entropy of the environment increases, in agreement with the second law of thermodynamics.

In mechanics, the second law in conjunction with the fundamental thermodynamic relation places limits on a system's ability to do useful work. The entropy change of a system at temperature T absorbing an infinitesimal amount of heat δq in a reversible way, is given by $\delta q/T$. More explicitly, an energy $T_R\, S$ is not available to do useful work, where T_R is the temperature of the coldest accessible reservoir or heat sink external to the system. Although this is possible, such an event has a small probability of occurring, making it unlikely.

The applicability of a second law of thermodynamics is limited to systems which are near or in equilibrium state. At the same time, laws governing systems which are far from equilibrium are still debatable. One of the guiding principles for such systems is the maximum entropy production principle. It claims that non-equilibrium systems evolve such as to maximize its entropy production.

Applications

The Fundamental Thermodynamic Relation

The entropy of a system depends on its internal energy and its external parameters, such as its volume. In the thermodynamic limit, this fact leads to an equation relating the change in the internal energy U to changes in the entropy and the external parameters. This relation is known as the *fundamental thermodynamic relation*. If external pressure P bears on the volume V as the only external parameter, this relation is:

$$dU = T\,dS - p\,dV$$

Since both internal energy and entropy are monotonic functions of temperature T, implying that the internal energy is fixed when one specifies the entropy and the volume, this relation is valid even if the change from one state of thermal equilibrium to another with infinitesimally larger entropy and volume happens in a non-quasistatic way (so during this change the system may be very far out of thermal equilibrium and then the entropy, pressure and temperature may not exist).

The fundamental thermodynamic relation implies many thermodynamic identities that are valid in general, independent of the microscopic details of the system. Important examples are the Maxwell relations and the relations between heat capacities.

Entropy in Chemical Thermodynamics

Thermodynamic entropy is central in chemical thermodynamics, enabling changes to be quantified and the outcome of reactions predicted. The second law of thermodynamics states that entropy in an isolated system – the combination of a subsystem under study and its surroundings – increases during all spontaneous chemical and physical processes. The Clausius equation of $\delta q_{rev}/T = \Delta S$ introduces the measurement of entropy change, ΔS. Entropy change describes the direction and quantifies the magnitude of simple changes such as heat transfer between systems – always from hotter to cooler spontaneously.

The thermodynamic entropy therefore has the dimension of energy divided by temperature, and the unit joule per kelvin (J/K) in the International System of Units (SI).

Thermodynamic entropy is an extensive property, meaning that it scales with the size or extent of a system. In many processes it is useful to specify the entropy as an intensive property independent of the size, as a specific entropy characteristic of the type of system studied. Specific entropy may be expressed relative to a unit of mass, typically the kilogram (unit: $Jkg^{-1}K^{-1}$). Alternatively, in chemistry, it is also referred to one mole of substance, in which case it is called the *molar entropy* with a unit of $Jmol^{-1}K^{-1}$.

Thus, when one mole of substance at about 0K is warmed by its surroundings to 298K, the sum of the incremental values of q_{rev}/T constitute each element's or compound's standard molar entropy, an indicator of the amount of energy stored by a substance at 298K. Entropy change also measures the mixing of substances as a summation of their relative quantities in the final mixture.

Entropy is equally essential in predicting the extent and direction of complex chemical reactions. For such applications, ΔS must be incorporated in an expression that includes both the system and its surroundings, $\Delta S_{universe} = \Delta S_{surroundings} + \Delta S_{system}$. This expression becomes, via some steps, the Gibbs free energy equation for reactants and products in the system: ΔG [the Gibbs free energy change of the system] = ΔH [the enthalpy change] $-T\Delta S$ [the entropy change].

Entropy Balance Equation for Open Systems

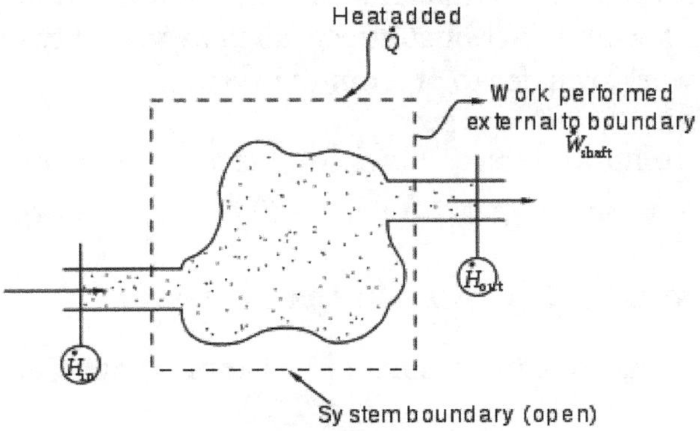

During steady-state continuous operation, an entropy balance applied to an open system accounts for system entropy changes related to heat flow and mass flow across the system boundary

In chemical engineering, the principles of thermodynamics are commonly applied to "open systems", i.e. those in which heat, work, and mass flow across the system boundary. Flows of both heat ($\dot{Q}$) and work, i.e. $\dot{W}_S$ (shaft work) and $P(dV/dt)$ (pressure-volume work), across the system boundaries, in general cause changes in the entropy of the system. Transfer as heat entails entropy transfer $\dot{Q}/T$, where T is the absolute thermodynamic temperature of the system at the point of the heat flow. If there are mass flows across the system boundaries, they also influence the total entropy of the system. This account, in terms of heat and work, is valid only for cases in which the work and heat transfers are by paths physically distinct from the paths of entry and exit of matter from the system.

To derive a generalized entropy balanced equation, we start with the general balance equation for the change in any extensive quantity Θ in a thermodynamic system, a quantity that may be either conserved, such as energy, or non-conserved, such as entropy. The basic generic balance expression states that $d\Theta/dt$, i.e. the rate of change of Θ in the system, equals the rate at which Θ enters the system at the boundaries, minus the rate at which Θ leaves the system across the system boundaries, plus the rate at which Θ is generated within the system. For an open thermodynamic system in which heat and work are transferred by paths separate from the paths for transfer of matter, using this generic balance equation, with respect to the rate of change with time t of the extensive quantity entropy S, the entropy balance equation is:

$$\frac{dS}{dt} = \sum_{k=1}^{K} \dot{M}_k \hat{S}_k + \frac{\dot{Q}}{T} + \dot{S}_{gen}$$

where

$\sum_{k=1}^{K} \dot{M}_k \hat{S}_k =$ the net rate of entropy flow due to the flows of mass into and out of the system (where $\hat{S} =$ entropy per unit mass).

$\frac{\dot{Q}}{T} =$ the rate of entropy flow due to the flow of heat across the system boundary.

$\dot{S}_{gen} =$ the rate of entropy production within the system. This entropy production arises from processes within the system, including chemical reactions, internal matter diffusion, internal heat transfer, and frictional effects such as viscosity occurring within the system from mechanical work transfer to or from the system.

Note, also, that if there are multiple heat flows, the term $\dot{Q}/T$ is replaced by $\sum \dot{Q}_j / T_j$, where $\dot{Q}_j$ is the heat flow and T_j is the temperature at the jth heat flow port into the system.

Entropy Change Formulas for Simple Processes

For certain simple transformations in systems of constant composition, the entropy changes are given by simple formulas.

Isothermal Expansion or Compression of an Ideal Gas

For the expansion (or compression) of an ideal gas from an initial volume V_0 and pressure P_0 to a final volume V and pressure P at any constant temperature, the change in entropy is given by:

$$\Delta S = nR\ln\frac{V}{V_0} = -nR\ln\frac{P}{P_0}.$$

Here n is the number of moles of gas and R is the ideal gas constant. These equations also apply for expansion into a finite vacuum or a throttling process, where the temperature, internal energy and enthalpy for an ideal gas remain constant.

Cooling and Heating

For heating or cooling of any system (gas, liquid or solid) at constant pressure from an initial temperature T_0 to a final temperature T, the entropy change is

$$\Delta S = nC_P\ln\frac{T}{T_0}.$$

provided that the constant-pressure molar heat capacity (or specific heat) C_P is constant and that no phase transition occurs in this temperature interval.

Similarly at constant volume, the entropy change is

$$\Delta S = nC_v\ln\frac{T}{T_0},$$

where the constant-volume heat capacity C_v is constant and there is no phase change.

At low temperatures near absolute zero, heat capacities of solids quickly drop off to near zero, so the assumption of constant heat capacity does not apply.

Since entropy is a state function, the entropy change of any process in which temperature and volume both vary is the same as for a path divided into two steps – heating at constant volume and expansion at constant temperature. For an ideal gas, the total entropy change is

$$\Delta S = nC_v\ln\frac{T}{T_0} + nR\ln\frac{V}{V_0}.$$

Similarly if the temperature and pressure of an ideal gas both vary,

$$\Delta S = nC_P\ln\frac{T}{T_0} - nR\ln\frac{P}{P_0}.$$

Phase Transitions

Reversible phase transitions occur at constant temperature and pressure. The reversible heat is the enthalpy change for the transition, and the entropy change is the enthalpy change divided by the thermodynamic temperature. For fusion (melting) of a solid to a liquid at the melting point T_m, the entropy of fusion is

$$\Delta S_{fus} = \frac{\Delta H_{fus}}{T_m}.$$

Similarly, for vaporization of a liquid to a gas at the boiling point T_b, the entropy of vaporization is

$$\Delta S_{vap} = \frac{\Delta H_{vap}}{T_b}.$$

Approaches to Understanding Entropy

As a fundamental aspect of thermodynamics and physics, several different approaches to entropy beyond that of Clausius and Boltzmann are valid.

Standard Textbook Definitions

The following is a list of additional definitions of entropy from a collection of textbooks:

- a measure of energy dispersal at a specific temperature.
- a measure of disorder in the universe or of the availability of the energy in a system to do work.
- a measure of a system's thermal energy per unit temperature that is unavailable for doing useful work.

In Boltzmann's definition, entropy is a measure of the number of possible microscopic states (or microstates) of a system in thermodynamic equilibrium. Consistent with the Boltzmann definition, the second law of thermodynamics needs to be re-worded as such that entropy increases over time, though the underlying principle remains the same.

Order and Disorder

Entropy has often been loosely associated with the amount of order or disorder, or of chaos, in a thermodynamic system. The traditional qualitative description of entropy is that it refers to changes in the status quo of the system and is a measure of "molecular disorder" and the amount of wasted energy in a dynamical energy transformation from one state or form to another. In this direction, several recent authors have derived exact entropy formulas to account for and measure disorder and order in atomic and molecular assemblies. One of the simpler entropy order/disorder formulas is that derived in 1984 by thermodynamic physicist Peter Landsberg, based on a combination of thermodynamics and information theory arguments. He argues that when constraints operate on a system, such that it is prevented from entering one or more of its possible or permitted states, as contrasted with its forbidden states, the measure of the total amount of "disorder" in the system is given by:

$$Disorder = \frac{C_D}{C_I}.$$

Similarly, the total amount of "order" in the system is given by:

$$Order = 1 - \frac{C_O}{C_I}.$$

In which C_D is the "disorder" capacity of the system, which is the entropy of the parts contained in the permitted ensemble, C_I is the "information" capacity of the system, an expression similar to Shannon's channel capacity, and C_O is the "order" capacity of the system.

Energy Dispersal

The concept of entropy can be described qualitatively as a measure of energy dispersal at a specific temperature. Similar terms have been in use from early in the history of classical thermodynamics, and with the development of statistical thermodynamics and quantum theory, entropy changes have been described in terms of the mixing or "spreading" of the total energy of each constituent of a system over its particular quantized energy levels.

Ambiguities in the terms *disorder* and *chaos*, which usually have meanings directly opposed to equilibrium, contribute to widespread confusion and hamper comprehension of entropy for most students. As the second law of thermodynamics shows, in an isolated system internal portions at different temperatures tend to adjust to a single uniform temperature and thus produce equilibrium. A recently developed educational approach avoids ambiguous terms and describes such spreading out of energy as dispersal, which leads to loss of the differentials required for work even though the total energy remains constant in accordance with the first law of thermodynamics. Physical chemist Peter Atkins, for example, who previously wrote of dispersal leading to a disordered state, now writes that "spontaneous changes are always accompanied by a dispersal of energy".

Relating Entropy to Energy Usefulness

Following on from the above, it is possible (in a thermal context) to regard entropy as an indicator or measure of the *effectiveness* or *usefulness* of a particular quantity of energy. This is because energy supplied at a high temperature (i.e. with low entropy) tends to be more useful than the same amount of energy available at room temperature. Mixing a hot parcel of a fluid with a cold one produces a parcel of intermediate temperature, in which the overall increase in entropy represents a "loss" which can never be replaced.

Thus, the fact that the entropy of the universe is steadily increasing, means that its total energy is becoming less useful: eventually, this will lead to the "heat death of the Universe".

Entropy and Adiabatic Accessibility

A definition of entropy based entirely on the relation of adiabatic accessibilitybetween equilibrium states was given by E.H.Lieb and J. Yngvason in 1999. This approach has several predecessors, including the pioneering work of Constantin Carathéodory from 1909 and the monograph by R. Giles. In the setting of Lieb and Yngvason one starts by picking, for a unit amount of the substance under consideration, two reference states X_0 and X_1 such that the latter is adiabatically accessible from the former but not vice versa. Defining the entropies of the reference states to be 0 and 1 respectively the entropy of a state X is defined as the largest number λ such that X is adiabatically accessible from a composite state consisting of an amount λ in the state X_1 and a complementary amount, $(1-\lambda)$, in the state X_0. A simple but important result within this setting is that entropy is uniquely determined, apart from a choice of unit and an additive constant for each chemical element, by the following properties: It is monotonic with respect to the relation of adiabatic accessibility, additive on composite systems, and extensive under scaling.

Entropy in Quantum Mechanics

In quantum statistical mechanics, the concept of entropy was developed by John von Neumann and is generally referred to as "von Neumann entropy",

$$S = -k_B \operatorname{Tr}(\rho \log \rho)$$

where ρ is the density matrix and Tr is the trace operator.

This upholds the correspondence principle, because in the classical limit, when the phases between the basis states used for the classical probabilities are purely random, this expression is equivalent to the familiar classical definition of entropy,

$$S = -k_B \sum_i p_i \log p_i,$$

i.e. in such a basis the density matrix is diagonal.

Von Neumann established a rigorous mathematical framework for quantum mechanics with his work *Mathematische Grundlagen der Quantenmechanik*. He provided in this work a theory of measurement, where the usual notion of wave function collapse is described as an irreversible process (the so-called von Neumann or projective measurement). Using this concept, in conjunction with the density matrix he extended the classical concept of entropy into the quantum domain.

Information Theory

When viewed in terms of information theory, the entropy state function is simply the amount of information (in the Shannon sense) that would be needed to specify the full microstate of the system. This is left unspecified by the macroscopic description.

In information theory, *entropy* is the measure of the amount of information that is missing before reception and is sometimes referred to as *Shannon entropy*. Shannon entropy is a broad and general concept which finds applications in information theory as well as thermodynamics. It was originally devised by Claude Shannon in 1948 to study the amount of information in a transmitted message. The definition of the information entropy is, however, quite general, and is expressed in terms of a discrete set of probabilities p_i *so that*

$$H(X) = -\sum_{i=1}^{n} p(x_i)\log p(x_i).$$

In the case of transmitted messages, these probabilities were the probabilities that a particular message was actually transmitted, and the entropy of the message system was a measure of the average amount of information in a message. For the case of equal probabilities (i.e. each message is equally probable), the Shannon entropy (in bits) is just the number of yes/no questions needed to determine the content of the message.

The question of the link between information entropy and thermodynamic entropy is a debated topic. While most authors argue that there is a link between the two, a few argue that they have nothing to do with each other.

The expressions for the two entropies are similar. If W is the number of microstates that can yield a given macrostate, and each microstate has the same *a priori* probability, then that probability is $p=1/W$. The Shannon entropy (in nats) is:

$$H = -\sum_{i=1}^{W} p\log(p) = \log(W)$$

and if entropy is measured in units of k per nat, then the entropy is given by:

$$H = k\log(W)$$

which is the famous Boltzmann entropy formula when k is Boltzmann's constant, which may be interpreted as the thermodynamic entropy per nat. There are many ways of demonstrating the equivalence of "information entropy" and "physics entropy", that is, the equivalence of "Shannon entropy" and "Boltzmann entropy". Nevertheless, some authors argue for dropping the word entropy for the H function of information theory and using Shannon's other term "uncertainty" instead.

Experimental Measurement of Entropy

Entropy of a substance can be measured, although in an indirect way. The measurement uses the *definition of temperature* in terms of entropy, while limiting energy exchange to heat ($dU \rightarrow dQ$).

$$T := \left(\frac{\partial U}{\partial S}\right)_{V,N} \Rightarrow \cdots \Rightarrow dS = dQ/T$$

The resulting relation describes how entropy changes dS when a small amount of energy dQ is introduced into the system at a certain temperature T.

The process of measurement goes as follows. First, a sample of the substance is cooled as close to absolute zero as possible. At such temperatures, the entropy approaches zero—due to the definition of temperature. Then, small amounts of heat are introduced into the sample and the change in temperature is recorded, until the temperature reaches a desired value (usually 25°C). The obtained data allows the user to integrate the equation above, yielding the absolute value of entropy of the substance at the final temperature. This value of entropy is called calorimetric entropy.

Causes of the Increase in Entropy of the System

In actual practice the reversible isentropic process never really occurs, it is only an ideal process. In actual practice whenever there is change in the state of the system the entropy of the system increases. Here are the various causes of the increase in entropy of the closed system are:

1) Due to external interaction: In closed system the mass of the system remains constant but it can exchange the heat with surroundings. Any change in the heat content of the system leads to disturbance in the system, which tends to increase the entropy of the system.

2) Internal changes in the system: Due to internal changes in the movements of the molecules of the system there is further disturbance inside the system. This causes irreversiblities inside the system and an increase in its entropy.

Misconceptions about Entropy

This view of the second law of thermodynamics is very popular, and it has been misused. Some argue that the second law of thermodynamics means that a system can never become more orderly. Not true. It just means that in order to become more orderly (for entropy to decrease), you must transfer energy from somewhere outside the system, such as when a pregnant woman draws energy from food to cause the fertilized egg to become a complete baby, completely in line with the second line's provisions.

Also known as: Disorder, Chaos, Randomness (all three imprecise synonyms).

References

- Taylor, Robert A.; Phelan, Patrick E.; Otanicar, Todd P.; Walker, Chad A.; Nguyen, Monica; Trimble, Steven; Prasher, Ravi (March 2011). "Applicability of nanofluids in high flux solar collectors". Journal of Renewable and Sustainable Energy. 3 (2): 023104. doi:10.1063/1.3571565
- Carey, V. P.; Chen, G.; Grigoropoulos, C.; Kaviany, M.; Majumdar, A. (2008). "A Review of Heat Transfer Physics". Nanoscale and Microscale Thermophysical Engineering. 12 (1): 1–60. doi:10.1080/15567260801917520
- Wilmore, Jack H.; Costill, David L.; Kenney, Larry (2008). Physiology of Sport and Exercise(6th ed.). Human Kinetics. p. 256. ISBN 9781450477673
- The scientific papers of J. Willard Gibbs in Two Volumes. 1. Longmans, Green, and Co. 1906. p. 11. Retrieved 2011-02-26
- Sakong, S.; Kratzer, P.; Han, X.; Laß, K.; Weingart, O.; Hasselbrink, E. (2008). "Density-functional theory study of vibrational relaxation of CO stretching excitation on Si(100)". The Journal of Chemical Physics. 129 (17): 174702. Bibcode:2008JChPh.129q4702S. doi:10.1063/1.2993254
- Oligschleger, C.; Schön, J. (1999). "Simulation of thermal conductivity and heat transport in solids". Physical Review B. 59 (6): 4125–4133. arXiv:cond-mat/9811156. Bibcode:1999PhRvB..59.4125O. doi:10.1103/PhysRevB.59.4125
- Edited by Tien, C.-L.; Majumdar, A.; Gerner, F. M. (1998). Microscale energy transport. Washington, D.C.: Taylor & Francis. ISBN 978-1560324591

- "Jaynes, E. T., "The Gibbs Paradox," In Maximum Entropy and Bayesian Methods; Smith, C. R; Erickson, G. J; Neudorfer, P. O., Eds; Kluwer Academic: Dordrecht, 1992, pp. 1–22" (PDF). Retrieved 2012-08-17
- Volz, S. (2010). Microscale and Nanoscale Heat Transfer (Topics in Applied Physics). Springer. ISBN 978-3642071584
- Greegor, R.; Lytle, F. (1979). "Extended x-ray absorption fine structure determination of thermal disorder in Cu: Comparison of theory and experiment". Physical Review B. 20 (12): 4902–4907. Bibcode:1979PhRvB..20.4902G. doi:10.1103/PhysRevB.20.4902
- Lambert, Frank L. "A Student's Approach to the Second Law and Entropy". entropysite.oxy.edu. Archived from the original on 17 July 2009. Retrieved 22 May 2016
- Chen, G. (2004). Nanoscale energy transport and conversion : a parallel treatment of electrones, molecules, phonons, and photons. New York: Oxford. ISBN 978-0195159424
- Bao, H.; Ruan, X. (2009). "Ab initio calculations of thermal radiative properties: The semiconductor GaAs". International Journal of Heat and Mass Transfer. 53 (7–8): 1308–1312. doi:10.1016/j.ijheatmasstransfer.2009.12.033
- Holland, M. (1963). "Analysis of Lattice Thermal Conductivity". Physical Review. 132 (6): 2461–2471. Bibcode:1963PhRv..132.2461H. doi:10.1103/PhysRev.132.2461
- Zhang, Z. M. (2007). Nano/microscale heat transfer ([Online-Ausg.]. ed.). New York: McGraw-Hill. ISBN 978-0071436748

4

Waves

A wave is a disturbance in space or matter, which involves the transfer of energy without any mass transport. It is generated by oscillations or vibrations in a field or physical medium. The different types of waves have been discussed in this chapter such as, mechanical waves, seismic waves, harmonic waves and standing waves. Other concepts discussed within the chapter include the phenomena of wave interference, Doppler effect and resonance.

A wave can be described as a disturbance that travels through a medium from one location to another location. Consider a slinky wave as an example of a wave. When the slinky is stretched from end to end and is held at rest, it assumes a natural position known as the equilibrium or rest position. The coils of the slinky naturally assume this position, spaced equally far apart. To introduce a wave into the slinky, the first particle is displaced or moved from its equilibrium or rest position. The particle might be moved upwards or downwards, forwards or backwards; but once moved, it is returned to its original equilibrium or rest position. The act of moving the first coil of the slinky in a given direction and then returning it to its equilibrium position creates a disturbance in the slinky. We can then observe this disturbance moving through the slinky from one end to the other. If the first coil of the slinky is given a single back-and-forth vibration, then we call the observed motion of the disturbance through the slinky a slinky pulse. A pulse is a single disturbance moving through a medium from one location to another location. However, if the first coil of the slinky is continuously and periodically vibrated in a back-and-forth manner, we would observe a repeating disturbance moving within the slinky that endures over some prolonged period of time. The repeating and periodic disturbance that moves through a medium from one location to another is referred to as a wave.

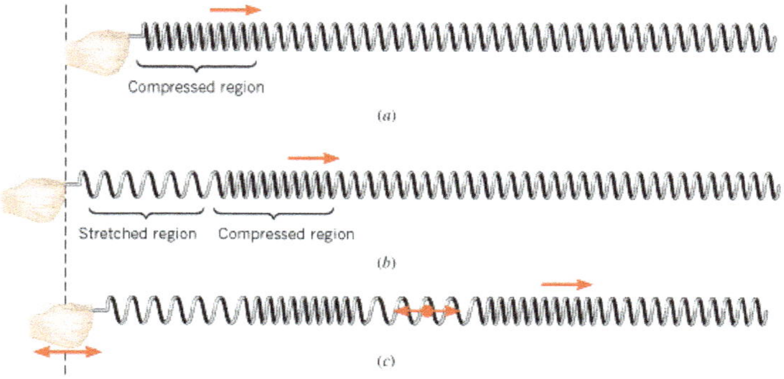

Particle-to-Particle Interaction

To fully understand the nature of a wave, it is important to consider the medium as a collection of interacting particles. In other words, the medium is composed of parts that are capable of interacting with each other. The interactions of one particle of the medium with the next adjacent particle allow the disturbance to travel through the medium. In the case of the slinky wave, the particles or interacting parts of the medium are the individual coils of the slinky. In the case of a sound wave in air, the particles or interacting parts of the medium are the individual molecules of air. And in the case of a stadium wave, the particles or interacting parts of the medium are the fans in the stadium.

Consider the presence of a wave in a slinky. The first coil becomes disturbed and begins to push or pull on the second coil; this push or pull on the second coil will displace the second coil from its equilibrium position. As the second coil becomes displaced, it begins to push or pull on the third coil; the push or pull on the third coil displaces it from its equilibrium position. As the third coil becomes displaced, it begins to push or pull on the fourth coil. This process continues in consecutive fashion, with each individual particle acting to displace the adjacent particle. Subsequently, the disturbance travels through the medium. The medium can be pictured as a series of particles connected by springs. As one particle moves, the spring connecting it to the next particle begins to stretch and apply a force to its adjacent neighbor. As this neighbor begins to move, the spring attaching this neighbor to its neighbor begins to stretch and apply a force on its adjacent neighbor.

A Wave Transports Energy and Not Matter

When a wave is present in a medium (that is, when there is a disturbance moving through a medium), the individual particles of the medium are only temporarily displaced from their rest position. There is always a force acting upon the particles that restores them to their original position. In a slinky wave, each coil of the slinky ultimately returns to its original position. In a water wave, each molecule of the water ultimately returns to its original position. And in a stadium wave, each fan in the bleacher ultimately returns to its original position. It is for this reason, that a wave is said to involve the movement of a disturbance without the movement of matter. The particles of the medium (water molecules, slinky coils, stadium fans) simply vibrate about a fixed position as the pattern of the disturbance moves from one location to another location.

Waves are said to be an energy transport phenomenon. As a disturbance moves through a medium from one particle to its adjacent particle, energy is being transported from one end of the medium to the other. In a slinky wave, a person imparts energy to the first coil by doing work upon it. The first coil receives a large amount of energy that it subsequently transfers to the second coil. When the first coil returns to its original position, it possesses the same amount of energy as it had before it was displaced. The first coil transferred its energy to the second coil. The second coil then has a large amount of energy that it subsequently transfers to the third coil. When the second coil returns to its original position, it possesses the same amount of energy as it had before it was displaced. The third coil has received the energy of the second coil. This process of energy transfer continues as each coil interacts with its neighbor. In this manner, energy is transported from one end of the slinky to the other, from its source to another location.

This characteristic of a wave as an energy transport phenomenon distinguishes waves from other types of phenomenon. Consider a common phenomenon observed at a softball game - the collision of a bat with a ball. A batter is able to transport energy from her to the softball by means of a bat. The batter applies a force to the bat, thus imparting energy to the bat in the form of kinetic energy. The bat then carries this energy to the softball and transports the energy to the softball upon collision. In this example, a bat is used to transport energy from the player to the softball. However, unlike wave phenomena, this phenomenon involves the transport of matter. The bat must move from its starting location to the contact location in order to transport energy. In a wave phenomenon, energy can move from one location to another, yet the particles of matter in the medium return to their fixed position. A wave transports its energy without transporting matter.

Waves are seen to move through an ocean or lake; yet the water always returns to its rest position. Energy is transported through the medium, yet the water molecules are not transported. Proof of this is the fact that there is still water in the middle of the ocean. The water has not moved from the middle of the ocean to the shore. If we were to observe a gull or duck at rest on the water, it would merely bob up-and-down in a somewhat circular fashion as the disturbance moves through the water. The gull or duck always returns to its original position. The gull or duck is not transported to the shore because the water on which it rests is not transported to the shore. In a water wave, energy is transported without the transport of water.

The same thing can be said about a stadium wave. In a stadium wave, the fans do not get out of their seats and walk around the stadium. We all recognize that it would be silly (and embarrassing) for any fan to even contemplate such a thought. In a stadium wave, each fan rises up and returns to the original seat. The disturbance moves through the stadium, yet the fans are not transported. Waves involve the transport of energy without the transport of matter.

Frequency and Period

A wave is an example of periodic motion. The particles of the medium oscillate back and forth about a fixed position. The time which it takes for such a particle to complete one full cycle is known as the period. Period units are simply seconds (or minutes or hours or some time unit). The frequency of a wave describes the number of complete cycles which are completed during a given period of time. As such, frequency is a rate quantity which describes the rate of oscillations or vibrations or cycles or waves on a per second basis. A common unit of frequency is the Hertz, abbreviated as Hz. A Hertz indicates the number of complete cyles per second.

There is a mathematical relationship between the frequency (f) of the wave and the period (T) of the wave as denoted by the following relationship:

$$f = 1 / T$$

Knowing frequency allows one to automatically calculate the period, and vice versa.

Wave Speed

Waves propagate through a medium. The familiar pattern of the sine wave consists of a crest and a trough (or in the case of longitudinal waves, a compression and a rarefaction). The crest can be seen moving along the medium from one location to another location. The speed (or velocity) of a wave is defined as the distance a crest on the wave travels per unit of time. Like the speed of a runner or a car, wave speed (v) is simply the ratio of the distance (d) traveled per time of travel (t) as shown by the following equation.

$$v = d / t$$

The Anatomy of a Wave

Imagine a wave moving through a medium such as a rope or snakey. And then imagine that a snapshot could be taken at an instant in time, thus freezing the motion of the wave and providing a picture of the shape of the medium. The result would be a diagram similar to that below (of course, without all the labels).

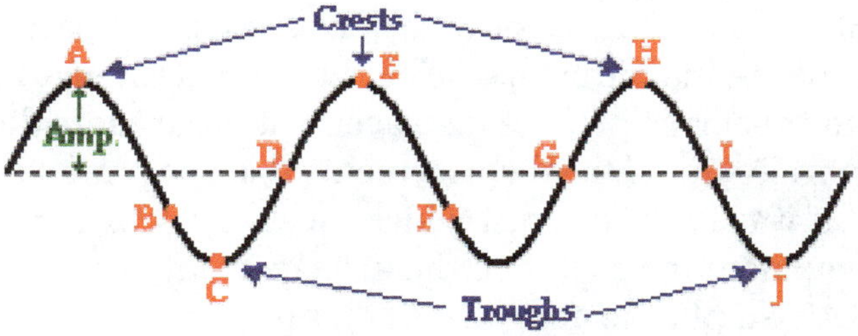

The labels have been added to indicate a few strategic points. The dashed line which stretches horizontally across the middle of the pattern is known as the rest position. The points labeled crests are at the points of maximum displacement above the rest position; the troughs are at the points of maximum displacement below the rest position. The vertical distance from a crest to the rest position is known as the amplitude. And the horizontal distance stretching between any two adjacent crests (or between adjacent troughs) is known as the wavelength.

Standing Wave Pattern Analysis

When a wave is introduced into a medium such as a rope or a snakey, it travels the length of the medium and then reflects back upon reaching the end of the medium. At certain frequencies, the reflected portion of the wave meets up with the original wave to create a pattern known as a standing wave pattern. In a standing wave pattern there are points along the medium which appear as if they are always standing still. These points are known as nodes and are easily remembered as the points of no displacement. There is always an antinode positioned between two adjacent nodes. Antinodes are points of maximum positive and negative displacement. In such standing wave patterns, there is a unique half-number relationship between the length of the medium and the wavelength of the waves which have established the pattern which is present in the medium. These relationships are shown below for the standing wave patterns having one node (first harmonic), two nodes (second harmonic) and three nodes (third harmonic).

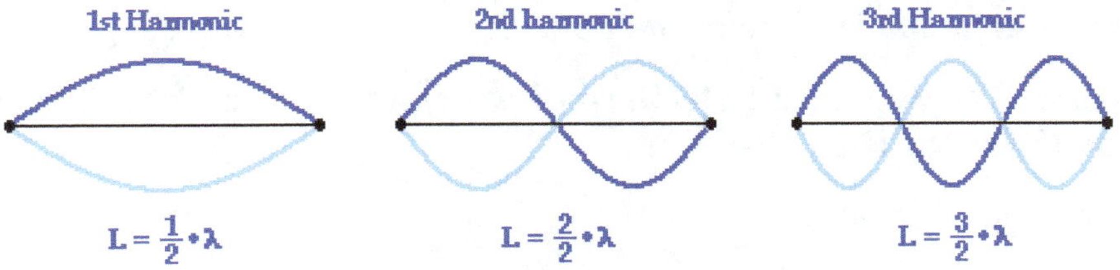

Other patterns for four nodes and more could be drawn by simply extending the principles which have been used for the first three patterns. From the pattern and knowledge of the length of the medium, the wavelength can be determined using the above equations. Similarly, the length of the medium can be determined from the wavelength.

Wave Interference

The meeting up of two waves while traveling in opposite directions along the same medium is known as interference. When two waves interfere, the medium takes on a new resultant shape which reflects the simultaneous influence of both waves upon the particles of the medium. The amount of displacement of the individual particles can be determined using the principle of superposition. According to this principle, the displacement of an individual particle at a moment in time is simply the algebraic sum of the displacements that the particle would have as a result of the individual waves. So if wave A has a crest which meets up with a trough of wave B at a given location and a specified moment in time, then the displacement of the medium at that location is simply the sum of wave A's amplitude and wave B's amplitude. In this case, Wave B would have a negative amplitude since it is a trough; and so a positive amplitude and a negative amplitude would be added together to determine the displacement of the medium at that location and moment in time.

Boundary Behavior

Mechanical waves travel through a medium (water, string, rope, snakey). The medium does not extend through space forever; eventually a medium ends. For instance, a rope may be terminated at wall to which it is tied, thus forming a boundary with a different medium (the wall). Or the end of a snakey may be tied to the end of another snakey, thus forming a boundary between two snakeys. As a wave reaches such a boundary, a portion of its energy is transmitted into the new medium and a portion of its energy is reflected off the boundary back into the original medium. This is known as boundary behavior. As a result of the reflection and transmission, changes in amplitude, wavelength and speed can often be observed. Most commonly, one observes that a transmitted wave (or pulse) has a different wavelength and speed than the incident wave (or pulse). While wavelength and speed may change as a wave is transmitted across a boundary, the frequency remains the same. Knowing this allows one to conduct a boundary analysis for the transmission of a wave across the boundary.

Mathematical Description of One-dimensional Waves

Wave Equation

Consider a traveling transverse wave (which may be a pulse) on a string (the medium). Consider the string to have a single spatial dimension. Consider this wave as traveling

- in the x direction in space. E.g., let the positive x direction be to the right, and the negative x direction be to the left.
- with constant amplitude u
- with constant velocity v, where v is
    - independent of wavelength (no dispersion)

- independent of amplitude (linear media, not nonlinear).
- with constant waveform, or shape

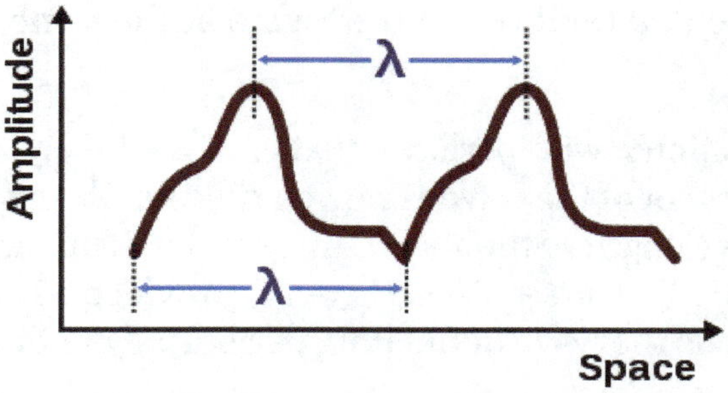

Wavelength λ, can be measured between any two corresponding points on a waveform

This wave can then be described by the two-dimensional functions

$u(x,t) = F(x - v\,t)$ (waveform F traveling to the right)

$u(x,t) = G(x + v\,t)$ (waveform G traveling to the left)

or, more generally, by d'Alembert's formula:

$$u(x,t) = F(x - vt) + G(x + vt).$$

representing two component waveforms F and G traveling through the medium in opposite directions. A generalized representation of this wave can be obtained as the partial differential equation

$$\frac{1}{v^2}\frac{\partial^2 u}{\partial t^2} = \frac{\partial^2 u}{\partial x^2}.$$

General solutions are based upon Duhamel's principle.

Wave Forms

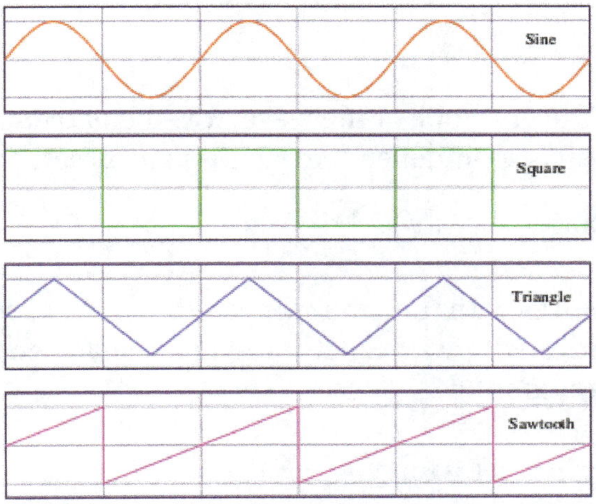

Sine, square, triangle and sawtooth waveforms

The form or shape of F in d'Alembert's formula involves the argument $x - vt$. Constant values of this argument correspond to constant values of F, and these constant values occur if x increases at the same rate that vt increases. That is, the wave shaped like the function F will move in the positive x-direction at velocity v (and G will propagate at the same speed in the negative x-direction).

In the case of a periodic function F with period λ, that is, $F(x + \lambda - vt) = F(x - vt)$, the periodicity of F in space means that a snapshot of the wave at a given time t finds the wave varying periodically in space with period λ (the wavelength of the wave). In a similar fashion, this periodicity of F implies a periodicity in time as well: $F(x - v(t + T)) = F(x - vt)$ provided $vT = \lambda$, so an observation of the wave at a fixed location x finds the wave undulating periodically in time with period $T = \lambda/v$.

Amplitude and Modulation

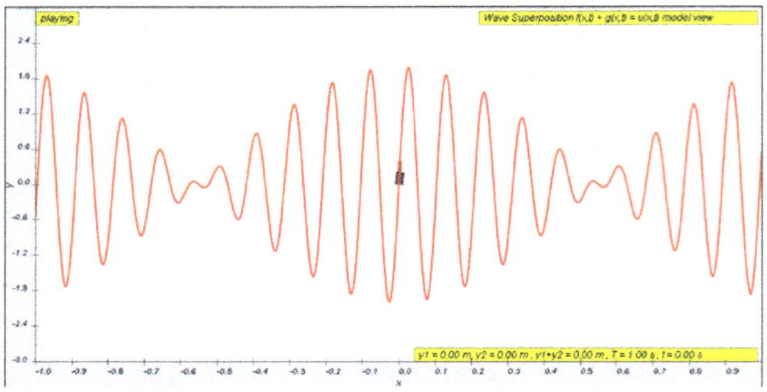

Amplitude modulation can be achieved through f(x,t) = 1.00*sin(2*pi/0.10*(x-1.00*t)) and g(x,t) = 1.00*sin (2*pi/0.11*(x-1.00*t))only the resultant is visible to improve clarity of waveform

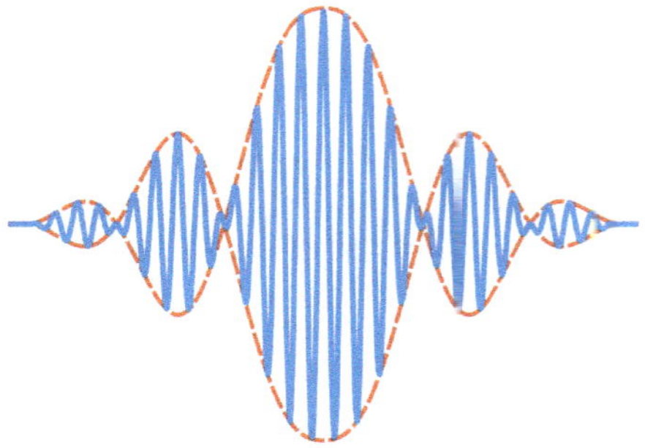

Illustration of the *envelope* (the slowly varying red curve) of an amplitude-modulated wave. The fast varying blue curve is the *carrier* wave, which is being modulated

The amplitude of a wave may be constant (in which case the wave is a *c.w.* or *continuous wave*), or may be *modulated* so as to vary with time and/or position. The outline of the variation in amplitude is called the *envelope* of the wave. Mathematically, the modulated wave can be written in the form:

$$u(x,t) = A(x,t)\sin(kx - \omega t + \phi),$$

where $A(x,t)$ is the amplitude envelope of the wave, k is the *wavenumber* and ϕ is the *phase*. If the group velocity v_g is wavelength-independent, this equation can be simplified as:

$$u(x,t) = A(x - v_g\, t)\sin(kx - \omega t + \phi),$$

showing that the envelope moves with the group velocity and retains its shape. Otherwise, in cases where the group velocity varies with wavelength, the pulse shape changes in a manner often described using an *envelope equation.*

Phase Velocity and Group Velocity

There are two velocities that are associated with waves, the phase velocity and the group velocity.

Phase velocity is the rate at which the phase of the wave propagates in space: any given phase of the wave (for example, the crest) will appear to travel at the phase velocity. The phase velocity is given in terms of the wavelength λ (lambda) and period T as

$$v_p = \frac{\lambda}{T}.$$

Group velocity is a property of waves that have a defined envelope, measuring propagation through space (i.e. phase velocity) of the overall shape of the waves' amplitudes—modulation or envelope of the wave.

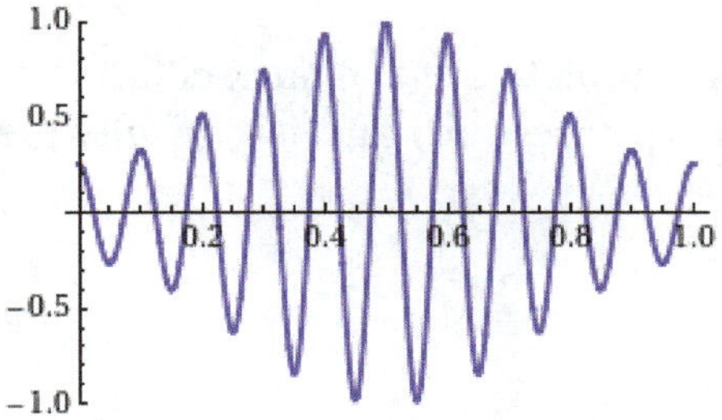

A wave with the group and phase velocities going in different directions

Sinusoidal Waves

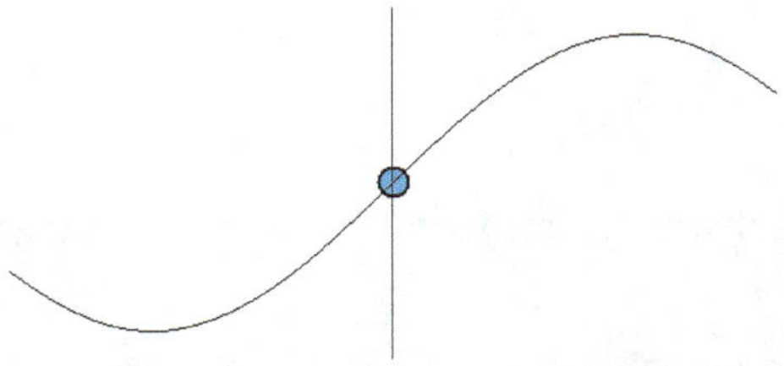
Sinusoidal waves correspond to simple harmonic motion

Mathematically, the most basic wave is the (spatially) one-dimensional sine wave (or *harmonic wave* or *sinusoid*) with an amplitude u described by the equation:

$$u(x,t) = A\sin(kx - \omega t + \phi),$$

where

- A is the maximum amplitude of the wave, maximum distance from the highest point of the disturbance in the medium (the crest) to the equilibrium point during one wave cycle. In the illustration to the right, this is the maximum vertical distance between the baseline and the wave.
- x is the space coordinate
- t is the time coordinate
- k is the wavenumber
- ω is the angular frequency
- ϕ is the phase constant.

The units of the amplitude depend on the type of wave. Transverse mechanical waves (e.g., a wave on a string) have an amplitude expressed as a distance (e.g., meters), longitudinal mechanical waves (e.g., sound waves) use units of pressure (e.g., pascals), and electromagnetic waves (a form of transverse vacuum wave) express the amplitude in terms of its electric field (e.g., volts/meter).

The wavelength λ is the distance between two sequential crests or troughs (or other equivalent points), generally is measured in meters. A wavenumber k, the spatial frequency of the wave in radians per unit distance (typically per meter), can be associated with the wavelength by the relation

$$k = \frac{2\pi}{\lambda}.$$

The period T is the time for one complete cycle of an oscillation of a wave. The frequency f is the number of periods per unit time (per second) and is typically measured in hertz denoted as Hz. These are related by:

$$f = \frac{1}{T}.$$

In other words, the frequency and period of a wave are reciprocals.

The angular frequency ω represents the frequency in radians per second. It is related to the frequency or period by

$$\omega = 2\pi f = \frac{2\pi}{T}.$$

The wavelength λ of a sinusoidal waveform traveling at constant speed v is given by:

$$\lambda = \frac{v}{f},$$

where v is called the phase speed (magnitude of the phase velocity) of the wave and f is the wave's frequency.

Wavelength can be a useful concept even if the wave is not periodic in space. For example, in an ocean wave approaching shore, the incoming wave undulates with a varying *local* wavelength that depends in part on the depth of the sea floor compared to the wave height. The analysis of the wave can be based upon comparison of the local wavelength with the local water depth.

Although arbitrary wave shapes will propagate unchanged in lossless linear time-invariant systems, in the presence of dispersion the sine wave is the unique shape that will propagate unchanged but for phase and amplitude, making it easy to analyze. Due to the Kramers–Kronig relations, a linear medium with dispersion also exhibits loss, so the sine wave propagating in a dispersive medium is attenuated in certain frequency ranges that depend upon the medium. The sine function is periodic, so the sine wave or sinusoid has a wavelength in space and a period in time.

The sinusoid is defined for all times and distances, whereas in physical situations we usually deal with waves that exist for a limited span in space and duration in time. Fortunately, an arbitrary wave shape can be decomposed into an infinite set of sinusoidal waves by the use of Fourier analysis.

As a result, the simple case of a single sinusoidal wave can be applied to more general cases. In particular, many media are linear, or nearly so, so the calculation of arbitrary wave behavior can be found by adding up responses to individual sinusoidal waves using the superposition principle to find the solution for a general waveform. When a medium is nonlinear, the response to complex waves cannot be determined from a sine-wave decomposition.

Plane Waves

Standing Waves

Standing wave in stationary medium. The red dots represent the wave nodes

A standing wave, also known as a *stationary wave*, is a wave that remains in a constant position. This phenomenon can occur because the medium is moving in the opposite direction to the wave, or it can arise in a stationary medium as a result of interference between two waves traveling in opposite directions.

The *sum* of two counter-propagating waves (of equal amplitude and frequency) creates a *standing wave*. Standing waves commonly arise when a boundary blocks further propagation of the wave, thus causing wave reflection, and therefore introducing a counter-propagating wave. For example, when a violin string is displaced, transverse waves propagate out to where the string is held in place at the bridge and the nut, where the waves are reflected back. At the bridge and nut, the two opposed waves are in antiphase and cancel each other, producing a node. Halfway between two nodes there is an antinode, where the two counter-propagating waves *enhance* each other maximally. There is no net propagation of energy over time.

Physical Properties

Waves exhibit common behaviors under a number of standard situations, e. g.

Transmission and Media

Waves normally move in a straight line (i.e. rectilinearly) through a *transmission medium*. Such media can be classified into one or more of the following categories:

- A *bounded medium* if it is finite in extent, otherwise an *unbounded medium*
- A *linear medium* if the amplitudes of different waves at any particular point in the medium can be added
- A *uniform medium* or *homogeneous medium* if its physical properties are unchanged at different locations in space
- An *anisotropic medium* if one or more of its physical properties differ in one or more directions
- An *isotropic medium* if its physical properties are the *same* in all directions

Light beam exhibiting reflection, refraction, transmission and dispersion when encountering a prism

Absorption

Absorption of waves means, if a kind of wave strikes a matter, it will be absorbed by the matter. When a wave with that same natural frequency impinges upon an atom, then the electrons of that atom will be set into vibrational motion. If a wave of a given frequency strikes a material with electrons having the same vibrational frequencies, then those electrons will absorb the energy of the wave and transform it into vibrational motion.

Reflection

When a wave strikes a reflective surface, it changes direction, such that the angle made by the incident wave and line normal to the surface equals the angle made by the reflected wave and the same normal line.

Refraction

Sinusoidal traveling plane wave entering a region of lower wave velocity at an angle, illustrating the decrease in wavelength and change of direction (refraction) that results

Refraction is the phenomenon of a wave changing its speed. Mathematically, this means that the size of the phase velocity changes. Typically, refraction occurs when a wave passes from one medium into another. The amount by which a wave is refracted by a material is given by the refractive index of the material. The directions of incidence and refraction are related to the refractive indices of the two materials by Snell's law.

Diffraction

A wave exhibits diffraction when it encounters an obstacle that bends the wave or when it spreads after emerging from an opening. Diffraction effects are more pronounced when the size of the obstacle or opening is comparable to the wavelength of the wave.

Interference

Waves that encounter each other combine through superposition to create a new wave called an interference pattern. Important interference patterns occur for waves that are in phase.

Polarization

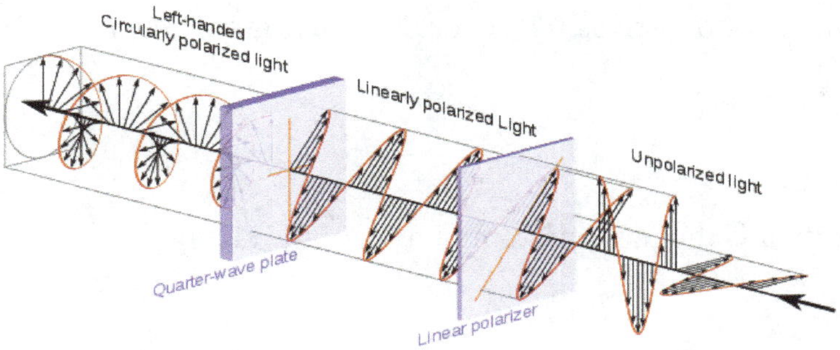

The phenomenon of polarization arises when wave motion can occur simultaneously in two orthogonal directions. Transverse waves can be polarized, for instance. When polarization is used as a descriptor without qualification, it usually refers to the special, simple case of linear polarization. A transverse wave is linearly polarized if it oscillates in only one direction or plane. In the case of linear polarization, it is often useful to add the relative orientation of that plane, perpendicular to the direction of travel, in which the oscillation occurs, such as "horizontal" for instance, if the plane of polarization is parallel to the ground. Electromagnetic waves propagating in free space, for instance, are transverse; they can be polarized by the use of a polarizing filter.

Longitudinal waves, such as sound waves, do not exhibit polarization. For these waves there is only one direction of oscillation, that is, along the direction of travel.

Dispersion

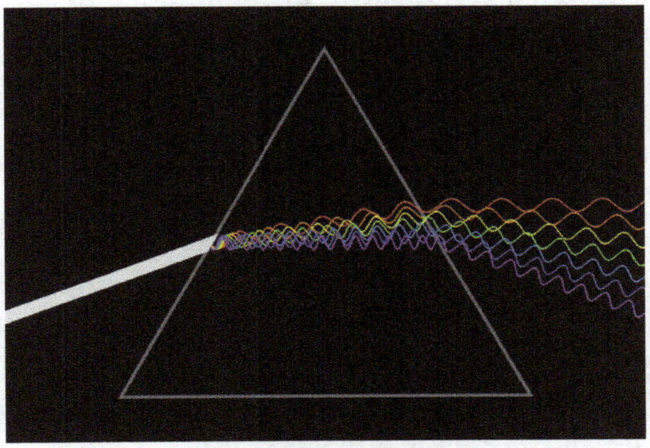

Schematic of light being dispersed by a prism

A wave undergoes dispersion when either the phase velocity or the group velocity depends on the wave frequency. Dispersion is most easily seen by letting white light pass through a prism, the result of which is to produce the spectrum of colours of the rainbow. Isaac Newton performed experiments with light and prisms, presenting his findings in the *Opticks* (1704) that white light consists of several colours and that these colours cannot be decomposed any further.

Mechanical Waves

Waves on Strings

The speed of a transverse wave traveling along a vibrating string (v) is directly proportional to the square root of the tension of the string (T) over the linear mass density (μ):

$$v = \sqrt{\frac{T}{\mu}},$$

where the linear density μ is the mass per unit length of the string.

Acoustic Waves

Acoustic or sound waves travel at speed given by

$$v = \sqrt{\frac{B}{\rho_0}},$$

or the square root of the adiabatic bulk modulus divided by the ambient fluid density.

Water waves

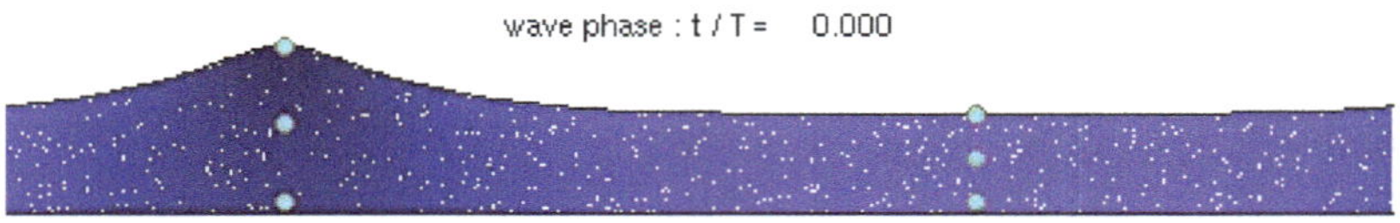

- Ripples on the surface of a pond are actually a combination of transverse and longitudinal waves; therefore, the points on the surface follow orbital paths.
- Sound—a mechanical wave that propagates through gases, liquids, solids and plasmas;
- Inertial waves, which occur in rotating fluids and are restored by the Coriolis effect;
- Ocean surface waves, which are perturbations that propagate through water.

Seismic Waves

Seismic waves are waves of energy that travel through the Earth's layers, and are a result of earthquakes, volcanic eruptions, magma movement, large landslides and large man-made explosions that give out low-frequency acoustic energy.

Shock Waves

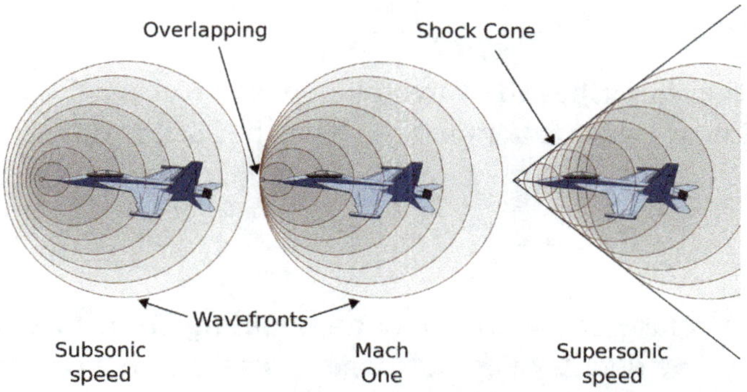

A shock wave is a type of propagating disturbance. When a wave moves faster than the local speed of sound in a fluid, it is a shock wave. Like an ordinary wave, a shock wave carries energy and can propagate through a medium; however, it is characterized by an abrupt, nearly discontinuous change in pressure, temperature and density of the medium.

Other

- Waves of traffic, that is, propagation of different densities of motor vehicles, and so forth, which can be modeled as kinematic waves.
- Metachronal wave refers to the appearance of a traveling wave produced by coordinated sequential actions.

Electromagnetic Waves

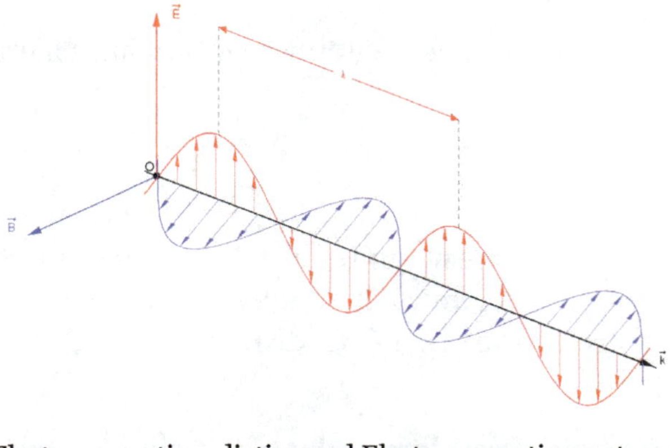

Electromagnetic radiation and Electromagnetic spectrum

An electromagnetic wave consists of two waves that are oscillations of the electric and magnetic fields. An electromagnetic wave travels in a direction that is at right angles to the oscillation direction of both fields. In the 19th century, James Clerk Maxwell showed that, in vacuum, the electric and magnetic fields satisfy the wave equation both with speed equal to that of the speed of light. From this emerged the idea that light is an electromagnetic wave. Electromagnetic waves can have different frequencies (and thus wavelengths), giving rise to various types of radiation such as radio waves, microwaves, infrared, visible light, ultraviolet, X-rays, and Gamma rays.

Quantum Mechanical Waves

Schrödinger Equation

The Schrödinger equation describes the wave-like behavior of particles in quantum mechanics. Solutions of this equation are wave functions which can be used to describe the probability density of a particle.

Dirac Equation

The Dirac equation is a relativistic wave equation detailing electromagnetic interactions. Dirac waves accounted for the fine details of the hydrogen spectrum in a completely rigorous way. The wave equation also implied the existence of a new form of matter, antimatter, previously unsuspected and unobserved and which was experimentally confirmed. In the context of quantum field theory, the Dirac equation is reinterpreted to describe quantum fields corresponding to spin-½ particles.

De Broglie Waves

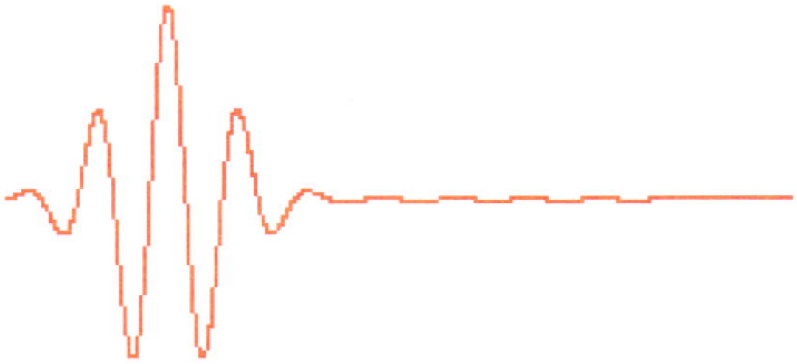

A propagating wave packet; in general, the *envelope* of the wave packet moves at a different speed than the constituent waves

Louis de Broglie postulated that all particles with momentum have a wavelength

$$\lambda = \frac{h}{p},$$

where h is Planck's constant, and p is the magnitude of the momentum of the particle. This hypothesis was at the basis of quantum mechanics. Nowadays, this wavelength is called the de Broglie wavelength. For example, the electrons in a CRT display have a de Broglie wavelength of about 10^{-13} m.

A wave representing such a particle traveling in the k-direction is expressed by the wave function as follows:

$$\psi(\mathbf{r}, t = 0) = A\, e^{i\mathbf{k}\cdot\mathbf{r}},$$

where the wavelength is determined by the wave vector k as:

$$\lambda = \frac{2\pi}{k},$$

and the momentum by:

$$\mathbf{p} = \hbar \mathbf{k} \, .$$

However, a wave like this with definite wavelength is not localized in space, and so cannot represent a particle localized in space. To localize a particle, de Broglie proposed a superposition of different wavelengths ranging around a central value in a wave packet, a waveform often used in quantum mechanics to describe the wave function of a particle. In a wave packet, the wavelength of the particle is not precise, and the local wavelength deviates on either side of the main wavelength value.

In representing the wave function of a localized particle, the wave packet is often taken to have a Gaussian shape and is called a *Gaussian wave packet*. Gaussian wave packets also are used to analyze water waves.

For example, a Gaussian wavefunction ψ might take the form:

$$\psi(x, t = 0) = A \exp\left(-\frac{x^2}{2\sigma^2} + ik_0 x \right),$$

at some initial time t = 0, where the central wavelength is related to the central wave vector k_o as $\lambda_o = 2\pi / k_o$. It is well known from the theory of Fourier analysis, or from the Heisenberg uncertainty principle (in the case of quantum mechanics) that a narrow range of wavelengths is necessary to produce a localized wave packet, and the more localized the envelope, the larger the spread in required wavelengths. The Fourier transform of a Gaussian is itself a Gaussian. Given the Gaussian:

$$f(x) = e^{-x^2/(2\sigma^2)} \, ,$$

the Fourier transform is:

$$\tilde{f}(k) = \sigma e^{-\sigma^2 k^2/2} \, .$$

The Gaussian in space therefore is made up of waves:

$$f(x) = \frac{1}{\sqrt{2\pi}} \int_{-\infty}^{\infty} \tilde{f}(k) e^{ikx} \, dk \, ;$$

that is, a number of waves of wavelengths λ such that $k\lambda = 2\,\pi$.

The parameter σ decides the spatial spread of the Gaussian along the x-axis, while the Fourier transform shows a spread in wave vector k determined by 1/σ. That is, the smaller the extent in space, the larger the extent in k, and hence in $\lambda = 2\pi/k$.

Gravity Waves

Gravity waves are waves generated in a fluid medium or at the interface between two media when the force of gravity or buoyancy tries to restore equilibrium. A ripple on a pond is one example.

Animation showing the effect of a cross-polarized gravitational wave on a ring of test particles

Gravitational Waves

Gravitational waves also travel through space. The first observation of gravitational waves was announced on 11 February 2016. Gravitational waves are disturbances in the curvature of spacetime, predicted by Einstein's theory of general relativity.

Mechanical Wave

Mechanical waves are a fundamental physical aspect of the movement of physical objects, whether these objects are solids, liquids, or gases. Every time an object moves under the influence of a force, waves are involved. In some cases the waves are more obvious, such as sound waves traveling through the air (which we can hear). But other times the presence of waves is not as obvious, such as in solid objects moving under the influence of a force.

A mechanical wave is really nothing more than a disturbance traveling through a medium (such as a solid, liquid, or gas). More precisely, a mechanical wave is the transmission of motion through matter, like a ripple moving across a lake. The speed, size, and shape of the wave depends on the mechanical properties of the medium, and on the force that caused the wave to occur in the first place.

A mechanical wave is caused by a force, and the movement of this wave is directly related to Newton's second law equation ($F = ma$). We are mostly familiar with this equation in the context of moving objects that we can see, such as a ball being acted on by a force and then moving as a result. But it also applies to matter in general.

For example, when a solid object is acted upon by a force, a wave emanates out from the point(s) of contact, and this wave travels through the body of the object. When it reaches the boundary of the object it reflects off of it and travels in the reverse direction. Different waves moving through the body may then combine with other waves moving through the body, and interfere with each other, either constructively or destructively. This behavior is fully predicted by applying Newton's second law equation, and combining it with equations that describe the mechanical properties of the solid object (these are called constitutive equations).

As the waves move through the solid object the object itself may begin to move as a whole. This will happen if the object is free to move and not constrained in some way. So for example, imagine you are picking up a ball, and for visualization purposes imagine this is occurring in extreme slow motion. At the instant that you touch it you cause waves to form inside the ball, which emanate out from the points of contact, and then these waves travel through the ball in the manner previously described. At the same time the ball begins to move and eventually lifts off the floor.

Wave speed through solids is very fast, much faster than wave speed through liquids and gases. For example, wave speed through air (the speed of sound) is about 340 m/s, wave speed through water is about 1500 m/s, and wave speed through a solid, such as iron, is 5100 m/s. Note that the 1500 m/s speed of waves traveling through water refers to the speed at which disturbances travel through the body of the water. The waves that we typically see on the surface of water do not travel at nearly this speed (far from it). These are called surface waves and are a different type of wave phenomenon. The motion of these waves occurs much more slowly than body wave motion. This is somewhat similar to how the lifting up of the ball, described in the previous example, occurs much more slowly than the speed of waves traveling through the inside of the ball. The ball motion occurs on a "bulk" (or macroscopic) level and as a result is subject to the entirety of whatever forces act on it, which will in general mean that it will move much, much slower than mechanical waves propagating through the ball interior.

It is interesting to note that the dynamics equations for rigid body motion are just a simplified mathematical representation for wave motion through bodies that are very stiff. If we were to solve Newton's second law equation (F = ma) along with the constitutive equations for a solid body that is very stiff, the result would be almost identical to the result we would get if we were to solve rigid body dynamics equations. The reason we apply rigid body dynamics equations to solid bodies instead of wave equations is because they are accurate enough and are much easier to solve.

When a solid body oscillates on its own, at its natural frequency, such as a vibrating ruler, it gives you a clue on how fast waves travel through it. The time it takes for a moving point on the ruler to undergo a complete oscillation cycle is the time that it takes for a mechanical wave to travel through the body of the ruler for one complete cycle. However, the ruler will not oscillate back and forth at the same linear speed as the wave speed. It will oscillate at a much slower speed than that.

The oscillation period of the ruler will be equal to the time it takes for a wave to travel through the body of the ruler for one complete cycle. Since the ruler oscillates due to the effect of mechanical waves moving through its body, one complete cycle of ruler oscillation must correspond to one complete travel cycle of the wave(s) moving through the body of the ruler.

Lastly, it is interesting to mention that no prior knowledge of mechanical waves is required to solve a problem involving mechanical waves, provided that the basic fundamental equations (Newton's second law and constitutive equations) are applied to the problem. Knowledge of wave behaviour for solving wave problems is only required when we don't start with the fundamental equations, in which case we have to apply known wave concepts to solve the problems. Sometimes this is best (and easiest) since the fundamental equations are quite complicated, and would be too time consuming to solve from scratch every time you want to solve a wave problem.

There are three types of mechanical waves:

- Transverse Wave

  Transverse waves cause the medium to vibrate at a right angle to the direction of the wave or energy being carried by the medium. Transverse waves have two parts—the crest and the trough. The crest is the highest point of the wave and the trough is the lowest. The distance between a subsequent crest and a trough is half of wavelength. The wavelength is the distance from crest to crest or from trough to trough.

  To see an example, move an end of a Slinky (whose other end is fixed) to the left-and-right of the Slinky, as opposed to to-and-fro. Light also has properties of a transverse wave, although it is an electromagnetic wave.

- Longitudinal Wave

  Longitudinal waves cause the medium to vibrate parallel to the direction of the wave. It consists of multiple compressions and rarefactions. The rarefaction is the farthest distance apart in the longitudinal wave and the compression is the closest distance together. The speed of the longitudinal wave is increased in higher index of refraction, due to the closer proximity of the atoms in the medium that is being compressed. Sound is considered a longitudinal wave.

- Surface Waves

  This type of wave travels along a surface that is between two media. An example of a surface wave would be waves in a pool, or in an ocean, lake, or any other type of water body. There are two types of surface waves, namely Rayleigh waves and Love waves.

  Rayleigh waves, also known as *ground roll*, are waves that travel as ripples with motion similar to those of waves on the surface of water. Rayleigh waves are much slower than body waves, roughly 90% of the velocity of body waves for a typical homogeneous elastic medium.

  A Love wave is a surface wave having horizontal waves that are shear or transverse to the direction of propagation. They usually travel slightly faster than Rayleigh waves, about 90% of the body wave velocity, and have the largest amplitude.

Examples

- Ocean waves
- Sound waves
- Seismic waves

Longitudinal Wave

Longitudinal wave, wave consisting of a periodic disturbance or vibration that takes place in the same direction as the advance of the wave. A coiled spring that is compressed at one end and then released experiences a wave of compression that travels its length, followed by a stretching; a point on any coil of the spring will move with the wave and return along the same path, passing through the neutral position and then reversing its motion again. Sound moving through air also compresses and rarefies the gas in the direction of travel of the sound wave as they vibrate back and forth. The P (primary) seismic waves are also longitudinal. In a longitudinal wave, each particle of matter vibrates about its normal rest position and along the axis of propagation, and all particles participating in the wave motion behave in the same manner, except that there is a progressive change in phase (q.v.) of vibration—i.e., each particle completes its cycle of reaction at a later time. The combined motions result in the advance of alternating regions of compression and rarefaction in the direction of propagation.

A mechanical model is helpful in explaining longitudinal waves. At the top of the figure, small masses A, B, C, etc. are joined together by coiled springs to represent a transmitting medium that has properties of both inertia and elasticity. Because mass B has inertia, motion of A toward the left (arrow 2) extends the spring it is attached to and motion to the right (arrow 1) compresses it. A corresponding motion will be communicated to B through the spring, except that there will be a slight lag in phase. Mass B will impart its motion to its partner C, and so on, the impulse travelling from A to K and the lag progressively increasing. At the instant shown, A leads J in phase by 360°; A is starting its second vibration, whereas J is just beginning its first.

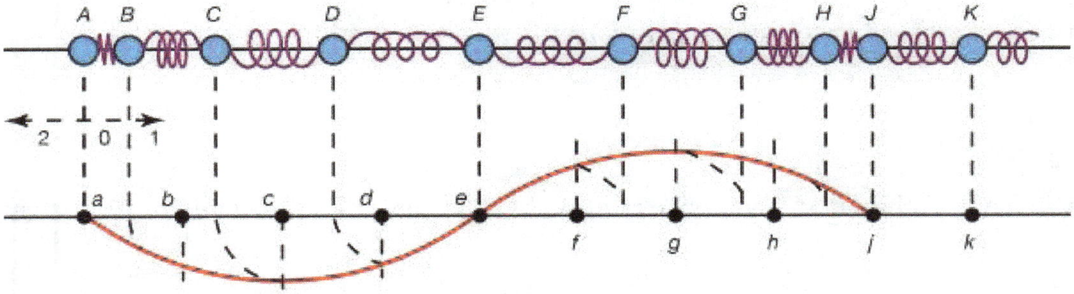

A longitudinal wave and its transverse representation

A transverse representation of a longitudinal wave is shown at the bottom of the figure. Here vertical lines are drawn through the rest positions (indicated by a,b,c, etc.), with lengths proportional to the distances that the masses have moved from equilibrium (their amplitudes). Lines are drawn upward from the axis when displacement is to the left and downward when to the right. A smooth curve drawn through the ends of the vertical lines gives a transverse curve. This transverse curve shows that there is one compression and one rarefaction per cycle, aj being one wavelength. Frequency would be represented by the number of complete cycles executed by any of the masses per second.

Examples

Longitudinal waves include sound waves (vibrations in pressure, particle of displacement, and particle velocity propagated in an elastic medium) and seismic P-waves (created by earthquakes and explosions).

In longitudinal waves, the displacement of the medium is parallel to the propagation of the wave, and waves can be either straight or round. A wave along the length of a stretched Slinky toy, where the distance between coils increases and decreases, is a good visualization.

Sound Waves

In the case of longitudinal harmonic sound waves, the frequency and wavelength can be described by the formula

$$y(x,t) = y_0 \cos\left(\omega\left(t - \frac{x}{c}\right)\right)$$

where:

- y is the displacement of the point on the traveling sound wave;
- x is the distance the point has traveled from the wave's source;
- t is the time elapsed;
- y_0 is the amplitude of the oscillations,
- c is the speed of the wave; and
- ω is the angular frequency of the wave.

The quantity x/c is the time that the wave takes to travel the distance x.

The ordinary frequency (f) of the wave is given by

$$f = \frac{\omega}{2\pi}.$$

The wavelength can be calculated as the relation between a wave's speed and ordinary frequency.

$$\lambda = \frac{c}{f}.$$

For sound waves, the amplitude of the wave is the difference between the pressure of the undisturbed air and the maximum pressure caused by the wave.

Sound's propagation speed depends on the type, temperature, and composition of the medium through which it propagates.

Pressure Waves

In an elastic medium with rigidity, a harmonic pressure wave oscillation has the form,

$$y(x,t) = y_0 \cos(kx - \omega t)$$

$$y(x,t) = y_0 \cos(kx - \omega t + \alpha)$$

where:

- y_0 is the amplitude of displacement,

- k is the angular wavenumber,
- x is the distance along the axis of propagation,
- ω is the angular frequency,
- t is the time, and
- ϕ is the phase difference.

The restoring force, which acts to return the medium to its original position, is provided by the medium's bulk modulus.

Electromagnetic

Maxwell's equations lead to the prediction of electromagnetic waves in a vacuum, which are transverse (in that the electric fields and magnetic fields vary perpendicularly to the direction of propagation). However, waves can exist in plasmas or confined spaces, called plasma waves, which can be longitudinal, transverse, or a mixture of both. Plasma waves can also occur in force-free magnetic fields.

In the early development of electromagnetism, there were some like Alexandru Proca (1897-1955) known for developing relativistic quantum field equations bearing his name (Proca's equations) for the massive, vector spin-1 mesons. In recent decades, some extended electromagnetic theorists, such as Jean-Pierre Vigier and Bo Lehnert of the Swedish Royal Society, have used the Proca equation in an attempt to demonstrate photon mass as a longitudinal electromagnetic component of Maxwell's equations, suggesting that longitudinal electromagnetic waves could exist in a Dirac polarized vacuum.

After Heaviside's attempts to generalize Maxwell's equations, Heaviside came to the conclusion that electromagnetic waves were not to be found as longitudinal waves in "*free space*" or homogeneous media. But Maxwell's equations do lead to the appearance of longitudinal waves under some circumstances, for example, in plasma waves or guided waves. Basically distinct from the "free-space" waves, such as those studied by Hertz in his UHF experiments, are Zenneck waves. The longitudinal modes of a resonant cavity are the particular standing wave patterns formed by waves confined in a cavity. The longitudinal modes correspond to those wavelengths of the wave which are reinforced by constructive interference after many reflections from the cavity's reflecting surfaces. Recently, Haifeng Wang et al. proposed a method that can generate a longitudinal electromagnetic (light) wave in free space, and this wave can propagate without divergence for a few wavelengths.

Surface Wave

In physics, a surface wave is a mechanical wave that propagates along the interface between differing media. A common example is gravity waves along the surface of liquids, such as ocean waves. Gravity waves can also occur within liquids, at the interface between two fluids with different densities. Elastic surface waves can travel along the surface of solids, such as *Rayleigh* or *Love* waves. Electromagnetic waves can also propagate as "surface waves" in that they can be guided along a refractive index gradient or along an interface between two media having different dielectric constants. In radio transmission, a *ground wave* is a guided wave that propagates close to the surface of the Earth.

A diving grebe creates surface waves

Mechanical Waves

In seismology, several types of surface waves are encountered. Surface waves, in this mechanical sense, are commonly known as either *Love waves* (L waves) or *Rayleigh waves*. A seismic wave is a wave that *travels through the Earth, often as the result of an earthquake or explosion*. Love waves have transverse motion (movement is perpendicular to the direction of travel, like light waves), whereas Rayleigh waves have both longitudinal (movement parallel to the direction of travel, like sound waves) and transverse motion. Seismic waves are studied by seismologists and measured by a seismograph or seismometer. Surface waves span a wide frequency range, and the period of waves that are most damaging is usually 10 seconds or longer. Surface waves can travel around the globe many times from the largest earthquakes. Surface waves are caused when P waves and S waves come to the surface.

Examples are the waves at the surface of water and air (ocean surface waves). Another example is internal waves, which can be transmitted along the interface of two water masses of different densities.

In theory of hearing physiology, the traveling wave (TW) of Von Bekesy, resulted from an acoustic surface wave of the basilar membrane into the cochlear duct. His theory pretended to explain every feature of the auditory sensation owing to these passive mechanical phenomena.

Electromagnetic Waves

Ground wave refers to the propagation of radio waves parallel to and adjacent to the surface of the Earth, following the curvature of the Earth. This radiative ground wave is known as the Norton surface wave. Other types of surface wave are the non-radiative Zenneck surface wave or Zenneck–Sommerfeld surface wave, the trapped surface wave and the gliding wave. Dyakonov surface waves (DSW) propagating at the interface of transparent materials with different symmetry.

Radio Propagation

Lower frequency radio waves, below 3 MHz, travel efficiently as ground waves. In ITU nomenclature, this includes (in order): medium frequency (MF), low frequency (LF), very low frequency (VLF), ultra low frequency (ULF), super low frequency (SLF), extremely low frequency (ELF) waves.

Ground propagation works because lower-frequency waves are more strongly diffracted around obstacles due to their long wavelengths, allowing them to follow the Earth's curvature. The Earth has one refractive index and the atmosphere has another, thus constituting an interface that supports the guided wave's transmission. Ground waves propagate in vertical polarization, with their magnetic field horizontal and electric field (close to) vertical. With VLF waves, the ionosphere and earth's surface act as a waveguide.

Conductivity of the surface affects the propagation of ground waves, with more conductive surfaces such as sea water providing better propagation. Increasing the conductivity in a surface results in less dissipation. The refractive indices are subject to spatial and temporal changes. Since the ground is not a perfect electrical conductor, ground waves are attenuated as they follow the earth's surface. The wavefronts initially are vertical, but the ground, acting as a lossy dielectric, causes the wave to tilt forward as it travels. This directs some of the energy into the earth where it is dissipated, so that the signal decreases exponentially.

Most long-distance LF "longwave" radio communication (between 30 kHz and 300 kHz) is a result of groundwave propagation. Mediumwave radio transmissions (frequencies between 300 kHz and 3000 kHz), including AM broadcast band, travel both as groundwaves and, for longer distances at night, as skywaves. Ground losses become lower at lower frequencies, greatly increasing the coverage of AM stations using the lower end of the band. The VLF and LF frequencies are mostly used for military communications, especially with ships and submarines. The lower the frequency the better the waves penetrate sea water. ELF waves (below 3 kHz) have even been used to communicate with deeply submerged submarines.

Ground waves have been used in over-the-horizon radar, which operates mainly at frequencies between 2–20 MHz over the sea, which has a sufficiently high conductivity to convey them to and from a reasonable distance (up to 100 km or more; over-horizon radar also uses skywave propagation at much greater distances). In the development of radio, ground waves were used extensively. Early commercial and professional radio services relied exclusively on long wave, low frequencies and ground-wave propagation. To prevent interference with these services, amateur and experimental transmitters were restricted to the high frequencies (HF), felt to be useless since their ground-wave range was limited. Upon discovery of the other propagation modes possible at medium wave and short wave frequencies, the advantages of HF for commercial and military purposes became apparent. Amateur experimentation was then confined only to authorized frequencies in the range.

Mediumwave and shortwave reflect off the ionosphere at night, which is known as skywave. During daylight hours, the lower D layer of the ionosphere forms and absorbs lower frequency energy. This prevents skywave propagation from being very effective on mediumwave frequencies in daylight hours. At night, when the D layer dissipates, mediumwave transmissions travel better by skywave. Ground waves *do not* include ionospheric and tropospheric waves.

The propagation of sound waves through the ground taking advantage of the Earth's ability to more efficiently transmit low frequency is known as audio ground wave (AGW).

Microwave Field Theory

Within microwave field theory, the interface of a dielectric and conductor supports "surface wave transmission". Surface waves have been studied as part of transmission lines and some may be considered as single-wire transmission lines.

Characteristics and utilizations of the electrical surface wave phenomenon include:

- The field components of the wave diminish with distance from the interface.
- Electromagnetic energy is not converted from the surface wave field to another form of energy (except in leaky or lossy surface waves) such that the wave does not transmit power normal to the interface, i.e. it is evanescent along that dimension.
- In optical fiber transmission, evanescent waves are surface waves.
- In coaxial cable in addition to the TEM mode there also exists a transverse-magnetic (TM) mode which propagates as a surface wave in the region around the central conductor. For coax of common impedance this mode is effectively suppressed but in high impedance coax and on a single central conductor without any outer shield, low attenuation and very broadband propagation is supported. Transmission line operation in this mode is called E-Line.

Surface Plasmon Polariton

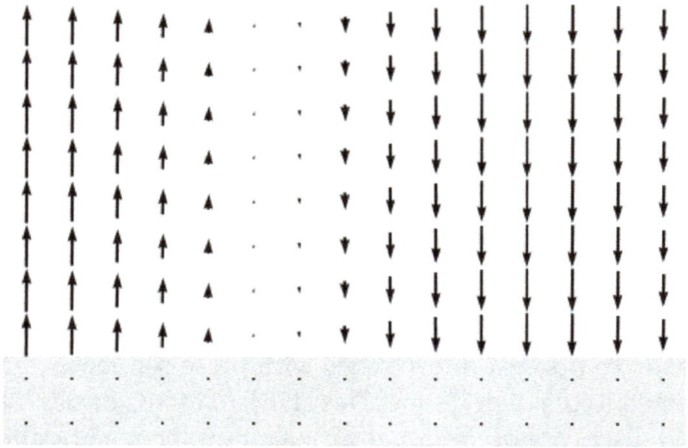

The E-field of a surface plasmon polariton at an silver–air interface, at a frequency corresponding to a free-space wavelength of 10μm. At this frequency, the silver behaves approximately as a perfect electric conductor, and the SPP is called a Sommerfeld–Zenneck wave

The surface plasmon polariton (SPP) is an electromagnetic surface wave that travels along an interface between two media with different dielectric constants. It exists under the condition that the permittivity of one of the materials forming the interface is negative, while the other one is positive, as is the case for the interface between air and a lossy conducting medium below the plasma frequency. The wave propagates parallel to the interface and decays exponentially vertical to it, a property called evanescence. Since the wave is on the boundary of a lossy conductor and a second medium, these oscillations can be sensitive to changes to the boundary, such as the adsorption of molecules by the conducting surface.

Sommerfeld–Zenneck Surface Wave

The Sommerfeld–Zenneck wave or Zenneck wave is a non-radiative guided electromagnetic wave that is supported by a planar or spherical interface between two homogeneous media having different dielectric constants. This surface wave propagates parallel to the interface and decays exponentially vertical to it, a property known as evanescence. It exists under the condition that the permittivity of one of the materials forming the interface is negative, while the other one is positive, as for example the interface between air and a lossy conducting medium such as the terrestrial transmission line, below the plasma frequency. Arising from original analysis by Arnold Sommerfeld and Jonathan Zenneck of the problem of wave propagation over a lossy earth, it exists as an exact solution to Maxwell's equations.

Transverse Wave

A wave motion, in which the particles of the medium oscillate about their mean positions at right angles to the direction of propagation of the wave, is called transverse wave.

In a transverse wave, the medium has particles that vibrate in a direction perpendicular to the direction of the propagation of wave. Here the formation of crest and trough takes place. The polarization of transverse wave is possible.

These waves can propagate through solids and liquids but not through gases, because gases do not possess elastic properties. Examples of these waves are: vibrations in strings, ripples on water surface and electromagnetic waves.

In a transverse wave, the particles of the medium oscillate in a direction perpendicular to the direction of propagation as shown in the figure. For example if a given transverse wave moves in x direction then the oscillations will occur in the y'z plane.

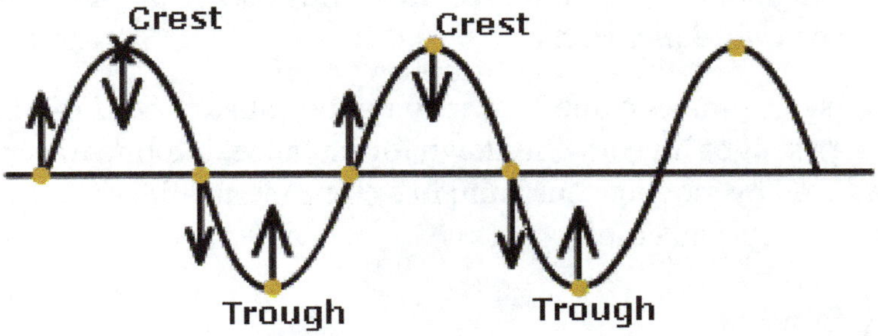

Particles of the medium oscillate in a direction perpendicular to the direction of propagation

Thus, during their oscillations, the particles may move upwards or downwards from the plane passing through their mean positions. The uppermost point of the wave, i.e., the position of maximum positive displacement is crest and the lowest point, i.e. the position of maximum displacement is called trough. Thus in a transverse wave crests and troughs appear alternatively.

For example: vibration of string, vibration of string, the surface wave produced on the surface of solid and liquid. Here the direction of propagation of energy is perpendicular to the direction of ossification.

There are always two directions that are independent of each other that can be used as the direction of wave. For example hold a ribbon in your hand and move it up and down can create transverse wave. Also moving the hand side by side can do the job.

A simple transverse wave can be represented by a sine or cosine curve, so called because the amplitude of any point on the curve—i.e., its distance from the axis—is proportional to the sine (or cosine) of an angle. In the Figure, sine curves of various amplitudes are shown. These curves represent how a standing transverse wave might look at consecutive (1, 2, 3, 4, and 5) intervals of time. The time required for a point on the wave to make a complete oscillation through the axis is called the period of the wave motion, and the number of oscillations executed per second is called the frequency. Wavelength is considered to be the distance between corresponding points on the wave—i.e., the distance between two adjacent peaks or troughs of the wave. Transverse waves may also be complex, in which the curves representing them are composed of two or more sine or cosine curves.

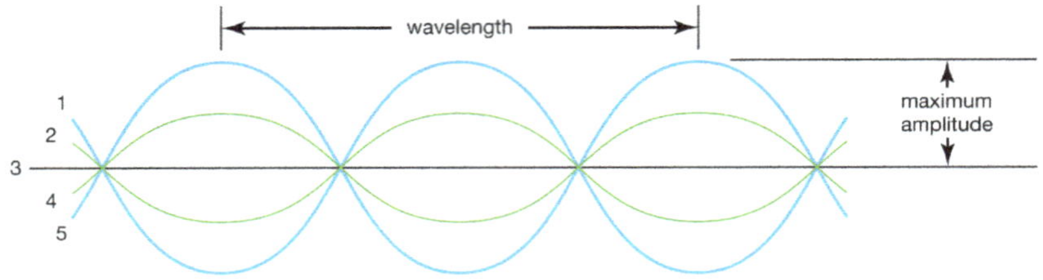

Polarized Waves

Continuing with the string example, if you move your hand in a clockwise circle, you will launch waves in the form of a left-handed helix as they propagate away. Similarly, if you move your hand in a counter-clockwise circle, a right-handed helix will form. These phenomena of *simultaneous* motion in two directions go beyond the kinds of waves we observe on the surface of water - in that a wave on a string can be *two-dimensional.*

Two-dimensional transverse waves exhibit a phenomenon called polarization. A wave produced by moving your hand in a straight line, up and down for instance, is a linearly polarized wave, a special case. A wave produced by moving your hand in a circle or an ellipse is a circularly or elliptically polarized wave, two other special cases.

Electromagnetic Waves

Electromagnetic waves behave in this same way. Electromagnetic waves are also two-dimensional transverse waves. Transverse waves are waves that travel perpendicular to the direction of the vibration. Ray theory does not describe phenomena such as interference and diffraction, which require wave theory (involving the phase of the wave). You can think of a ray of light, in optics, as an idealized narrow beam of electromagnetic radiation. Rays are used to model the propagation of light through an optical system, by dividing the real light field up into discrete rays that can be computationally propagated through the system by the techniques of ray tracing. A light ray is a line or curve that is perpendicular to the light's wavefronts (and is therefore collinear with the wave vector). Light rays bend at the interface between two dissimilar media and may be curved in a medium in which the refractive index changes. Geometric optics describes how rays propagate through an optical system.

This two-dimensional nature should not be confused with the two components of an electromagnetic wave, the electric and magnetic field components, which are shown in the light wave diagram here. Each of these fields, the electric and the magnetic, exhibits two-dimensional transverse wave behavior, just like the waves on a string.

Speed of a Transverse Wave (ESACQ)

Wave Speed

Wave speed is the distance a wave travels per unit time.

Quantity: Wave speed (v) Unit name: metre per second Unit symbol: $m \cdot s^{-1}$

The distance between two successive crests is 1 wavelength, λ. Thus in a time of 1 period, the wave will travel 1 wavelength in distance. Thus the speed of the wave, v, is:

$$v = \frac{\text{distance travelled}}{\text{time taken}} = \frac{\lambda}{T}$$

However, $f = \frac{1}{T}$ Therefore, we can also write:

$$v = \frac{\lambda}{T}$$

$$= \lambda \cdot \frac{1}{T}$$

$$= \lambda \cdot f$$

We call this equation the *wave equation*. To summarise, we have that $v = \lambda \cdot f$ where

- v = speed in $m.s^{-1}$
- λ = wavelength in m m
- f = frequency in Hz Hz

Wave equation:

$$v = f \cdot \lambda$$

or

$$v = \frac{\lambda}{T}$$

Seismic Waves

Seismic wave, vibration generated by an earthquake, explosion, or similar energetic source and propagated within the Earth or along its surface. Earthquakes generate four principal types of elastic waves; two, known as body waves, travel within the Earth, whereas the other two, called surface waves, travel along its surface. Seismographs record the amplitude and frequency of seismic waves and yield information about the Earth and its subsurface structure. Artificially generated seismic waves recorded during seismic surveys are used to collect data in oil and gas prospecting and engineering.

Of the body waves, the primary, or P, wave has the higher speed of propagation and so reaches a seismic recording station faster than the secondary, or S, wave. P waves, also called compressional or longitudinal waves, give the transmitting medium—whether liquid, solid, or gas—a back-and-forth motion in the direction of the path of propagation, thus stretching or compressing the medium as the wave passes any one point in a manner similar to that of sound waves in air. In the Earth, P waves travel at speeds from about 6 km (3.7 miles) per second in surface rock to about 10.4 km (6.5 miles) per second near the Earth's core some 2,900 km (1,800 miles) below the surface. As the waves enter the core, the velocity drops to about 8 km (5 miles) per second. It increases to about 11 km (6.8 miles) per second near the centre of the Earth. The speed increase with depth results from increased hydrostatic pressure as well as from changes in rock composition; in general, the increase causes P waves to travel in curved paths that are concave upward.

S waves, also called shear or transverse waves, cause points of solid media to move back and forth perpendicular to the direction of propagation; as the wave passes, the medium is sheared first in one direction and then in another. In the Earth the speed of S waves increases from about 3.4 km (2.1 miles) per second at the surface to 7.2 km (4.5 miles) per second near the boundary of the core, which, being liquid, cannot transmit them; indeed, their observed absence is a compelling argument for the liquid nature of the outer core. Like P waves, S waves travel in curved paths that are concave upward.

Of the two surface seismic waves, Love waves—named after the British seismologist A.E.H. Love, who first predicted their existence—travel faster. They are propagated when the solid medium near the surface has varying vertical elastic properties. Displacement of the medium by the wave is entirely perpendicular to the direction of propagation and has no vertical or longitudinal components. The energy of Love waves, like that of other surface waves, spreads from the source in two directions rather than in three, and so these waves produce a strong record at seismic stations even when originating from distant earthquakes.

The other principal surface waves are called Rayleigh waves after the British physicist Lord Rayleigh, who first mathematically demonstrated their existence. Rayleigh waves travel along the free surface of an elastic solid such as the Earth. Their motion is a combination of longitudinal compression and dilation that results in an elliptical motion of points on the surface. Of all seismic waves, Rayleigh waves spread out most in time, producing a long wave duration on seismographs.

The are many different seismic waves, but all of basically of four types:

- Compressional or P (for primary)

- Transverse or S (for secondary)
- Love
- Rayleigh

An earthquake radiates P and S waves in all directions and the interaction of the P and S waves with Earth's surface and shallow structure produces surface waves.

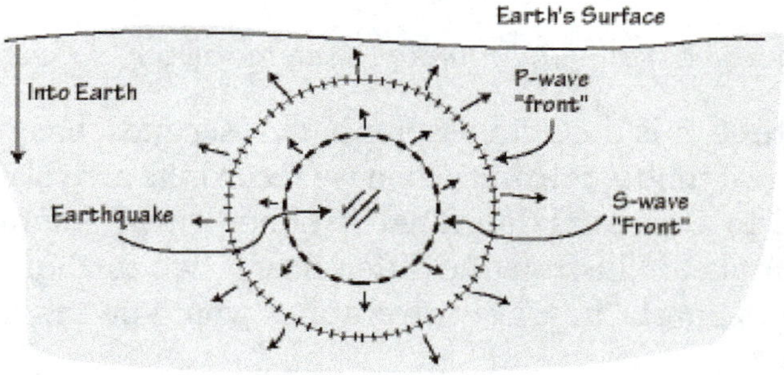

Near an earthquake the shaking is large and dominated by shear-waves and short-period surface waves. These are the waves that do the most damage to our buildings, highways, etc. Even in large earthquakes the intense shaking generally lasts only a few tens of seconds, but it can last for minutes in the greatest earthquakes. At farther distances the amplitude of the seismic waves decreases as the energy released by the earthquake spreads throughout a larger volume of Earth. Also with increasing distance from the earthquake, the waves are separated apart in time and dispersed because P, S, and surface waves travel at different speeds.

Seismic waves can be distinguished by a number of properties including the speed the waves travel, the direction that the waves move particles as they pass by, where and where they don't propagate. We'll go through each wave type individually to expound upon the differences.

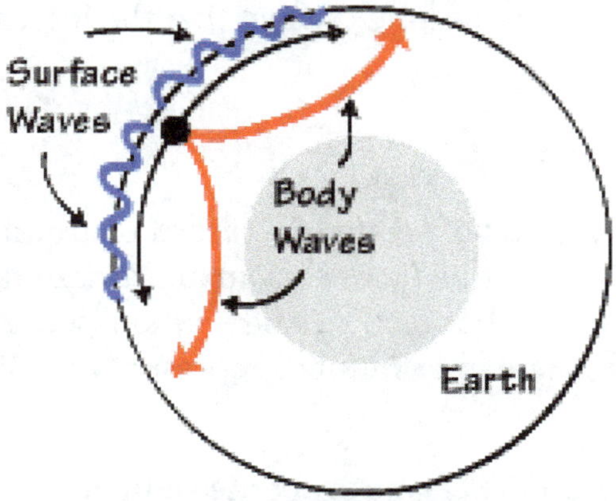

The first two wave types, P and S , are called body waves because they travel or propagate through the body of Earth. The latter two are called surface waves they the travel along Earth's surface and their amplitude decreases with depth into Earth

Wave Travel Times

Travel times are best conceptualized of with an analogy of an auto trip. If you have to travel 120 miles and you drive 60 mph, you'll get to your destination in two hours, if you are forced to drive at a speed of 30 mph, it will take you twice as long to arrive at your destination. The mathematical formula we use in this problem is

driving time = (distance of trip) / (driving speed)

To apply those ideas to earthquake studies, think of the earthquake location as the starting point for the trip and the seismometer as the place where the trip concludes. Faster waves will travel the distance quicker and show up on the seismogram first.

travel time = (distance from earthquake to seismometer) / (seismic wave speed)

Travel time is a relative time, it is the number of minutes, seconds, etc. that the wave took to complete its journey. The arrival time is the time when we record the arrival of a wave - it is an absolute time, usually referenced to Universal Coordinated Time (a 24-hour time system used in many sciences). Here's an example to illustrate the difference: if two earthquakes occurred at the same place but exactly 24 hours apart, the wave travel times would be the same but the arrival times would differ by one day.

Seismic Wave Speed

Seismic waves travel fast, on the order of kilometers per second (km/s). The precise speed that a seismic wave travels depends on several factors, most important is the composition of the rock. We are fortunate that the speed depends on the rock type because it allows us to use observations recorded on seismograms to infer the composition or range of compositions of the planet. But the process isn't always simple, because sometimes different rock types have the same seismic-wave velocity, and other factors also affect the speed, particularly temperature and pressure. Temperature tends to lower the speed of seismic waves and pressure tends to increase the speed. Pressure increases with depth in Earth because the weight of the rocks above gets larger with increasing depth. Usually, the effect of pressure is the larger and in regions of uniform composition, the velocity generally increases with depth, despite the fact that the increase of temperature with depth works to lower the wave velocity.

Compressional or P-Waves

P-waves are the first waves to arrive on a complete record of ground shaking because they travel the fastest (their name derives from this fact - P is an abbreviation for primary, first wave to arrive). They typically travel at speeds between ~1 and ~14 km/sec. The slower values corresponds to a P-wave traveling in water, the higher number represents the P-wave speed near the base of Earth's mantle.

The velocity of a wave depends on the elastic properties and density of a material. If we let κ represent the bulk modulus of a material, μ the shear-modulus, and ρ the density, then the P-wave velocity, which we represent by α, is defined by:

$$\alpha = \sqrt{\frac{\kappa + \frac{4}{3}\mu}{\rho}}$$

A modulus is a measure of how easy or difficulty it is to deforms a material. For example, the bulk modulus is a measure of how a material changes volume when pressure is applied and is a characteristic of a material. For example, foam rubber has a lower bulk modulus than steel.

P-waves are sound waves, it's just that in seismology we are interested in frequencies that are lower than humans' range of hearing (the speed of sound in air is about 0.3 km/sec). The vibration caused by P waves is a volume change, alternating from compression to expansion in the direction that the wave is traveling. P-waves travel through all types of media - solid, liquid, or gas.

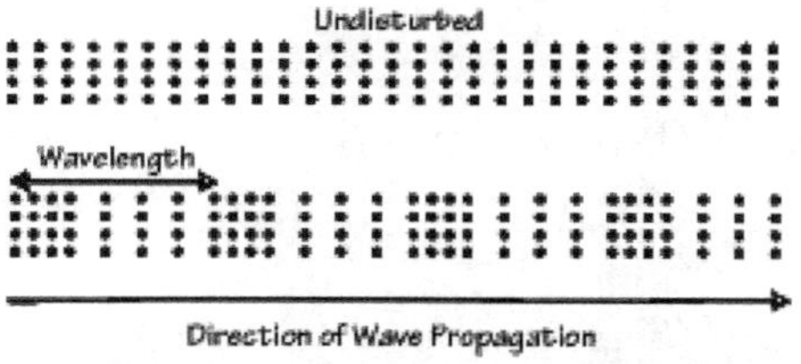

As a P-wave passes the ground is vibrated in the direction that the wave is propagating

S-Waves

Secondary , or S waves, travel slower than P waves and are also called "shear" waves because they don't change the volume of the material through which they propagate, they shear it. S-waves are transverse waves because they vibrate the ground in a the direction "transverse", or perpendicular, to the direction that the wave is traveling.

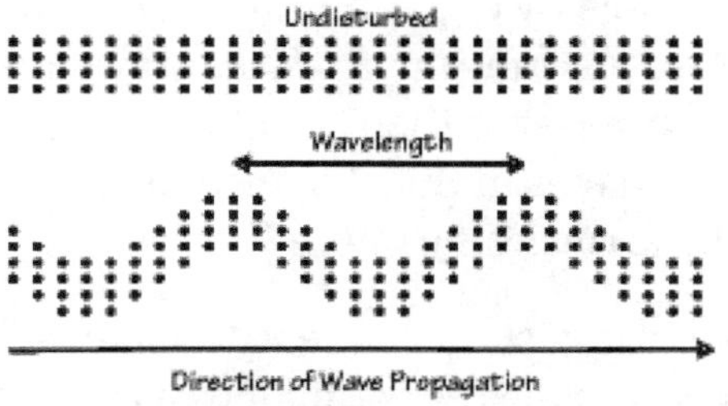

As a transverse wave passes the ground perpendicular to the direction that the wave is propagating.
S-waves are transverse waves

The S-wave speed, call it β, depends on the shear modulus and the density

$$\beta = \sqrt{\frac{\mu}{\rho}}$$

Even though they are slower than P-waves, the S-waves move quickly. Typical S-wave propagation speeds are on the order of 1 to 8 km/sec. The lower value corresponds to the wave speed in loose, unconsolidated sediment, the higher value is near the base of Earth's mantle.

An important distinguishing characteristic of an S-wave is its inability to propagate through a fluid or a gas because a fluids and gasses cannot transmit a shear stress and S-waves are waves that shear the material.

In general, earthquakes generate larger shear waves than compressional waves and much of the damage close to an earthquake is the result of strong shaking caused by shear waves.

Using P and S-waves to Locate Earthquakes

We can use the fact that P and S waves travel at different speeds to locate earthquakes. Assume a seismometer are is far enough from the earthquake that the waves travel roughly horizontally, which is about 50 to 500 km for shallow earthquakes. When an earthquake occurs the P and S waves travel outward from the region of the fault that ruptured and the P waves arrive at the seismometer first, followed by the S-wave. Once the S-wave arrives we can measure the time interval between the onset of P-wave and the onset of S-wave shaking.

The travel time of the P wave is

distance from earthquake / (P-wave speed)

The travel time of the S wave is

distance from earthquake / (S-wave speed)

The difference in the arrival times of the waves is

distance from earthquake / (S-wave speed) - distance from earthquake / (P-wave speed)

which equals

*distance from earthquake * (1/ (S-wave speed) - 1 / (P-wave speed))*

We can measure that difference from a seismogram and if we also know the speed that the waves travel, we could calculate the distance by equating the measured time difference and the expression. For the distance range 50 to 500 km, the S-waves travel about 3.45 km/s and the P-waves around 8 km/s. The value in parentheses is then equal to about (1/3.45 - 1/8) or about 1/8. Thus the simple rule of thumb for earthquakes in this distance range is the distance is about eight times the arrival time of S-wave less the arrival time of the P-wave.

That means that we can estimate the distance an earthquake is from a seismometer. The earthquake can be in any direction, but must be the estimated distance away. Geometrically that means that the earthquake must be located on a circle surrounding the seismometer, and the radius of the circle is about eight times the observed wave travel-time difference (in kilometers).

If we have two other seismometers which recorded the same earthquake, we could make a similar measurement and construct a circle of possible locations for each seismometer. Since the earthquake location since it must lie on each circle centered on a seismometer, if we plot three or more circles on a map we could find that the three circles will intersect at a single location - the earthquake's epicenter.

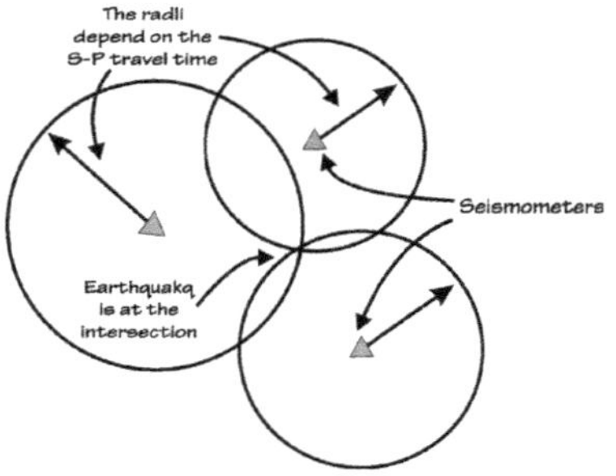

Using the "S minus P arrival time" to locate an earthquake. You need at least three stations and some idea of the P and S velocities between the earthquake and the seismometers

In practice we use better estimates of the speed than our simple rule of thumb and solve the problem using algebra instead of geometry. We also can include the earthquake depth and the time that earthquake rupture initiated (called the "origin time") into the problem.

Love Waves

Love waves are transverse waves that vibrate the ground in the horizontal direction perpendicular to the direction that the waves are traveling. They are formed by the interaction of S waves with Earth's surface and shallow structure and are dispersive waves. The speed at which a dispersive wave travels depends on the wave's period. In general, earthquakes generate Love waves over a range of periods from 1000 to a fraction of a second, and each period travels at a different velocity but the typical range of velocities is between 2 and 6 km/second.

Love waves are transverse and restricted to horizontal movement - they are recorded only on seismometers that measure the horizontal ground motion

Another important characteristic of Love waves is that the amplitude of ground vibration caused by a Love wave decreases with depth - they're surface waves. Like the velocity the rate of amplitude decrease with depth also depends on the period.

Rayleigh Waves

Rayleigh waves are the slowest of all the seismic wave types and in some ways the most complicated. Like Love waves they are dispersive so the particular speed at which they travel depends on the wave period and the near-surface geologic structure, and they also decrease in amplitude with depth. Typical speeds for Rayleigh waves are on the order of 1 to 5 km/s.

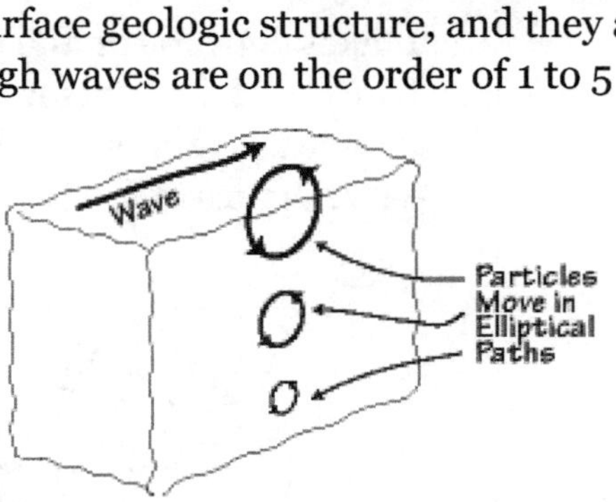

Rayleigh waves are similar to water waves in the ocean (before they "break" at the surf line). As a Rayleigh wave passes, a particle moves in an elliptical trajectory that is counterclockwise (if the wave is traveling to your right). The amplitude of Rayleigh-wave shaking decreases with depth

Seismic Wave Propagation

Waves on a Seismogram

As you might expect, the difference in wave speed has a profound influence on the nature of seismograms. Since the travel time of a wave is equal to the distance the wave has traveled, divided by the average speed the wave moved during the transit, we expect that the fastest waves arrive at a seismometer first. Thus, if we look at a seismogram, we expect to see the first wave to arrive to be a P-wave (the fastest), then the S-wave, and finally, the Love and Rayleigh (the slowest) waves. Although we have neglected differences in the travel path (which correspond to differences in travel distance) and the abundance waves that reverberate within Earth, the overall character is as we have described.

The fact that the waves travel at speeds which depend on the material properties (elastic moduli and density) allows us to use seismic wave observations to investigate the interior structure of the planet. We can look at the travel times, or the travel times and the amplitudes of waves to infer the existence of features within the planet, and this is a active area of seismological research. To understand how we "see" into Earth using vibrations, we must study how waves interact with the rocks that make up Earth.

Several types of interaction between waves and the subsurface geology (i.e. the rocks) are commonly observable on seismograms

- Refraction
- Reflection
- Dispersion
- Diffraction
- Attenuation

We'll examine the two simplest types of interaction refraction and reflection.

Refraction

As a wave travels through Earth, the path it takes depends on the velocity. Perhaps you recall from high school a principle called Snell's law, which is the mathematical expression that allows us to determine the path a wave takes as it is transmitted from one rock layer into another. The change in direction depends on the ratio of the wave velocities of the two different rocks.

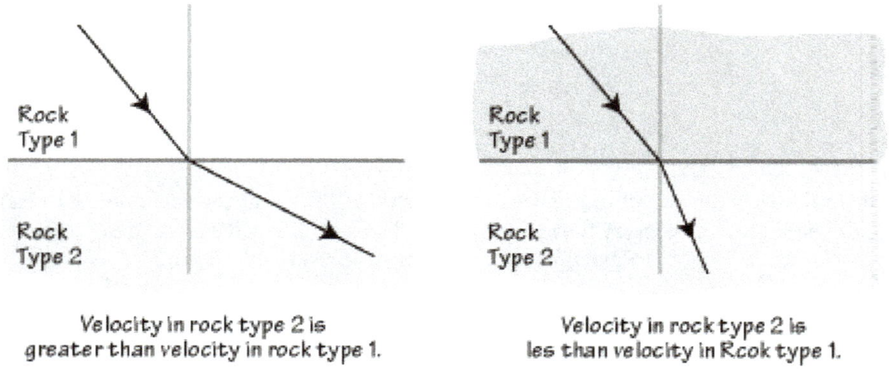

When waves reach a boundary between different rock types, part of the energy is transmitted across the boundary. The transmitted wave travels in a different direction which depends on the ratio of velocities of the two rock types. Part of the energy is also reflected backwards into the region with Rock Type 1

Refraction has an important affect on waves that travel through Earth. In general, the seismic velocity in Earth increases with depth (there are some important exceptions to this trend) and refraction of waves causes the path followed by body waves to curve upward.

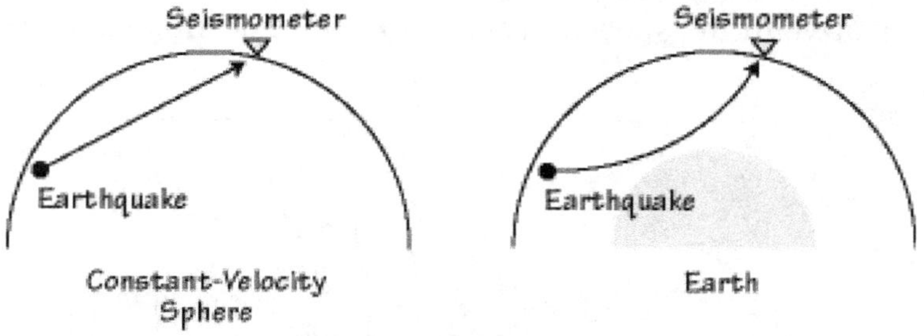

The overall increase in seismic wave speed with depth into Earth produces an upward curvature to rays that pass through the mantle. A notable exception is caused by the decrease in velocity from the mantle to the core. This speed decrease bends waves backwards and creates a "P-wave Shadow Zone" between about 100° and 140° distance (1° = 111.19 km)

Reflection

The second wave interaction with variations in rock type is reflection. I am sure that you are familiar with reflected sound waves; we call them echoes. And your reflection in a mirror or pool of water is composed of reflected light waves. In seismology, reflections are used to prospect for petroleum and investigate Earth's internal structure. In some instances reflections from the boundary between the mantle and crust may induce strong shaking that causes damage about 100 km from an earthquake (we call that boundary the "Moho" in honor of Mohorovicic, the scientist who discovered it).

A seismic reflection occurs when a wave impinges on a change in rock type (which usually is accompanied by a change in seismic wave speed). Part of the energy carried by the incident wave is transmitted through the material (that's the refracted wave described above) and part is reflected back into the medium that contained the incident wave.

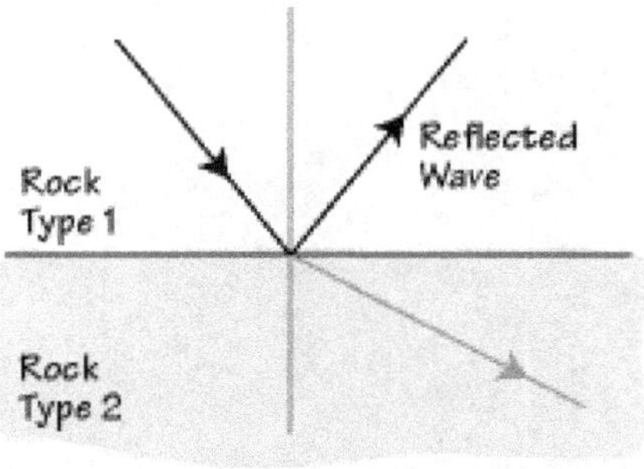

When a wave encounters a change in material properties
(seismic velocities and or density) its energy is split into reflected and refracted waves

The amplitude of the reflection depends strongly on the angle that the incidence wave makes with the boundary and the contrast in material properties across the boundary. For some angles all the energy can be returned into the medium containing the incident wave.

The actual interaction between a seismic wave and a contrast in rock properties is more complicated because an incident P wave generates transmitted and reflected P- *and* S-waves and so five waves are involved. Likewise, when an S-wave interacts with a boundary in rock properties, it too generates reflected and refracted P- and S-waves.

Dispersion

The effects of dispersion become more noticeable with increasing distance because the longer travel distance spreads the energy out (it disperses the energy). Usually, the long periods arrive first since they are sensitive to the speeds deeper in Earth, and the deeper regions are generally faster.

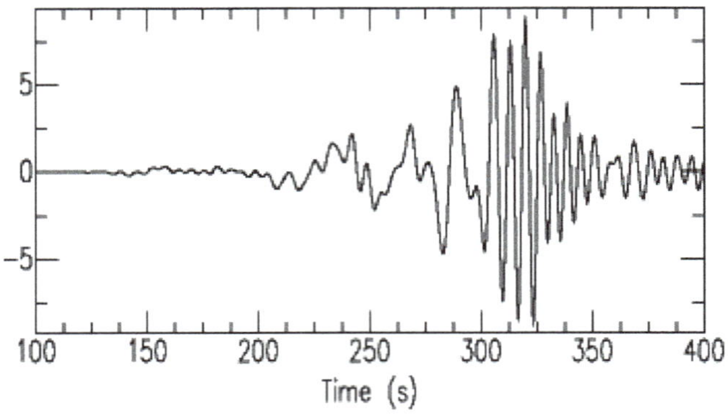

A dispersed Rayleigh wave generated by an earthquake in Alabama near the Gulf coast, and recorded in Missouri

P-Waves in Earth

The mathematics behind wave propagation is elegant and relatively simple, considering the fact that similar mathematical tools are useful for studying light, sound, and seismic waves. We can solve these equations or an appropriate approximation to them to compute the paths that seismic waves follow in Earth. The diagram below is an example of the paths P-waves generated by an earthquake near Earth's surface would follow.

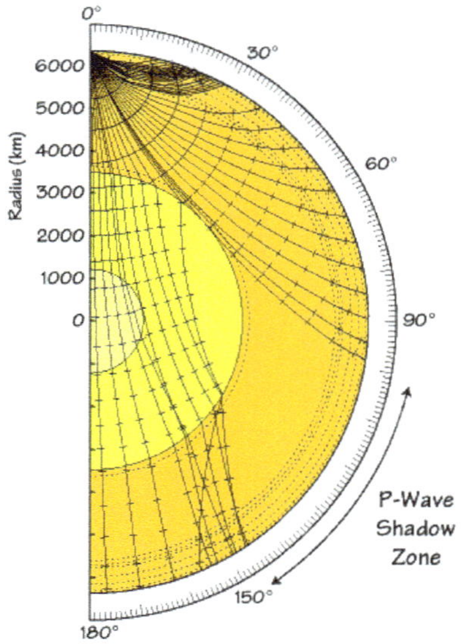

The main chemical shells of Earth are shown by different colors and regions with relatively abrupt velocity changes are shown by dashed lines. The curves show the paths of waves, and the lines crossing the rays show mark the wavefront at one minute intervals

Note the curvature of the rays in the mantle, the complexities in the upper mantle, and the dramatic impact of the core on the wavefronts. The decrease in velocity from the lower mantle to the outer core casts a "shadow" on the P-waves that extends from about 100° to 140° distance. Other waves such as surface waves and body waves reflecting off the surface are recorded in the "shadow" region, but the P-wave "dies out" near 100°. Since the outer core is fluid, and S-waves cannot travel through a fluid, the "S-wave shadow zone" is even larger, extending from about 100° to 180°.

P Wave

P waves are formed when energy is applied exactly at right angles to a medium. Particle motion under the influence of the wave is then in the direction of propagation of the wave. As a result of the particle motion, the rock particles are alternatively compressed and rarefracted or pulled apart as the waves propagate. P waves are also known as compressional waves, because of the pushing and pulling they do. They are the fastest kind of seismic waves.

In seismic processing, one generally assumes that we are dealing with P waves (Figure) only, travelling through isotropic and homogeneous material.

If the elastic moduli (K) and the bulk density (ρ) of the rocks are known, P-wave velocity can be calculated using the following equation:

$Vp=K+4/3\mu\rho$

Characteristics of P Waves

- It is a sound wave or pressure wave.
- It is always the first arriving seismic wave.
- Its velocity is the fastest of all seismic waves.
- Its particle motion is in the direction of propagation of the wave.
- It can propagate in solids or fluids (oil, gas and water).

Nomenclature

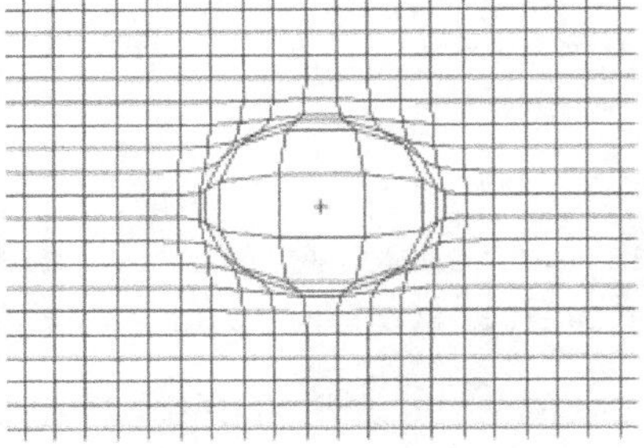

Representation of the propagation of a P-wave on a 2D grid (empirical shape)

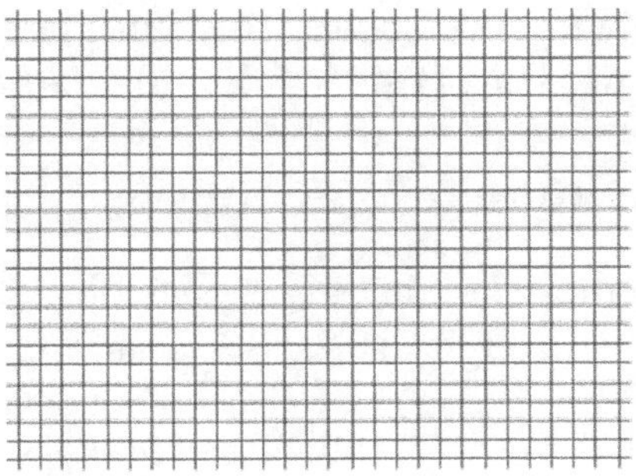

Plane P-wave

The name P-wave can stand for either pressure wave (as it is formed from alternating compressions and rarefactions) or primary wave (as it has high velocity and is therefore the first wave to be recorded by a seismograph).

In isotropic and homogeneous solids, the mode of propagation of a P-wave is always longitudinal; thus, the particles in the solid vibrate along the axis of propagation (the direction of motion) of the wave energy.

Seismic Waves in the Earth

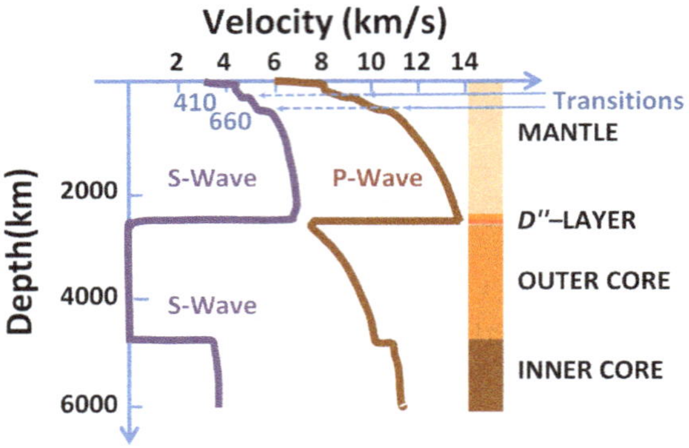

Velocity of seismic waves in the Earth versus depth. The negligible *S*-wave velocity in the outer core occurs because it is liquid, while in the solid inner core the *S*-wave velocity is non-zero

Primary and secondary waves are body waves that travel within the Earth. The motion and behavior of both *P*-type and *S*-type in the Earth are monitored to probe the interior structure of the Earth. Discontinuities in velocity as a function of depth are indicative of changes in phase or composition. Differences in arrival times of waves originating in a seismic event like an earthquake as a result of waves taking different paths allow mapping of the Earth's inner structure.

P-wave Shadow Zone

Almost all the information available on the structure of the Earth's deep interior is derived from observations of the travel times, reflections, refractions and phase transitions of seismic body waves, or normal modes. P-waves travel through the fluid layers of the Earth's interior, and yet they are refracted slightly when they pass through the transition between the semisolid mantle and the liquid outer core. As a result, there is a P-wave "shadow zone" between 103° and 142° from the earthquake's focus, where the initial P-waves are not registered on seismometers. In contrast, S-waves do not travel through liquids, rather, they are attenuated.

As an Earthquake Warning

Advance earthquake warning is possible by detecting the nondestructive primary waves that travel more quickly through the Earth's crust than do the destructive secondary and Rayleigh waves.

The amount of advance warning depends on the delay between the arrival of the P-wave and other destructive waves, generally on the order of seconds up to about 60 to 90 seconds for deep, distant, large quakes such as the 2011 Tohoku earthquake. The effectiveness of advance warning depends on accurate detection of the P-waves and rejection of ground vibrations caused by local activity (such as trucks or construction). Earthquake early warning systems can be automated to allow for immediate safety actions, such as issuing alerts, stopping elevators at the nearest floors and switching off utilities.

Velocity

The velocity of P-waves in a homogeneous isotropic medium is given by

$$v_p = \sqrt{\frac{K + \frac{4}{3}\mu}{\rho}} = \sqrt{\frac{\lambda + 2\mu}{\rho}}$$

where K is the bulk modulus (the modulus of incompressibility), μ is the shear modulus (modulus of rigidity, sometimes denoted as G and also called the second Lamé parameter), ρ is the density of the material through which the wave propagates, and λ is the first Lamé parameter.

Of these, density shows the least variation, so the velocity is mostly *controlled* by K and μ.

The elastic moduli P-wave modulus, M, is defined so that $M = K + 4\mu / 3$ and thereby

$$v_p = \sqrt{M / \rho}$$

Typical values for P-wave velocity in earthquakes are in the range 5 to 8 km/s. The precise speed varies according to the region of the Earth's interior, from less than 6 km/s in the Earth's crust to 13.5 km/s in the lower mantle, and 11 km/s through the inner core.

Velocity of Common Rock Typeshide		
Rock Type	**Velocity [m/s]**	**Velocity [ft/s]**
Unconsolidated Sandstone	4600 - 5200	15000 - 17000
Consolidated Sandstone	5800	19000
Shale	1800 - 4900	6000 -16000
Limestone	5800 - 6400	19000 - 21000
Dolomite	6400 - 7300	21000 - 24000
Anhydrite	6100	20000
Granite	5800 - 6100	19000 - 20000
Gabbro	7200	23600

Geologist Francis Birch discovered a relationship between the velocity of P waves and the density of the material the waves are traveling in:

$$V_p = a(\bar{M}) + b\rho$$

which later became known as Birch's law

S Wave

The second type of body wave is the S wave or secondary wave, which is the second wave you feel in an earthquake. An S wave is slower than a P wave and can only move through solid rock, not through any liquid medium. It is this property of S waves that led seismologists to conclude that the Earth's outer core is a liquid. S waves move rock particles up and down, or side-to-side--perpindicular to the direction that the wave is traveling in (the direction of wave propagation).

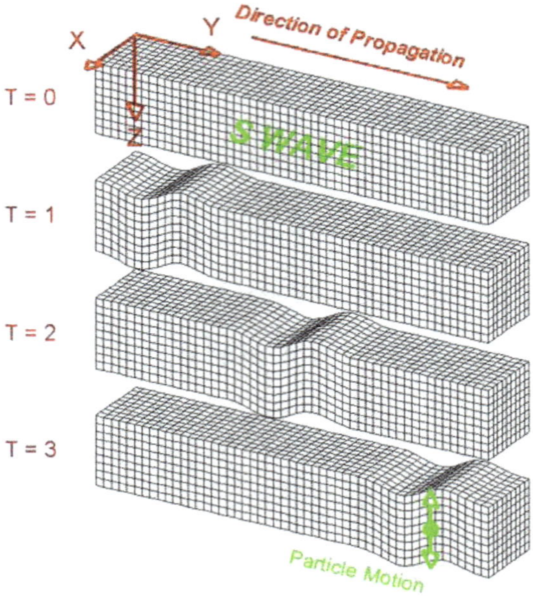

Figure- An S wave travels through a medium.
Particles are represented by cubes in this model

Theory

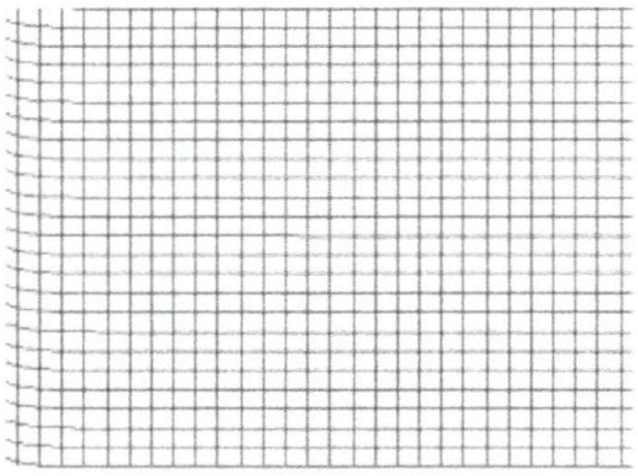

Plane shear wave

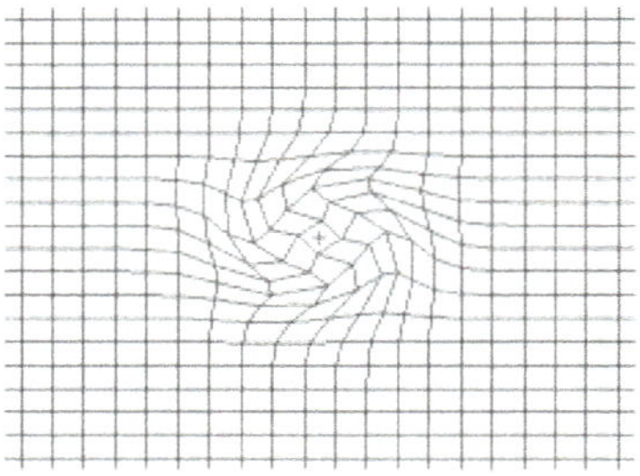

Propagation of a spherical S-wave in a 2d grid (empirical model)

The prediction of S-waves came out of theory in the 1800s. Starting with the stress-strain relationship for an isotropic solid in Einstein notation:

$$\tau_{ij} = \lambda \delta_{ij} e_{kk} + 2\mu e_{ij}$$

where τ is the stress, λ and μ are the Lamé parameters (with μ as the shear modulus), δ_{ij} is the Kronecker delta, and the strain tensor is defined

$$e_{ij} = \frac{1}{2}\left(\partial_i u_j + \partial_j u_i\right)$$

for strain displacement u. Plugging the latter into the former yields:

$$\tau_{ij} = \lambda \delta_{ij} \partial_k u_k + \mu\left(\partial_i u_j + \partial_j u_i\right)$$

Newton's 2nd law in this situation gives the *homogeneous equation of motion* for seismic wave propagation:

$$\rho \frac{\partial^2 u_i}{\partial t^2} = \partial_j \tau_{ij}$$

where ρ is the mass density. Plugging in the above stress tensor gives:

$$\begin{aligned} \rho \frac{\partial^2 u_i}{\partial t^2} &= \partial_i \lambda \partial_k u_k + \partial_j \mu\left(\partial_i u_j + \partial_j u_i\right) \\ &= \lambda \partial_i \partial_k u_k + \mu \partial_i \partial_j u_j + \mu \partial_j \partial_j u_i \end{aligned}$$

Applying vector identities and making certain approximations gives the seismic wave equation in homogeneous media:

$$\rho \ddot{u} = \left(\lambda + 2\mu\right) \nabla(\nabla \cdot u) - \mu \nabla \times (\nabla \times u)$$

where Newton's notation has been used for the time derivative. Taking the curl of this equation and applying vector identities eventually gives:

$$\nabla^2 (\nabla \times u) - \frac{1}{\beta^2} \frac{\partial^2}{\partial t^2} \left(\nabla \times u\right) = 0$$

which is simply the wave equation applied to the curl of u with a velocity β satisfying

$$\beta^2 = \frac{\mu}{\rho}$$

This describes S-wave propagation. Taking the divergence of seismic wave equation in homogeneous media, instead of the curl, yields an equation describing P-wave propagation. The steady-state SH waves are defined by the Helmholtz equation

$$(\nabla^2 + k^2) u = 0$$

where k is the wave number.

Reason P-waves Travel Faster than S-waves

P-waves and S-waves are body waves that propagate through the planet. P-waves travel 60% faster than S-waves on average because the interior of the Earth does not react the same way to both of them.

P-waves are compression waves that apply a force in the direction of propagation. As the interior of the Earth is almost incompressible, P-waves transmit their energy quite easily through the medium and thus travel quickly.

On the other hand, S-waves are shear waves, which means that the motion of the medium is perpendicular to the direction of propagation of the wave. The energy is thus less easily transmitted through the medium, and S-waves are slower.

Standing Wave

Standing wave, also called stationary wave, combination of two waves moving in opposite directions, each having the same amplitude and frequency. The phenomenon is the result of interference—that is, when waves are superimposed, their energies are either added together or cancelled out. In the case of waves moving in the same direction, interference produces a travelling wave; for oppositely moving waves, interference produces an oscillating wave fixed in space. A vibrating rope tied at one end will produce a standing wave, as shown in the Figure; the wave train (line *B*), after arriving at the fixed end of the rope, will be reflected back and superimposed on itself as another train of waves (line *C*) in the same plane. Because of interference between the two waves, the resultant amplitude (*R*) of the two waves will be the sum of their individual amplitudes. Part I of the Figure shows the wave trains *B* and C coinciding so that standing wave R has twice their amplitude. In part II, $^1/_8$ period later, *B* and C have each shifted $^1/_8$ wavelength. Part III represents the case $^1/_8$ period still later, when the amplitudes of the component waves *B* and *C* are oppositely directed. At all times there are positions (*N*) along the rope, called nodes, at which there is no movement at all; there the two wave trains are always in opposition. On either side of a node is a vibrating antinode (*A*). The antinodes alternate in the direction of displacement so that the rope at any instant resembles a graph of the mathematical function called the sine, as represented by line R. Both longitudinal (e.g., sound) waves and transverse (e.g., water) waves can form standing waves.

Formation of Standing Waves

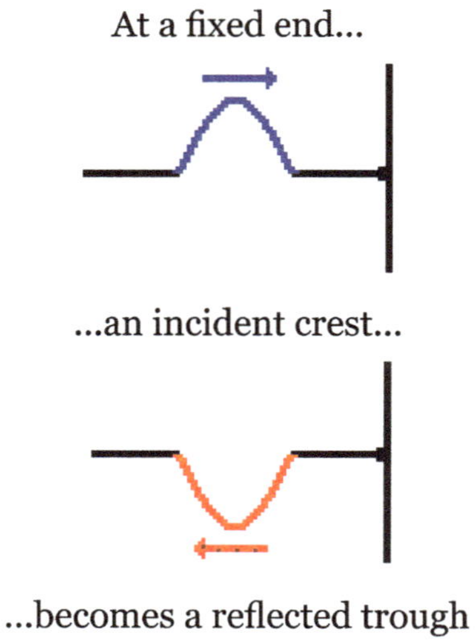

A standing wave pattern is a vibrational pattern created within a medium when the vibrational frequency of the source causes reflected waves from one end of the medium to interfere with incident waves from the source. This interference occurs in such a manner that specific points along the medium appear to be standing still. Because the observed wave pattern is characterized by points that appear to be standing still, the pattern is often called a standing wave pattern. Such patterns are only created within the medium at specific frequencies of vibration. These frequencies are known as harmonic frequencies, or merely harmonics. At any frequency other than a harmonic frequency, the interference of reflected and incident waves leads to a resulting disturbance of the medium that is irregular and non-repeating.

But how are standing wave formations formed? And why are they only formed when the medium is vibrated at specific frequencies? And what makes these so-called harmonic frequencies so special and magical? To answer these questions, let's consider a snakey stretched across the room, approximately 4-meters from end to end. (A "snakey" is a slinky-like device that consists of a large concentration of small-diameter metal coils.) If an upward displaced pulse is introduced at the left end of the snakey, it will travel rightward across the snakey until it reaches the fixed end on the right side of the snakey. Upon reaching the fixed end, the single pulse will reflect and undergo inversion. That is, the upward displaced pulse will become a downward displaced pulse. Now suppose that a second upward displaced pulse is introduced into the snakey at the precise moment that the first crest undergoes its fixed end reflection. If this is done with perfect timing, a rightward moving, upward displaced pulse will meet up with a leftward moving, downward displaced pulse in the exact middle of the snakey. As the two pulses pass through each other, they will undergo destructive interference. Thus, a point of no displacement in the exact middle of the snakey will be produced. The animation below shows several snapshots of the meeting of the two pulses at various stages in their interference. The individual pulses are drawn in blue and red; the resulting shape of the medium (as found by the principle of superposition) is shown in green. Note that there is a point on the diagram in the exact middle of the medium that never experiences any displacement from the equilibrium position.

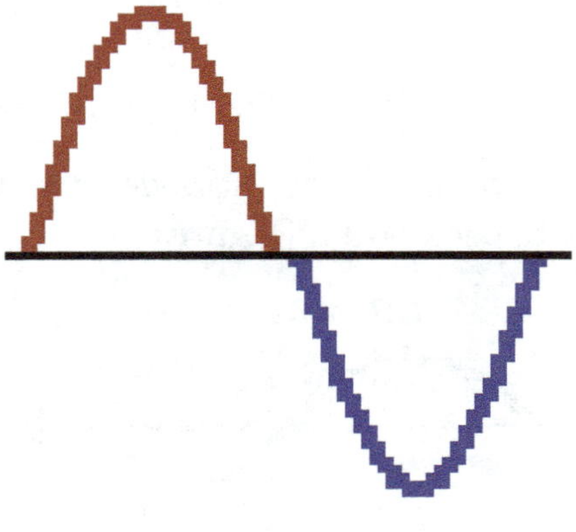

An upward displaced pulse introduced at one end will destructively interfere in the exact middle of the snakey with a second upward displaced pulse introduced from the same end if the introduction of the second pulse is performed with perfect timing. The same rationale could be applied to two downward displaced pulses introduced from the same end. If the second pulse is introduced at precisely the moment that the first pulse is reflecting from the fixed end, then destructive interference will occur in the exact middle of the snakey.

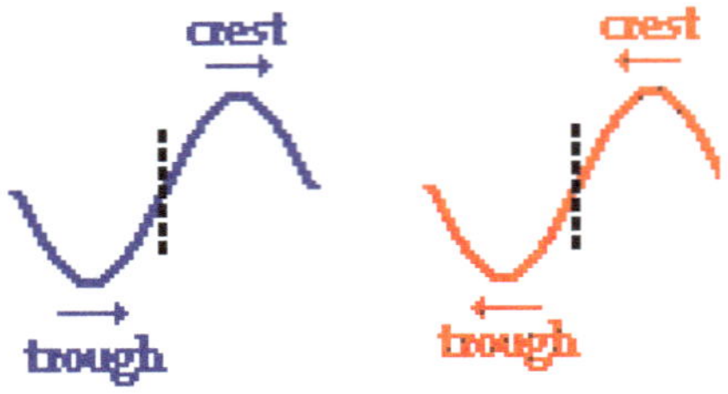

An incident wave and a reflected wave can be thought of as merely a repeating pattern of crests and throughs

The above discussion only explains why two pulses might interfere destructively to produce a point of no displacement in the middle of the snakey. A wave is certainly different than a pulse. What if there are two waves traveling in the medium? Understanding why two waves introduced into a medium with perfect timing might produce a point of displacement in the middle of the medium is a mere extension of the above discussion. While a pulse is a single disturbance that moves through a medium, a wave is a repeating pattern of crests and troughs. Thus, a wave can be thought of as an upward displaced pulse (crest) followed by a downward displaced pulse (trough) followed by an upward displaced pulse (crest) followed by a downward displaced pulse (trough) followed by. Since the introduction of a crest is followed by the introduction of a trough, every crest and trough will destructively interfere in such a way that the middle of the medium is a point of no displacement.

Parts of Standing Wave

(1) Nodes: Points which are stationary and with zero amplitude. At nodes, two waves travel in exactly same amplitude and out of phrase offset with each other. (Figure).

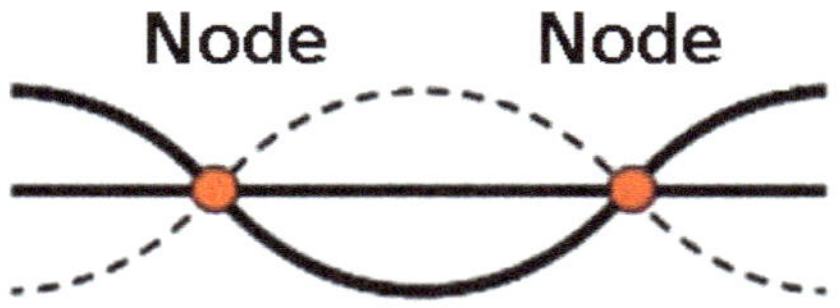

Figure: Nodes of Standing Wave

(2) Antinodes: Points at maximum amplitude. At antinodes, two waves travel in exactly same amplitude and in phrase reinforce with each other. (Figure).

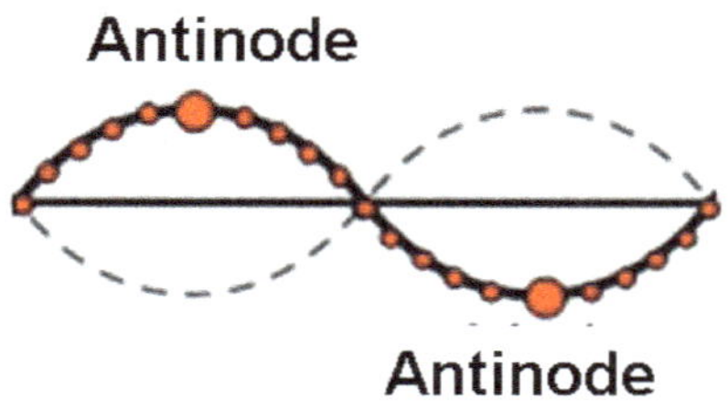

Figure: Antinodes of Standing Wave

A Standing Wave has the Following Properties

Maximum displacement of antinodes is ±2A. All points (except nodes) pass their mean position twice in one time period.

(a) Since antinodes have always maximum displacement, their velocity is also maximum compared to other points and velocity at nodes is zero.

(b) Standing waves can be transverse or longitudinal. In case of longitudinal stationary waves nodes are points of maximum pressure (minimum displacement) because phase difference of $\pi/2$ between pressure wave and displacement wave. And antinodal points have minimum pressure (maximum displacement). In organ pipes longitudinal stationary waves are obtained.

(c) In stationary waves, amplitudes of different points are not equal.

$A_{node} = 0$; $A_{antinode} =$ maximum $= 2A$.

For other points As increases from 0 to A and then decreases for A to 0.

(d) As in stationary waves nodes are permanently at rest; so no energy can be transmitted across them. I.e. energy of one region (segment) is confined in that region. However this energy oscillates between elastic PE and KE of the particles of the medium.

When all the particles are at their extreme positions the KE is minimum. while elastic PE is maximum and when all the particles (simultaneously) pass through their mean position KE will be maximum while elastic PE is a minimum. The total energy confined in a segment (Elastic PE & KE) always remains the same

(e) In standing wave if the amplitude of component waves are not equal then. $A_{min} \neq 0$ i.e. node will not be permanently at rest and so some energy will pass across the node and wave will be partially standing.

One Dimension: Two Fixed Ends

If a medium is bounded such that its opposite ends can be considered fixed, nodes will then be found at the ends. The simplest standing wave that can form under these circumstances has one antinode in the middle. This is half a wavelength. To make the next possible standing wave, place a node in the center. We now have one whole wavelength. To make the third possible standing wave, divide the length into thirds by adding another node. This gives us one and a half wavelengths. It should become obvious that to continue all that is needed is to keep adding nodes, dividing the medium into fourths, then fifths, sixths, etc. There are an infinite number of harmonics for this system, but no matter how many times we divide the medium up, we always get a whole number of half wavelengths $(^1/_2\lambda, ^2/_2\lambda, ^3/_2\lambda, \ldots, ^n/_2\lambda)$.

There are important relations among the harmonics themselves in this sequence. The wavelengths of the harmonics are simple fractions of the fundamental wavelength. If the fundamental wavelength were 1 m the wavelength of the second harmonic would be ½ m, the third harmonic would be ⅓ m, the fourth ¼ m, and so on. Since frequency is inversely proportional to wavelength, the frequencies are also related. The frequencies of the harmonics are whole-number multiples of the fundamental frequency. If the fundamental frequency were 1 Hz the frequency of the second harmonic would be 2 Hz, the third harmonic would be 3 Hz, the fourth 4 Hz, and so on.

One Dimension: Two Free Ends

If a medium is bounded such that its opposite ends can be considered free, antinodes will then be found at the ends. The simplest standing wave that can form under these circumstances has one node in the middle. This is half a wavelength. To make the next possible standing wave, place another antinode in the center. We now have one whole wavelength. To make the third possible standing wave, divide the length into thirds by adding another antinode. This gives us one and a half wavelengths. It should become obvious that we will get the same relationships for the standing waves formed between two free ends that we have for two fixed ends. The only difference is that the nodes have been replaced with antinodes and vice versa. Thus when standing waves form in a linear medium that has two free ends a whole number of half wavelengths fit inside the medium and the overtones are whole number multiples of the fundamental frequency

One Dimension: One Fixed End — One Free End

When the medium has one fixed end and one free end the situation changes in an interesting way. A node will always form at the fixed end while an antinode will always form at the free end. The simplest standing wave that can form under these circumstances is one-quarter wavelength long. To make the next possible standing wave add both a node and an antinode, dividing the drawing up into thirds. We now have three-quarters of a wavelength. Repeating this procedure we get five-quarters of a wavelength, then seven-quarters, etc. In this arrangement, there are always an odd number of quarter wavelengths present. Thus the wavelengths of the harmonics are always fractional multiples of the fundamental wavelength with an odd number in the denominator. Likewise, the frequencies of the harmonics are always odd multiples of the fundamental frequency.

The three cases above show that, although not all frequencies will result in standing waves, a simple, one-dimensional system possesses an infinite number of natural frequencies that will. It also shows that these frequencies are simple multiples of some fundamental frequency. For any real-world system, however, the higher frequency standing waves are difficult if not impossible to produce. Tuning forks, for example, vibrate strongly at the fundamental frequency, very little at the second harmonic, and effectively not at all at the higher harmonics.

Filtering

The best part of a standing wave is not that it appears to stand still, but that the amplitude of a standing wave is much larger that the amplitude of the disturbance driving it. It seems like getting something for nothing. Put a little bit of energy in at the right rate and watch it accumulate into something with a lot of energy. This ability to amplify a wave of one particular frequency over those of any other frequency has numerous applications.

- Basically, all non-digital musical instruments work directly on this principle. What gets put into a musical instrument is vibrations or waves covering a spread of frequencies (for brass, it's the buzzing of the lips; for reeds, it's the raucous squawk of the reed; for percussion, it's the relatively indiscriminate pounding; for strings, it's plucking or scraping; for flutes and organ pipes, it's blowing induced turbulence). What gets amplified is the fundamental frequency plus its multiples. These frequencies are louder than the rest and are heard. All the other frequencies keep their original amplitudes while some are even de-amplified. These other frequencies are quieter in comparison and are not heard.

- You don't need a musical instrument to illustrate this principle. Cup your hands together loosely and hold them next to your ear forming a little chamber. You will notice that one frequency gets amplified out of the background noise in the space around you. Vary the size and shape of this chamber. The amplified pitch changes in response. This is what people hear when the hold a seashell up to their ears. It's not "the ocean" but a few select frequencies amplified out of the noise that always surrounds us.
- During speech, human vocal cords tend to vibrate within a much smaller range that they would while singing. How is it then possible to distinguish the sound of one vowel from another? English is not a tonal language (unlike Chinese and many African languages). There is very little difference in the fundamental frequency of the vocal cords for English speakers during a declarative sentence. (Interrogative sentences rise in pitch near the end. Don't they?) Vocal cords don't vibrate with just one frequency, but with all the harmonic frequencies. Different arrangements of the parts of the mouth (teeth, lips, front and back of tongue, etc.) favor different harmonics in a complicated manner. This amplifies some of the frequencies and de-amplifies others. This makes "EE" sound like "EE" and "OO" sound like "OO".
- The filtering effect of resonance is not always useful or beneficial. People that work around machinery are exposed to a variety of frequencies. (This is what noise is.) Due to resonance in the ear canal, sounds near 4000 Hz are amplified and are thus louder than the other sounds entering the ear. Everyone should know that loud sounds can damage one's hearing. What everyone may not know is that exposure to loud sounds of just one frequency will damage one's hearing at that frequency. People exposed to noise are often experience 4000 Hz hearing loss. Those afflicted with this condition do not hear sounds near this frequency with the same acuity that unafflicted people do. It is often a precursor to more serious forms of hearing loss.

Two Dimensions

The type of reasoning used in the discussion so far can also be applied to two-dimensional and three-dimensional systems. As you would expect, the descriptions are a bit more complex. Standing waves in two dimensions have numerous applications in music. A circular drum head is a reasonably simple system on which standing waves can be studied. Instead of having nodes at opposite ends, as was the case for guitar and piano strings, the entire rim of the drum is a node. Other nodes are straight lines and circles. The harmonic frequencies are not simple multiples of the fundamental frequency.

Standing waves in two dimensions have been applied extensively to the study of violin bodies. Violins manufactured by the Italian violin maker Antonio Stradivari (1644–1737) are renowned for their clarity of tone over a wide dynamic range. Acoustic physicists have been working on reproducing violins equal in quality to those produced by Stradivarius for quite some time. One technique developed by the German physicist Ernst Chladni (1756–1794) involves spreading grains of fine sand on a plate from a dismantled violin that is then clamped and set vibrating with a bow. The sand grains bounce away from the lively antinodes and accumulate at the relatively quiet nodes. The resulting Chladni patterns from different violins could then be compared. Presumably, the patterns from better sounding violins would be similar in some way. Through trial and error, a violin designer should be able to produce components whose behavior mimicked those of the legendary master. This is, of course, just one factor in the design of a violin.

Chladni patterns on violin plates in order of increasing frequency			
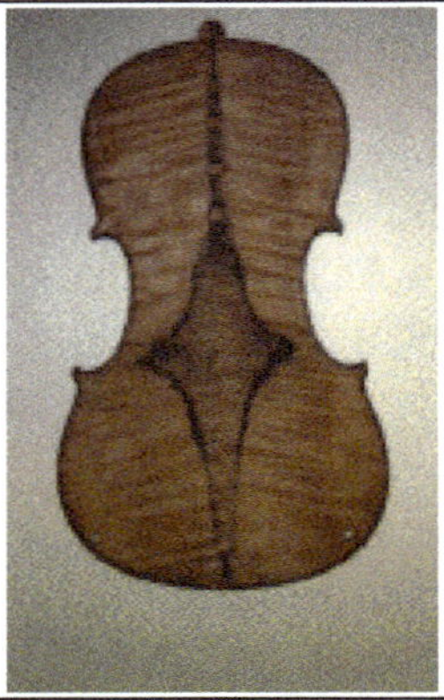	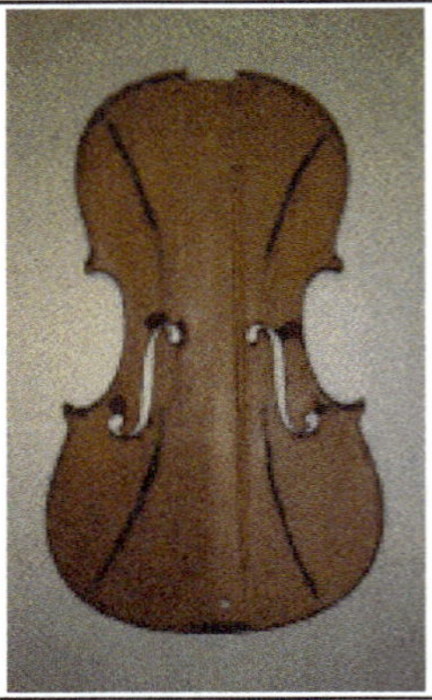	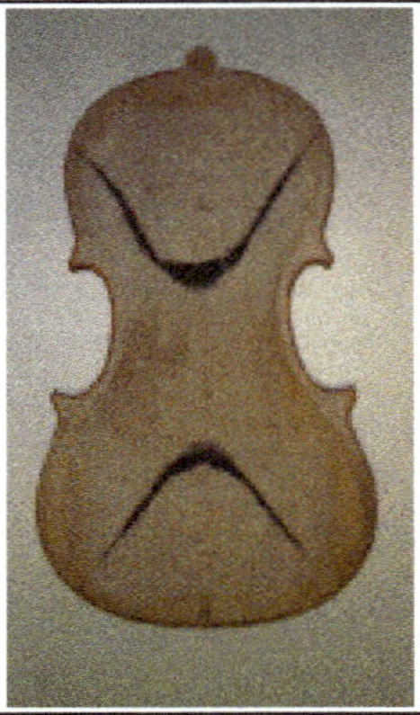	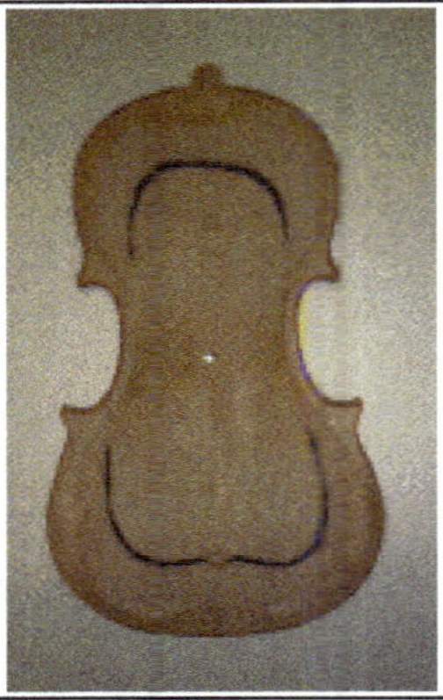
91 Hz	145 Hz	170 Hz	384 Hz

Three Dimensions

Probability density of a ground state electron |1,0,0> in a hydrogen atom

In the one-dimensional case the nodes were points (zero-dimensional). In the two-dimensional case the nodes were curves (one-dimensional). The dimension of the nodes is always one less than the dimension of the system. Thus, in a three-dimensional system the nodes would be two-dimensional surfaces. The most important example of standing waves in three dimensions are the orbitals of an electron in an atom. On the atomic scale, it is usually more appropriate to describe the electron as a wave than as a particle. The square of an electron's wave equation gives the probability function for locating the electron in any particular region. The orbitals used by chemists describe the shape of the region where there is a high probability of finding a particular electron. Electrons are confined to the space surrounding a nucleus in much the same manner that the waves in a guitar string are constrained within the string. The constraint of a string in a guitar forces the string to vibrate with specific frequencies. Likewise, an electron can only vibrate with specific frequencies. In the case of an electron, these frequencies are called eigenfrequencies and the states associated with these frequencies are called eigenstates or eigenfunctions. The set of all eigenfunctions for an electron form a mathematical set called the spherical harmonics.

There are an infinite number of these spherical harmonics, but they are specific and discrete. That is, there are no in-between states. Thus an atomic electron can only absorb and emit energy in specific in small packets called quanta. It does this by making a quantum leap from one eigenstate to another. This term has been perverted in popular culture to mean any sudden, large change. In physics, quite the opposite is true. A quantum leap is the smallest possible change of system, not the largest.

Some probability densities for excited electrons in a hydrogen atom		
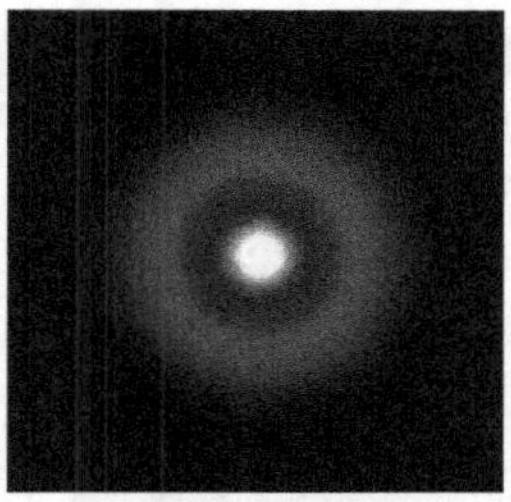 $\vert 2,0,0\rangle$	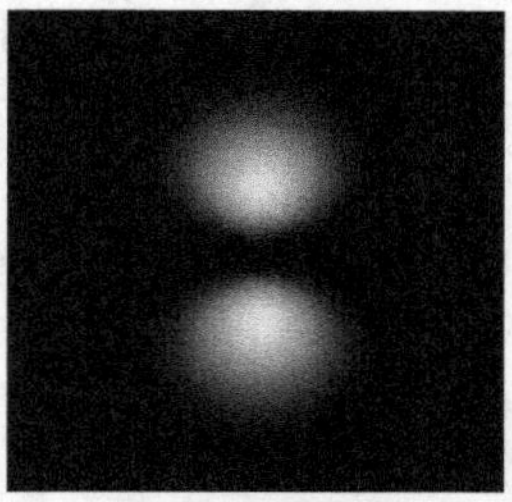 $\vert 2,1,0\rangle$	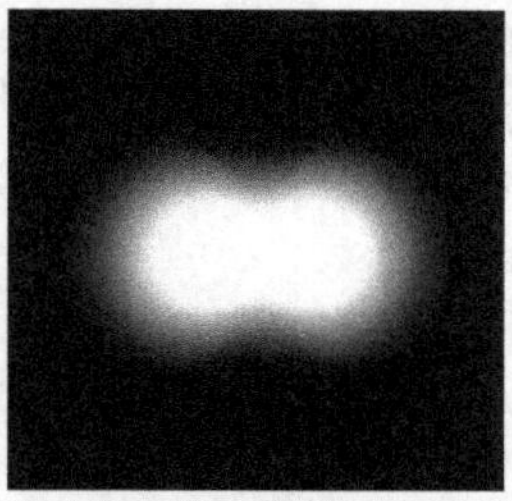 $\vert 2,1,1\rangle$
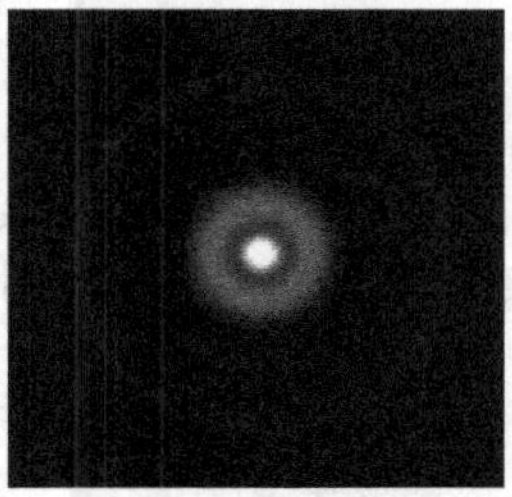 $\vert 3,0,0\rangle$	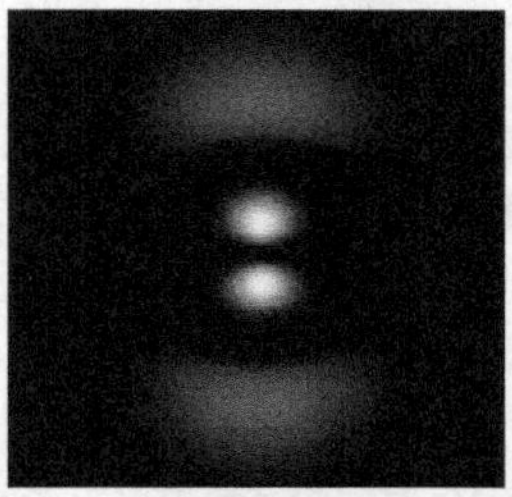 $\vert 3,1,0\rangle$	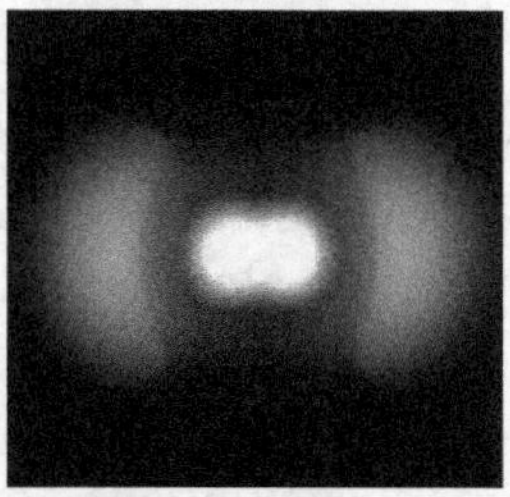 $\vert 3,1,1\rangle$
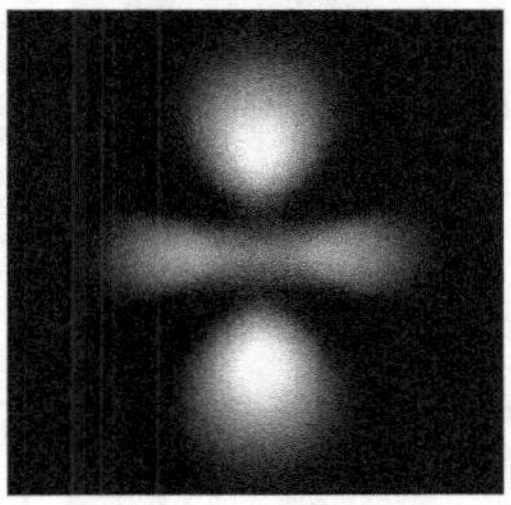 $\vert 3,2,0\rangle$	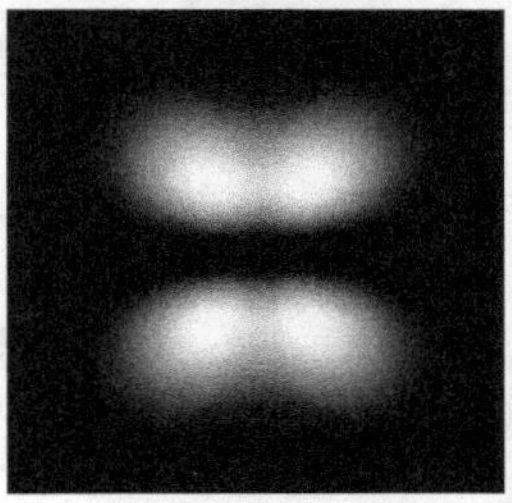 $\vert 3,2,1\rangle$	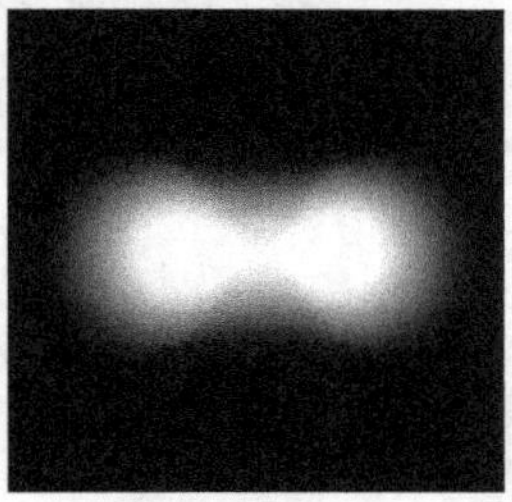 $\vert 3,2,2\rangle$

Mathematics

In mathematics, the infinite sequence of fractions $\frac{1}{1}, \frac{1}{2}, \frac{1}{3}, \frac{1}{4}, \ldots$ is called the harmonic sequence. Surprisingly, there are exactly the same number of harmonics described by the harmonic sequence as there are harmonics described by the "odds only" sequence: $\frac{1}{1}, \frac{1}{3}, \frac{1}{5}, \frac{1}{7}, \ldots$ "What? Obviously there are more numbers in the harmonic sequence than there are in the 'odds only' sequence." Nope. There are exactly the same number. Here's the proof. We can set up a one-to-one correspondence between the whole numbers and the odd numbers and we observe.

1, 2, 3, 4, 5, 6, 7, 8, 9,...

1, 3, 5, 7, 9, 11, 13, 15, 17,...

This can go on forever. Which means there are exactly the same number of odd numbers as there are whole numbers. Both the whole numbers and the odd numbers are examples of countable infinite sets.

There are an infinite number of possible wavelengths that can form standing waves under all of the circumstances described above, but there are an even greater number of wavelengths that can't form standing waves. "What? How can you have more than an infinite amount of something?" Well I don't want to prove that right now so you'll have to trust me, but there are more real numbers between 0 and 1 than there are whole numbers between zero and infinity. Not only do we have all the rational numbers less than one ($^1/_2$, $^3/_5$, $^{733}/_{2741}$, etc.) we also have all the possible algebraic numbers (square root 2, 7 minus square root 13, etc.) and the whole host of bizarre transcendental numbers (π, e, eπ, Feigenbaum's number, etc.). All of these numbers together form an uncountable infinite set called the real numbers. The number of whole numbers is an infinity called aleph null (0ℵ) the number of real numbers is an infinity called c (for continuum). The study of infinitely large numbers is known as transfinite mathematics. In this field, it is possible to prove that aleph null is less than c. There is no one-to-one correspondence between the real numbers and the whole numbers. Thus, there are more frequencies that won't form standing waves than there are frequencies that will form standing waves.

Mathematical Description

In one dimension, two waves with the same frequency, wavelength and amplitude traveling in opposite directions will interfere and produce a standing wave or stationary wave. For example: a wave traveling to the right along a taut string held stationary at its right end will reflect back in the other direction along the string, and the two waves will superpose to produce a standing wave. The reflected wave has to have the same amplitude and frequency as the incoming wave. The phenomenon can be demonstrated mathematically by deriving the equation for the sum of two oppositely moving waves:

A harmonic wave traveling to the right along the x-axis is described by the equation

$$y_1(x,t) = A\sin(\frac{2\pi x}{\lambda} - \omega t)$$

An identical harmonic wave traveling to the left is described by the equation

$$y_2(x,t) = A\sin(\frac{2\pi x}{\lambda} + \omega t)$$

where:

- A is the amplitude of the wave,
- ω (called the angular frequency and measured in *radians per second*) is 2π times the frequency (in *hertz*)
- λ is the wavelength of the wave (in *metres*)
- x and t are variables for longitudinal position and time, respectively.

So the equation of the resultant wave y will be the sum of y_1 and y_2:

$$y(x,t) = y_1 + y_2 = A\sin(\frac{2\pi x}{\lambda} - \omega t) + A\sin(\frac{2\pi x}{\lambda} + \omega t)$$

Using the trigonometric sum-to-product identity

$$\sin a + \sin b = 2\sin\left(\frac{a+b}{2}\right)\cos\left(\frac{a-b}{2}\right) \text{ to simplify:}$$

$$y = 2A\sin\left(\frac{2\pi x}{\lambda}\right)\cos(\omega t)$$

This describes a wave that oscillates in time, but has a spatial dependence that is stationary; at any point x the amplitude of the oscillations is constant with value $2A\sin\left(\frac{2\pi x}{\lambda}\right)$.. At locations which are *even* multiples of a quarter wavelength

$$x = \ldots -\frac{3\lambda}{2}, -\lambda, -\frac{\lambda}{2}, 0, \frac{\lambda}{2}, \lambda, \frac{3\lambda}{2}, \ldots$$

called the nodes, the amplitude is always zero, whereas at locations which are *odd* multiples of a quarter wavelength

$$x = \ldots -\frac{5\lambda}{4}, -\frac{3\lambda}{4}, -\frac{\lambda}{4}, \frac{\lambda}{4}, \frac{3\lambda}{4}, \frac{5\lambda}{4}, \ldots$$

called the anti-nodes, the amplitude is maximum, with a value of twice the amplitude of the original waves. The distance between two consecutive nodes or anti-nodes is $\lambda/2$.

Standing waves can also occur in two- or three-dimensional resonators. With standing waves on two-dimensional membranes such as drumheads, illustrated in the animations above, the nodes become nodal lines, lines on the surface at which there is no movement, that separate regions vibrating with opposite phase. These nodal line patterns are called Chladni figures. In three-dimensional resonators, such as musical instrument sound boxes and microwave cavity resonators, there are nodal surfaces.

Opposing Waves

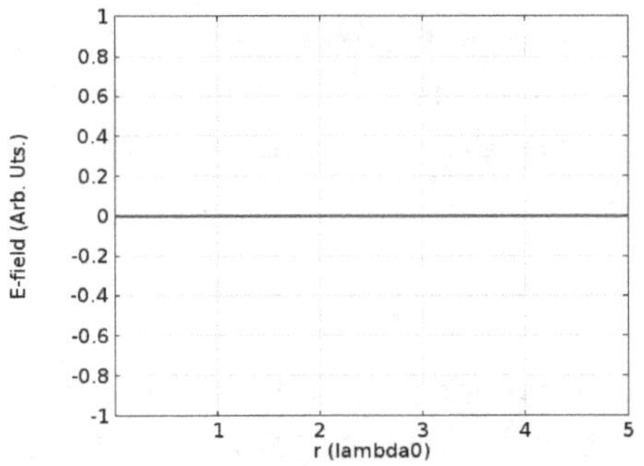

Transient analysis of a damped traveling wave reflecting at a boundary.

As an example of the second type, a *standing wave* in a transmission line is a wave in which the distribution of current, voltage, or field strength is formed by the superposition of two waves of the same frequency propagating in opposite directions. The effect is a series of nodes (zero displacement) and anti-nodes (maximum displacement) at fixed points along the transmission line. Such a standing wave may be formed when a wave is transmitted into one end of a transmission line and is reflected from the other end by an impedance mismatch, *i.e.*, discontinuity, such as an open circuit or a short. The failure of the line to transfer power at the standing wave frequency will usually result in attenuation distortion.

In practice, losses in the transmission line and other components mean that a perfect reflection and a pure standing wave are never achieved. The result is a *partial standing wave*, which is a superposition of a standing wave and a traveling wave. The degree to which the wave resembles either a pure standing wave or a pure traveling wave is measured by the standing wave ratio (SWR).

Another example is standing waves in the open ocean formed by waves with the same wave period moving in opposite directions. These may form near storm centres, or from reflection of a swell at the shore, and are the source of microbaroms and microseisms.

Standing Wave Ratio, Phase, and Energy Transfer

Standing wave ratio (SWR) is the ratio of the amplitude at the antinode (maximum) to the amplitude at the node (minimum). A pure standing wave will have an infinite SWR. It will also have a constant phase at any point in space (but it may undergo a 180° inversion every half cycle). A finite, non-zero SWR indicates a wave that is partially stationary and partially travelling. Such waves can be decomposed into two linearly superpositional components (assuming the medium is linear) of a travelling wave component and a stationary wave component. A zero SWR indicates that the wave does not have a stationary component – it is purely a travelling wave.

A pure standing wave does not transfer energy from the source to the destination. However, the wave is still subject to losses in the medium. Such losses will manifest as a finite SWR, indicating a travelling wave component leaving the source to supply the losses. Even though the SWR is now finite, it may still be the case that no energy reaches the destination because the travelling component is purely supplying the losses. However, in a lossless medium, a finite SWR implies a definite transfer of energy to the destination.

Examples

One easy example to understand standing waves is two people shaking either end of a jump rope. If they shake in sync the rope can form a regular pattern of waves oscillating up and down, with stationary points along the rope where the rope is almost still (nodes) and points where the arc of the rope is maximum (antinodes)

Sound Waves

Standing waves are also observed in physical media such as strings and columns of air. Any waves traveling along the medium will reflect back when they reach the end. This effect is most noticeable in musical instruments where, at various multiples of a vibrating string or air column's natural frequency, a standing wave is created, allowing harmonics to be identified. Nodes occur at fixed ends and anti-nodes at open ends. If fixed at only one end, only odd-numbered harmonics are available. At the open end of a pipe the anti-node will not be exactly at the end as it is altered by its contact with the air and so end correctionis used to place it exactly. The density of a string will affect the frequency at which harmonics will be produced; the greater the density the lower the frequency needs to be to produce a standing wave of the same harmonic.

Visible Light

Standing waves are also observed in optical media such as optical wave guides, optical cavities, etc. Lasers use optical cavities in the form of a pair of facing mirrors. The gain medium in the cavity (such as a crystal) emits light coherently, exciting standing waves of light in the cavity. The wavelength of light is very short (in the range of nanometers, 10^{-9} m) so the standing waves are microscopic in size. One use for standing light waves is to measure small distances, using optical flats.

X-rays

Interference between X-ray beams can form an X-ray standing wave(XSW) field. Because of the short wavelength of X-rays (less than 1 nanometer), this phenomenon can be exploited for measuring atomic-scale events at material surfaces. The XSW is generated in the region where an X-ray beam interferes with a diffracted beam from a nearly perfect single crystal surface or a reflection from an X-ray mirror. By tuning the crystal geometry or X-ray wavelength, the XSW can be translated in space, causing a shift in the X-ray fluorescence or photoelectron yield from the atoms near the surface. This shift can be analyzed to pinpoint the location of a particular atomic species relative to the underlying crystal structure or mirror surface. The XSW method has been used to clarify the atomic-scale details of dopants in semiconductors, atomic and molecular adsorption on surfaces, and chemical transformations involved in catalysis.

Mechanical Waves

Standing waves can be mechanically induced into a solid medium using resonance. One easy to understand example is two people shaking either end of a jump rope. If they shake in sync, the rope will form a regular pattern with nodes and antinodes and appear to be stationary, hence the name standing wave. Similarly a cantilever beam can have a standing wave imposed on it by applying a base excitation. In this case the free end moves the greatest distance laterally compared to any location along the beam. Such a device can be used as a sensor to track changes in frequency or phase of the resonance of the fiber. One application is as a measurement device for dimensional metrology.

Seismic Waves

Standing surface waves on the Earth are observed as free oscillations of the Earth.

Faraday Waves

The Faraday wave is a non-linear standing wave at the air-liquid interface induced by hydrodynamic instability. It can be used as a liquid-based template to assemble microscale materials.

Wave Interference

Wave interference is the phenomenon that occurs when two waves meet while traveling along the same medium. The interference of waves causes the medium to take on a shape that results from the net effect of the two individual waves upon the particles of the medium. To begin our exploration of wave interference, consider two pulses of the same amplitude traveling in different directions along the same medium. Let's suppose that each displaced upward 1 unit at its crest and has the shape of a sine wave. As the sine pulses move towards each other, there will eventually be a moment in time when they are completely overlapped. At that moment, the resulting shape of the medium would be an upward displaced sine pulse with an amplitude of 2 units. The diagrams below depict the before and during interference snapshots of the medium for two such pulses. The individual sine pulses are drawn in red and blue and the resulting displacement of the medium is drawn in green.

Before Interferance

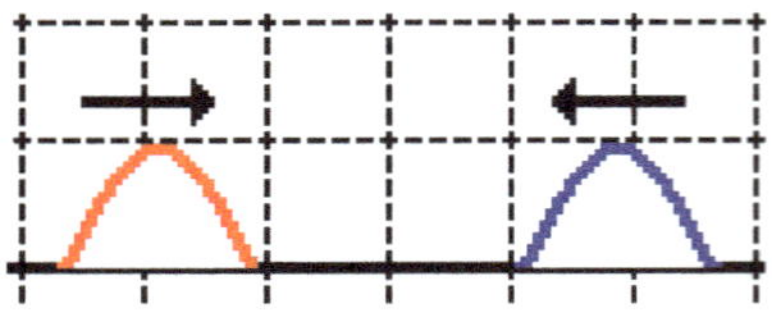

During Interferance

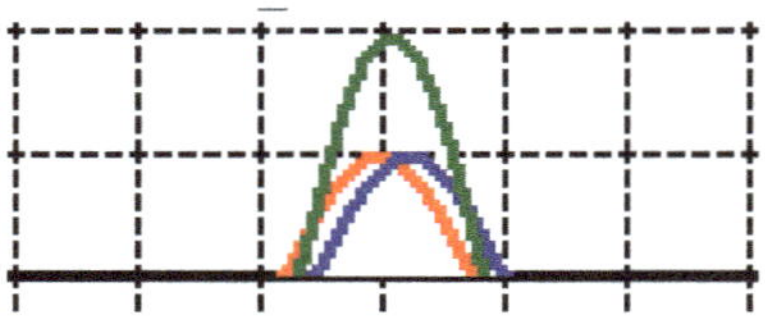

The Principle of Superposition

The task of determining the shape of the resultant demands that the principle of superposition is applied. The principle of superposition is sometimes stated as follows:

> When two waves interfere, the resulting displacement of the medium at any location is the algebraic sum of the displacements of the individual waves at that same location.

In the cases above, the summing the individual displacements for locations of complete overlap was made out to be an easy task - as easy as simple arithmetic:

Displacement of Pulse 1	Displacement of Pulse 2		Resulting Displacement
+1	+1	=	+2
-1	-1	=	-2
+1	-1	=	0
+1	-2	=	-1

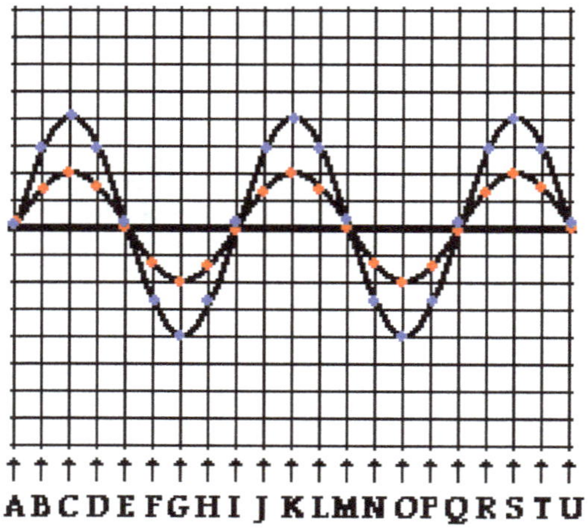

In actuality, the task of determining the complete shape of the entire medium during interference demands that the principle of superposition be applied for every point (or nearly every point) along the medium. As an example of the complexity of this task, consider the two interfering waves at the right. A snapshot of the shape of each individual wave at a particular instant in time is shown. To determine the precise shape of the medium at this given instant in time, the principle of superposition must be applied to several locations along the medium. A short cut involves measuring the displacement from equilibrium at a few strategic locations. Thus, approximately 20 locations have been picked and labeled as A, B, C, D, etc. The actual displacement of each individual wave can be counted by measuring from the equilibrium position up to the particular wave. At position A, there is no displacement for either individual wave; thus, the resulting displacement of the medium at position will be 0 units. At position B, the smaller wave has a displacement of approximately 1.4 units (indicated by the red dot); the larger wave has a displacement of approximately 2 units (indicated by the blue dot). Thus, the resulting displacement of the medium will be approximately 3.4 units. At position C, the smaller wave has a displacement of approximately 2 units; the larger wave has a displacement of approximately 4 units; thus, the resulting displacement of the medium will be approximately 6 units. At position D, the smaller wave has a displacement of approximately 1.4 units; the larger wave has a displacement of approximately 2 units; thus, the resulting displacement of the medium will be approximately 3.4 units. This process can be repeated for every position. When finished, a dot (done in green below) can be marked on the graph to note the displacement of the medium at each given location. The actual shape of the medium can then be sketched by estimating the position between the various marked points and sketching the wave. This is shown as the green line in the diagram below.

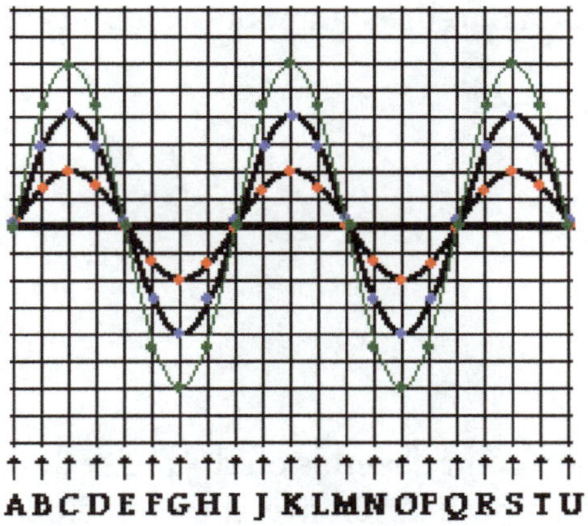

Let us consider two simple harmonic progressive waves of the same frequency travelling in the same direction. Let a_1 and a_2 be the amplitudes of the waves and ϕ the phase difference between them at any point in the medium. If the displacements due to these waves at that point be y_1 and y_2 then

$$Y_1 = a_1 \sin \omega t \text{............}(a)$$

and

$$Y_2 = a_2 \sin(\omega t + \phi) \text{...........}(b)$$

And $\frac{\omega}{2\pi}$ is the frequency of each wave. By the principle of superposition, the net displacement at the given point is given by:

$$\begin{aligned} Y &= Y_1 + Y_2 \\ &= a_1 \sin \omega t + a_2 \sin(\omega t + \phi) \\ &= a_1 \sin \omega t + a_2 \sin \omega t \cos \phi + a_2 \cos \omega t \sin \phi \\ &= \sin \omega t(a_1 + a_2 \cos \phi) + \cos \omega t(a_2 \sin \phi) \end{aligned}$$

Now let

$$a_1 + a_2 \cos \phi = R \cos \theta \ldots\ldots\ldots..(c)$$
$$a_1 \sin \phi = R \sin \theta \ldots\ldots\ldots..(d)$$

Here R and θ are the newly generated constants. Then,

$$Y = R \sin \omega t \cos \theta + R \cos \omega t \sin \theta$$

or

$$Y = R \sin(\omega t + \theta)$$

This equation is similar to the previous equations. Hence the resultant displacement at that point is due to a similar wave of amplitude R. To determine R we square (c) and (d) and add.

The intensity is directly proportional to the square of the amplitude. So the resultant intensity I is represented by:

$$I = a_1^2 + a_2^2 + 2a_1a_2 \cos \phi$$

And phase angle,

$$\phi = \tan^{-1}\left(\frac{a2 \sin \phi}{a_1 + a_2 \cos \phi}\right)$$

Thus the resultant intensity at any point depends on the phase difference ϕ between the two waves at that point. If ϕ is equivalent to x path difference, then remember

$$\frac{\Phi}{2\pi} = \frac{x}{\lambda} = \frac{\Delta t}{T}$$

The above equations are called Wave Interference Equation.

Mechanisms

The principle of superposition of waves states that when two or more propagating waves of same type are incident on the same point, the resultant amplitude at that point is equal to the vector sum of the amplitudes of the individual waves. If a crest of a wave meets a crest of another wave of the same frequency at the same point, then the amplitude is the sum of the individual amplitudes—this is constructive interference. If a crest of one wave meets a trough of another wave, then the amplitude is equal to the difference in the individual amplitudes—this is known as destructive interference.

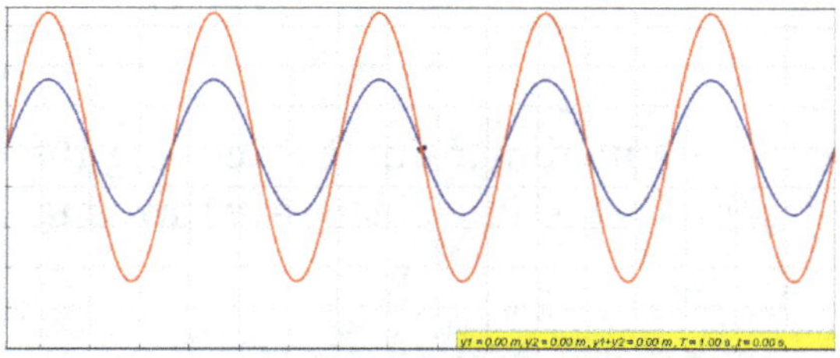

Interference of left traveling (green) and right traveling (blue) waves in one dimension, resulting in final (red) wave

Constructive interference occurs when the phase difference between the waves is an even multiple of π (180°) , whereas destructive interference occurs when the difference is an odd multiple of π. If the difference between the phases is intermediate between these two extremes, then the magnitude of the displacement of the summed waves lies between the minimum and maximum values.

Interference of waves from two point sources.

A magnified image of a coloured interference pattern in a soap film. The "black holes" are areas of almost total destructive interference (antiphase).

Consider, for example, what happens when two identical stones are dropped into a still pool of water at different locations. Each stone generates a circular wave propagating outwards from the point where the stone was dropped. When the two waves overlap, the net displacement at a particular point is the sum of the displacements of the individual waves. At some points, these will be in phase, and will produce a maximum displacement. In other places, the waves will be in anti-phase, and there will be no net displacement at these points. Thus, parts of the surface will be stationary—these are seen in the figure above and to the right as stationary blue-green lines radiating from the centre.

Interference of light is a common phenomenon that can be explained classically by the superposition of waves, however a deeper understanding of light interference requires knowledge of wave-particle duality of light which is due to quantum mechanics. Prime examples of light interference are the famous double-slit experiment, laser speckle, anti-reflective coatings and interferometers. Traditionally the classical wave model is taught as a basis for understanding optical interference, based the Huygens–Fresnel principle.

Derivation

The above can be demonstrated in one dimension by deriving the formula for the sum of two waves. The equation for the amplitude of a sinusoidal wave traveling to the right along the x-axis is

$$W_1(x,t) = A\cos(kx - \omega t)$$

where A is the peak amplitude, $k = 2\pi / \lambda$ is the wavenumber and $\omega = 2\pi f$ is the angular frequency of the wave. Suppose a second wave of the same frequency and amplitude but with a different phase is also traveling to the right

$$W_2(x,t) = A\cos(kx - \omega t + \phi)$$

where ϕ is the phase difference between the waves in radians. The two waves will superpose and add: the sum of the two waves is

$$W_1 + W_2 = A[\cos(kx - \omega t) + \cos(kx - \omega t + \phi)]$$

Using the trigonometric identity for the sum of two cosines:

$$\cos a + \cos b = 2\cos\left(\frac{a-b}{2}\right)\cos\left(\frac{a+b}{2}\right),$$

this can be written

$$W_1 + W_2 = 2A\cos\left(\frac{\phi}{2}\right)\cos\left(kx - \omega t + \frac{\phi}{2}\right)$$

This represents a wave at the original frequency, traveling to the right like the components, whose amplitude is proportional to the cosine of $\phi / 2$.

- *Constructive interference*: If the phase difference is an even multiple of pi: $\phi = \ldots -4\pi, -2\pi, 0, 2\pi, 4\pi, \ldots$ then $|\cos(\phi/2)| = 1$,, so the sum of the two waves is a wave with twice the amplitude

  $$W_1 + W_2 = 2A\cos(kx - \omega t)$$

- *Destructive interference*: If the phase difference is an odd multiple of pi: $\phi = \ldots -3\pi, -\pi, \pi, 3\pi, 5\pi, \ldots$ then $\cos(\phi/2) = 0$, so the sum of the two waves is zero

  $W_1 + W_2 = 0$.

Between Two Plane Waves

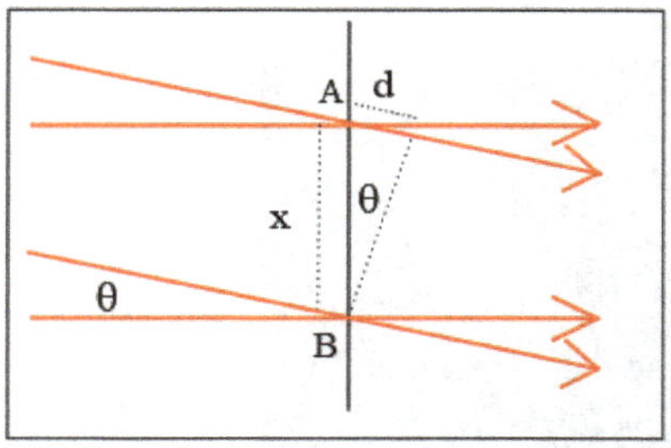

Geometrical arrangement for two plane wave interference

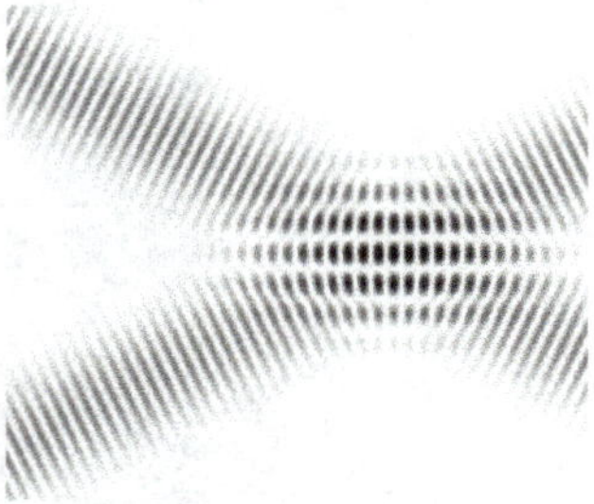

Interference fringes in overlapping plane waves

A simple form of interference pattern is obtained if two plane waves of the same frequency intersect at an angle. Interference is essentially an energy redistribution process. The energy which is lost at the destructive interference is regained at the constructive interference. One wave is travelling horizontally, and the other is travelling downwards at an angle θ to the first wave. Assuming that the two waves are in phase at the point B, then the relative phase changes along the x-axis. The phase difference at the point A is given by

$$\Delta\varphi = \frac{2\pi d}{\lambda} = \frac{2\pi x \sin\theta}{\lambda}$$

It can be seen that the two waves are in phase when

$$\frac{x\sin\theta}{\lambda} = 0, \pm 1, \pm 2, \ldots,$$

and are half a cycle out of phase when

$$\frac{x\sin\theta}{\lambda} = \pm\frac{1}{2}, \pm\frac{3}{2}, \ldots$$

Constructive interference occurs when the waves are in phase, and destructive interference when they are half a cycle out of phase. Thus, an interference fringe pattern is produced, where the separation of the maxima is

$$d_f = \frac{\lambda}{\sin\theta}$$

and d_f is known as the fringe spacing. The fringe spacing increases with increase in wavelength, and with decreasing angle θ. The fringes are observed wherever the two waves overlap and the fringe spacing is uniform throughout.

Between Two Spherical Waves

A point source produces a spherical wave. If the light from two point sources overlaps, the interference pattern maps out the way in which the phase difference between the two waves varies in space. This depends on the wavelength and on the separation of the point sources. The figure to the right shows interference between two spherical waves. The wavelength increases from top to bottom, and the distance between the sources increases from left to right.

Optical interference between two point sources that have different wavelengths and separations of sources

When the plane of observation is far enough away, the fringe pattern will be a series of almost straight lines, since the waves will then be almost planar.

Multiple Beams

Interference occurs when several waves are added together provided that the phase differences between them remain constant over the observation time.

It is sometimes desirable for several waves of the same frequency and amplitude to sum to zero (that is, interfere destructively, cancel). This is the principle behind, for example, 3-phase power and the diffraction grating. In both of these cases, the result is achieved by uniform spacing of the phases.

It is easy to see that a set of waves will cancel if they have the same amplitude and their phases are spaced equally in angle. Using phasors, each wave can be represented as $Ae^{i\varphi_n}$ for N waves from $n = 0$ to , where $n = N - 1$

$$\varphi_n - \varphi_{n-1} = \frac{2\pi}{N}.$$

To show that

$$\sum_{n=0}^{N-1} Ae^{i\varphi_n} = 0$$

one merely assumes the converse, then multiplies both sides by $e^{i\frac{2\pi}{N}}$

The Fabry–Pérot interferometer uses interference between multiple reflections.

A diffraction grating can be considered to be a multiple-beam interferometer; since the peaks which it produces are generated by interference between the light transmitted by each of the elements in the grating.

Optical Interference

Because the frequency of light waves (~10^{14} Hz) is too high to be detected by currently available detectors, it is possible to observe only the intensity of an optical interference pattern. The intensity of the light at a given point is proportional to the square of the average amplitude of the wave. This can be expressed mathematically as follows. The displacement of the two waves at a point r is:

$$U_2(\mathbf{r},t) = A_2(\mathbf{r})e^{i[\varphi_2(\mathbf{r})-\omega t]}$$

$$U_2(\mathbf{r},t) = A_2(\mathbf{r})e^{i[\varphi_2(\mathbf{r})-\omega t]}$$

where A represents the magnitude of the displacement, ϕ represents the phase and ω represents the angular frequency.

The displacement of the summed waves is

$$U(\mathbf{r},t) = A_1(\mathbf{r})e^{i[\varphi_1(\mathbf{r})-\omega t]} + A_2(\mathbf{r})e^{i[\varphi_2(\mathbf{r})-\omega t]}$$

The intensity of the light at r is given by

$$I(\mathbf{r}) = \int U(\mathbf{r},t)U^*(\mathbf{r},t)dt \propto A_1^2(\mathbf{r}) + A_2^2(\mathbf{r}) + 2A_1(\mathbf{r})A_2(\mathbf{r})\cos[\varphi_1(\mathbf{r}) - \varphi_2(\mathbf{r})]$$

This can be expressed in terms of the intensities of the individual waves as

$$I(\mathbf{r}) = I_1(\mathbf{r}) + I_2(\mathbf{r}) + 2\sqrt{I_1(\mathbf{r})I_2(\mathbf{r})}\cos[\varphi_1(\mathbf{r}) - \varphi_2(\mathbf{r})]$$

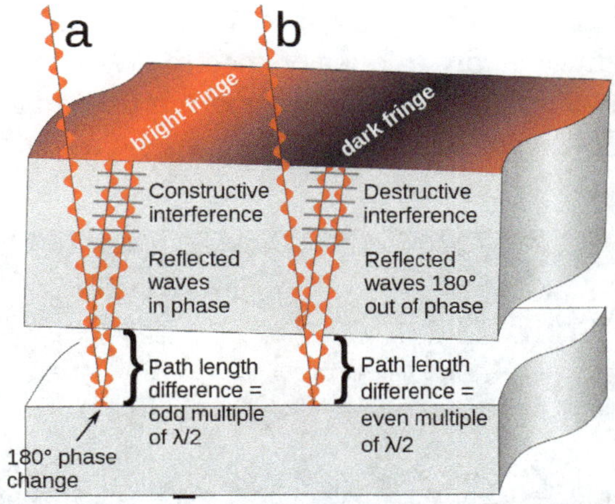

Creation of interference fringes by an optical flat on a reflective surface

Light rays from a monochromatic source pass through the glass and reflect off both the bottom surface of the flat and the supporting surface. The tiny gap between the surfaces means the two reflected rays have different path lengths. In addition the ray reflected from the bottom plate undergoes a 180° phase reversal. As a result, at locations *(a)* where the path difference is an odd multiple of $\lambda/2$, the waves reinforce. At locations *(b)* where the path difference is an even multiple of $\lambda/2$ the waves cancel. Since the gap between the surfaces varies slightly in width at different points, a series of alternating bright and dark bands, *interference fringes*, are seen.

Thus, the interference pattern maps out the difference in phase between the two waves, with maxima occurring when the phase difference is a multiple of 2π. If the two beams are of equal intensity, the maxima are four times as bright as the individual beams, and the minima have zero intensity.

The two waves must have the same polarization to give rise to interference fringes since it is not possible for waves of different polarizations to cancel one another out or add together. Instead, when waves of different polarization are added together, they give rise to a wave of a different polarization state.

Light Source Requirements

The discussion above assumes that the waves which interfere with one another are monochromatic, i.e. have a single frequency—this requires that they are infinite in time. This is not, however, either practical or necessary. Two identical waves of finite duration whose frequency is fixed over that period will give rise to an interference pattern while they overlap. Two identical waves which consist of a narrow spectrum of frequency waves of finite duration, will give a series of fringe patterns of slightly differing spacings, and provided the spread of spacings is significantly less than the average fringe spacing, a fringe pattern will again be observed during the time when the two waves overlap.

Conventional light sources emit waves of differing frequencies and at different times from different points in the source. If the light is split into two waves and then re-combined, each individual light wave may generate an interference pattern with its other half, but the individual fringe patterns generated will have different phases and spacings, and normally no overall fringe pattern will be observable. However, single-element light sources, such as sodium- or mercury-vapor lamps have emission lines with quite narrow frequency spectra. When these are spatially and colour filtered, and then split into two waves, they can be superimposed to generate interference fringes. All interferometry prior to the invention of the laser was done using such sources and had a wide range of successful applications.

White light interference in a soap bubble. The iridescence is due to thin-film interference

A laser beam generally approximates much more closely to a monochromatic source, and it is much more straightforward to generate interference fringes using a laser. The ease with which interference fringes can be observed with a laser beam can sometimes cause problems in that stray reflections may give spurious interference fringes which can result in errors.

Normally, a single laser beam is used in interferometry, though interference has been observed using two independent lasers whose frequencies were sufficiently matched to satisfy the phase requirements.

It is also possible to observe interference fringes using white light. A white light fringe pattern can be considered to be made up of a 'spectrum' of fringe patterns each of slightly different spacing. If all the fringe patterns are in phase in the centre, then the fringes will increase in size as the wavelength decreases and the summed intensity will show three to four fringes of varying colour. Young describes this very elegantly in his discussion of two slit interference. Since white light fringes are obtained only when the two waves have travelled equal distances from the light source, they can be very useful in interferometry, as they allow the zero path difference fringe to be identified.

Optical Arrangements

To generate interference fringes, light from the source has to be divided into two waves which have then to be re-combined. Traditionally, interferometers have been classified as either amplitude-division or wavefront-division systems.

In an amplitude-division system, a beam splitter is used to divide the light into two beams travelling in different directions, which are then superimposed to produce the interference pattern. The Michelson interferometer and the Mach-Zehnder interferometer are examples of amplitude-division systems.

In wavefront-division systems, the wave is divided in space—examples are Young's double slit interferometer and Lloyd's mirror.

Interference can also be seen in everyday phenomena such as iridescence and structural coloration. For example, the colours seen in a soap bubble arise from interference of light reflecting off the front and back surfaces of the thin soap film. Depending on the thickness of the film, different colours interfere constructively and destructively.

Applications

Optical Interferometry

Interferometry has played an important role in the advancement of physics, and also has a wide range of applications in physical and engineering measurement.

Thomas Young's double slit interferometer in 1803 demonstrated interference fringes when two small holes were illuminated by light from another small hole which was illuminated by sunlight. Young was able to estimate the wavelength of different colours in the spectrum from the spacing of the fringes. The experiment played a major role in the general acceptance of the wave theory of light. In quantum mechanics, this experiment is considered to demonstrate the inseparability of the wave and particle natures of light and other quantum particles (wave–particle duality). Richard Feynman was fond of saying that all of quantum mechanics can be gleaned from carefully thinking through the implications of this single experiment.

The results of the Michelson–Morley experiment are generally considered to be the first strong evidence against the theory of a luminiferous aether and in favor of special relativity.

Interferometry has been used in defining and calibrating length standards. When the metre was defined as the distance between two marks on a platinum-iridium bar, Michelson and Benoît used interferometry to measure the wavelength of the red cadmium line in the new standard, and also showed that it could be used as a length standard. Sixty years later, in 1960, the metre in the new SI system was defined to be equal to 1,650,763.73 wavelengths of the orange-red emission line in the electromagnetic spectrum of the krypton-86 atom in a vacuum. This definition was replaced in 1983 by defining the metre as the distance travelled by light in vacuum during a specific time interval. Interferometry is still fundamental in establishing the calibration chain in length measurement.

Interferometry is used in the calibration of slip gauges (called gauge blocks in the US) and in coordinate-measuring machines. It is also used in the testing of optical components.

Radio Interferometry

The Very Large Array, an interferometric array formed from many smaller telescopes, like many larger radio telescopes

In 1946, a technique called astronomical interferometry was developed. Astronomical radio interferometers usually consist either of arrays of parabolic dishes or two-dimensional arrays of omni-directional antennas. All of the telescopes in the array are widely separated and are usually connected together using coaxial cable, waveguide, optical fiber, or other type of transmission line. Interferometry increases the total signal collected, but its primary purpose is to vastly increase the resolution through a process called Aperture synthesis. This technique works by superposing (interfering) the signal waves from the different telescopes on the principle that waves that coincide with the same phase will add to each other while two waves that have opposite phases will cancel each other out. This creates a combined telescope that is equivalent in resolution (though not in sensitivity) to a single antenna whose diameter is equal to the spacing of the antennas furthest apart in the array.

Acoustic Interferometry

An acoustic interferometer is an instrument for measuring the physical characteristics of sound wave in a gas or liquid. It may be used to measure velocity, wavelength, absorption, or impedance. A vibrating crystal creates the ultrasonic waves that are radiated into the medium. The waves strike a reflector placed parallel to the crystal. The waves are then reflected back to the source and measured.

Quantum Interference

If a system is in state ψ, its wavefunction is described in Dirac or bra–ket notation as:

$$|\psi\rangle = \sum_i |i\rangle\langle i|\psi\rangle = \sum_i |i\rangle\psi_i$$

where the $|i\rangle$ s specify the different quantum "alternatives" available (technically, they form an eigenvector basis) and the ψ_i are the probability amplitude coefficients, which are complex numbers.

The probability of observing the system making a transition or quantum leap from state ψ to a new state φ is the square of the modulus of the scalar or inner product of the two states:

$$\text{prob}(\psi \Rightarrow \varphi) = |\langle\psi|\varphi\rangle|^2 = |\sum_i \psi_i^* \varphi_i|^2$$

$$= \sum_{ij} \psi_i^* \psi_j \varphi_j^* \varphi_i = \sum_i |\psi_i|^2 |\varphi_i|^2 + \sum_{ij;i\neq j} \psi_i^* \psi_j \varphi_j^* \varphi_i$$

where $\psi_i = \langle i|\psi\rangle$ (as defined above) and similarly $\varphi_i = \langle i|\varphi\rangle$ are the coefficients of the final state of the system. * is the complex conjugate so that $\psi_i^* = \langle\psi|i\rangle$, , etc

Now let's consider the situation classically and imagine that the system transited from $|\psi\rangle$ to $|\varphi\rangle$ via an intermediate state $\rangle$. Then we would *classically* expect the probability of the two-step transition to be the sum of all the possible intermediate steps. So we would have

$$\text{prob}(\psi \Rightarrow \varphi) = \sum_i \text{prob}(\psi \Rightarrow i \Rightarrow \varphi)$$

$$= \sum_i |\langle\psi|i\rangle|^2 |\langle i|\varphi\rangle|^2 = \sum_i |\psi_i|^2 |\varphi_i|^2,$$

The classical and quantum derivations for the transition probability differ by the presence, in the quantum case, of the extra terms $\sum_{ij;i\neq j} \psi_i^* \psi_j \varphi_j^* \varphi_i$; these extra quantum terms represent *interference* between the different $i \neq j$ intermediate "alternatives". These are consequently known as the quantum interference terms, or cross terms. This is a purely quantum effect and is a consequence of the non-additivity of the probabilities of quantum alternatives.

The interference terms vanish, via the mechanism of quantum decoherence, if the intermediate state $|i\rangle$ is measured or coupled with the environment.

Sound Wave Interference

When two coherent sound waves superpose in certain region interference takes place. For waves to be coherent, the phase difference between the two waves should not change with time. This requires that the wavelength of the two waves must be the same. Furthermore to have a better contrast between the maximum and minimum intensities, the two waves producing interference and should propagate in the same direction.

Consider two waves,

$$Y_1 = a_1 \sin \omega t$$
$$Y_2 = a_2 \sin(\omega t + \phi)$$

Here ϕ is the phase difference between the two waves. This arises because of the path difference or may be some initial phase difference or may be both. Then the resultant is obtained from superposition and is given by:

$$Y = Y_1 + Y_2$$

Sound wave interference can be of two types: constructive and destructive interference.

Constructive Interference

When two waves of same phase difference or constant phase difference superimpose at some points, where the crust of one wave falls on the crust of the other, a resultant maximum amplitude is obtained. This is called Constructive Wave Interference.

At these points, the maximum intensity of light is produced because intensity is directly proportional to the square of the amplitude of the wave.

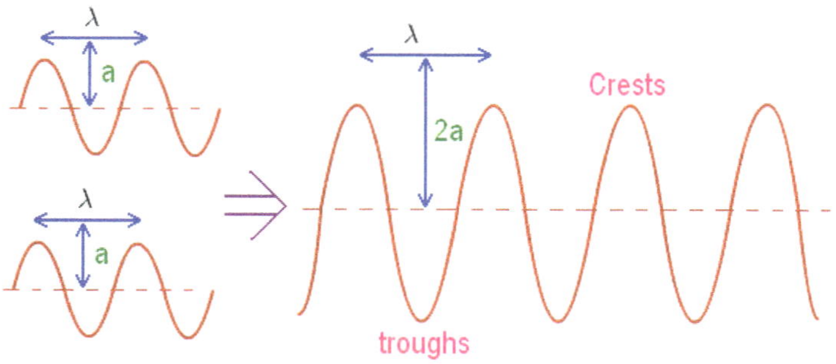

Crests and trough reinforce

The mathematical representation of the above statement is

$$I \,\alpha\, A^2$$

Constructive Interference Equation

The constructive interference is seen when the phase difference between the two waves is

$\phi = 0, 2\pi, 4\pi, 6\pi ....2n\pi$.

Then $\cos\phi = +1$

$A_{max} = a_1 + a_2$

The phase difference between two waves an even multiple of $\phi = 2n\pi$. Here $n = 0, 1, 2..........$

The path difference between two waves is an even multiple of $\frac{\lambda}{2}$.

$\delta = 2n\frac{\lambda}{2}$.

The time interval between two waves is even multiple of $\frac{T}{2}$.

$\theta = 2n\frac{T}{2}$, where $n = 0, 1, 2, 3.........$

The resultant amplitude of wave is equal to the sum of amplitudes of individual waves

$A = a_1 + a_2$

The resultant intensity is more than the sum of intensities of individual waves

$$1_{max} = 1_1 + 1_2 + 2\sqrt{I_1 I_2})$$
$$= (\sqrt{I_1} + \sqrt{I_2})^2$$

Intensity is more than the given intensity $I > I_1 + I_2$.

Constructive Interference Example

These are the few examples of constructive interference:

1. Constructive interference takes place in light.
2. Constructive interference takes place in sound.
3. The rogue waves in the ocean are the example of constructive interference.
4. When two speakers are put one meter apart and run we get constructive interference between the sound waves.
5. Beats are the example of constructive interference.

Destructive Interference

When two waves of same frequency travel in a medium simultaneously in the one and the same direction then, the resultant at any point of the medium is different from the sum of the intensities of the two waves, due to their superposition. At some point it is very less and negligible and this is called Destructive Interference.

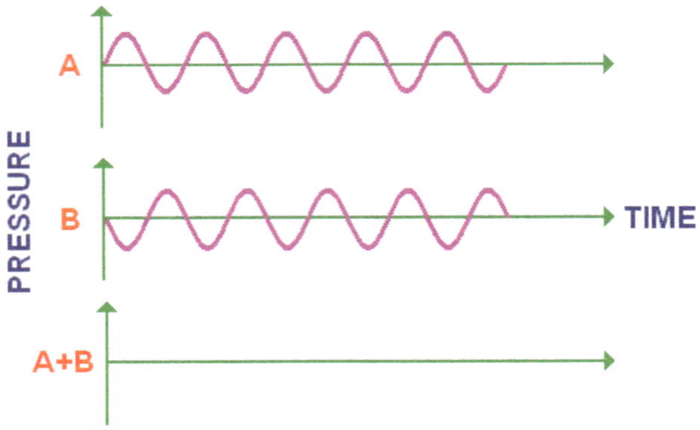

The above figure shows Destructive Interference

In the figure, A and B be the two waves of same frequency travel in a medium in the same direction. And A + B be the resultant.

Destructive Interference Equation

1. The phase difference between two waves an odd multiple of π that is, $(2n-1)\pi$, Here n = 1, 2, 3
2. The path difference between two waves is an odd multiple of $\frac{\lambda}{2}$

   $\Delta=(2n-1)\frac{\lambda}{2}$, Here n = 1, 2
3. The time interval between two waves is odd multiple of $\frac{T}{2}$

   $\theta=(2n-1)\frac{T}{2}$, n = 1, 2, 3
4. The resultant amplitude of wave is equal to the difference of amplitudes of individual waves Amplitude = A - B if $\phi=\pi,3\pi,5\pi.......(2n-1)\pi$
5. The resultant intensity is less than the sum of intensities of individual waves

   $I_{\min}=I_1+I_2-2\sqrt{I_1I_2}==(\sqrt{\ }I_1-\sqrt{I_2})^2$

Destructive Interference Examples

These are the few Examples of Destructive Interference:

1. Gravitational waves are the example of Destructive Interference.
2. Light beams may show Destructive Interference.
3. Acoustics is another example.

4. Moving electrons also display this phenomenon.
5. Can be observed in radio waves.
6. Surface water waves show Destructive Interference.

Complete Destructive Interference

Complete Destructive Interference will never be possible. There would always be some constructive interference associated with it. At some places the energy would be zero (Destructive Interference) but at other places it would appear as the case for constructive interference.

Doppler Effect

The Doppler effect is observed whenever the source of waves is moving with respect to an observer. The Doppler effect can be described as the effect produced by a moving source of waves in which there is an apparent upward shift in frequency for observers towards whom the source is approaching and an apparent downward shift in frequency for observers from whom the source is receding. It is important to note that the effect does not result because of an actual change in the frequency of the source. Using the example above, the bug is still producing disturbances at a rate of 2 disturbances per second; it just appears to the observer whom the bug is approaching that the disturbances are being produced at a frequency greater than 2 disturbances/second. The effect is only observed because the distance between observer B and the bug is decreasing and the distance between observer A and the bug is increasing.

The Doppler effect can be observed for any type of wave - water wave, sound wave, light wave, etc. We are most familiar with the Doppler effect because of our experiences with sound waves. Perhaps you recall an instance in which a police car or emergency vehicle was traveling towards you on the highway. As the car approached with its siren blasting, the pitch of the siren sound (a measure of the siren's frequency) was high; and then suddenly after the car passed by, the pitch of the siren sound was low. That was the Doppler effect - an apparent shift in frequency for a sound wave produced by a moving source.

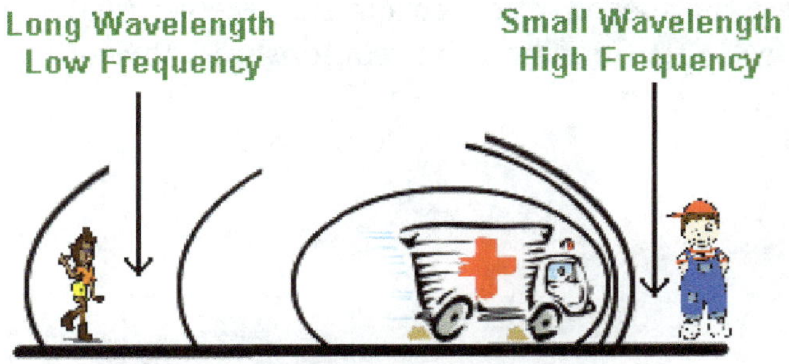

The doppler effect for a moving sound source

Suppose that there is a happy bug in the center of a circular water puddle. The bug is periodically shaking its legs in order to produce disturbances that travel through the water. If these disturbances originate at a point, then they would travel outward from that point in all directions. Since each disturbance is traveling in the same medium, they would all travel in every direction at the same speed. The pattern produced by the bug's *shaking* would be a series of concentric circles as shown in the diagram at the right. These circles would reach the edges of the water puddle at the same frequency. An observer at point A (the left edge of the puddle) would observe the disturbances to strike the puddle's edge at the same frequency that would be observed by an observer at point B (at the right edge of the puddle). In fact, the frequency at which disturbances reach the edge of the puddle would be the same as the frequency at which the bug produces the disturbances. If the bug produces disturbances at a frequency of 2 per second, then each observer would observe them approaching at a frequency of 2 per second.

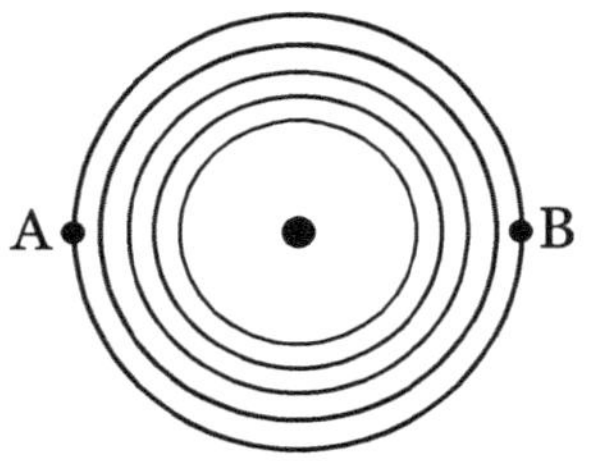

A stationary bug producing disturbance in water

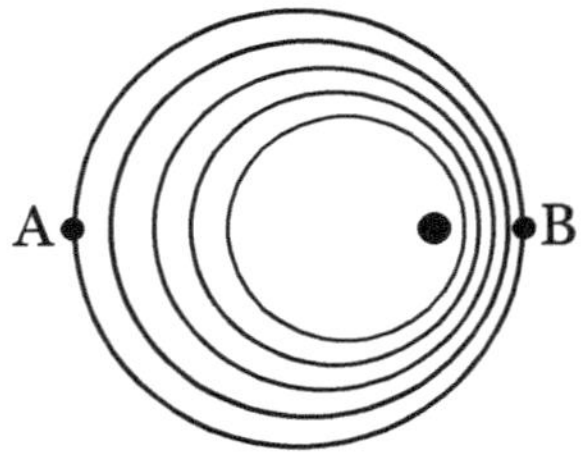

A bug moving to right and producing disturbances

Now suppose that our bug is moving to the right across the puddle of water and producing disturbances at the same frequency of 2 disturbances per second. Since the bug is moving towards the right, each consecutive disturbance originates from a position that is closer to observer B and farther from observer A. Subsequently, each consecutive disturbance has a shorter distance to travel before reaching observer B and thus takes less time to reach observer B. Thus, observer B observes that the frequency of arrival of the disturbances is higher than the frequency at which disturbances are produced. On the other hand, each consecutive disturbance has a further distance to travel before reaching observer A. For this reason, observer A observes a frequency of arrival that is less than the frequency at which the disturbances are produced. The net effect of the motion of the bug (the source of waves) is that the observer towards whom the bug is moving observes a frequency that is higher than 2 disturbances/second; and the observer away from whom the bug is moving observes a frequency that is less than 2 disturbances/second. This effect is known as the Doppler effect.

General

In classical physics, where the speeds of source and the receiver relative to the medium are lower than the velocity of waves in the medium, the relationship between observed frequency f and emitted frequency f_0 is given by:

$$f = \left(\frac{c \pm v_r}{c \pm v_s}\right) f_0$$

where

c is the velocity of waves in the medium;

v_r is the velocity of the receiver relative to the medium; positive if the receiver is moving towards the source (and negative in the other direction);

v_s is the velocity of the source relative to the medium; positive if the source is moving away from the receiver (and negative in the other direction).

The frequency is decreased if either is moving away from the other.

The above formula assumes that the source is either directly approaching or receding from the observer. If the source approaches the observer at an angle (but still with a constant velocity), the observed frequency that is first heard is higher than the object's emitted frequency. Thereafter, there is a monotonic decrease in the observed frequency as it gets closer to the observer, through equality when it is coming from a direction perpendicular to the relative motion (and was emitted at the point of closest approach; but when the wave is received, the source and observer will no longer be at their closest), and a continued monotonic decrease as it recedes from the observer. When the observer is very close to the path of the object, the transition from high to low frequency is very abrupt. When the observer is far from the path of the object, the transition from high to low frequency is gradual.

If the speeds v_s and v_r are small compared to the speed of the wave, the relationship between observed frequency f and emitted frequency f_0 is approximately

Observed frequency	Change in frequency
$f = \left(1 + \frac{\Delta v}{c}\right) f_0$	$\Delta f = \frac{\Delta v}{c} f_0$

where

$\Delta f = f - f_0$

$\Delta v = v_r - v_s$ is the velocity of the receiver relative to the source: it is positive when the source and the receiver are moving towards each other.

Analysis

To understand what happens, consider the following analogy. Someone throws one ball every second at a man. Assume that balls travel with constant velocity. If the thrower is stationary, the man will receive one ball every second. However, if the thrower is moving towards the man, he will receive balls more frequently because the balls will be less spaced out. The inverse is true if the thrower is moving away from the man. So it is actually the *wavelength* which is affected; as a consequence, the received frequency is also affected. It may also be said that the velocity of the wave remains constant whereas wavelength changes; hence frequency also changes.

With an observer stationary relative to the medium, if a moving source is emitting waves with an actual frequency f_0 (in this case, the wavelength is changed, the transmission velocity of the wave keeps constant -- note that the *transmission velocity* of the wave does not depend on the *velocity of the source*), then the observer detects waves with a frequency f given by:

$$f = \left(\frac{c}{c \pm v_s}\right) f_0$$

A similar analysis for a moving *observer* and a stationary source (in this case, the wavelength keeps constant, but due to the motion, the rate at which the observer receives waves -- and hence the *transmission velocity* of the wave [with respect to the observer] -- is changed) yields the observed frequency:

$$f = \left(\frac{c \pm v_r}{c}\right) f_0$$

These can be generalized into the equation that was presented in the previous section.

$$f = \left(\frac{c \pm v_r}{c \pm v_s}\right) f_0$$

An interesting effect was predicted by Lord Rayleigh in his classic book on sound: if the source is moving toward the observer at twice the speed of sound, a musical piece emitted by that source would be heard in correct time and tune, but *backwards*. The Doppler effect with sound is only clearly heard with objects moving at high speed, as change in frequency of musical tone involves a speed of around 40 meters per second, and smaller changes in frequency can easily be confused by changes in the amplitude of the sounds from moving emitters. Neil A Downie has demonstrated how the Doppler effect can be made much more easily audible by using an ultrasonic (e.g. 40 kHz) emitter on the moving object. The observer then uses a heterodyne frequency converter, as used in many bat detectors, to listen to a band around 40 kHz. In this case, with the bat detector tuned to give frequency for the stationary emitter of 2000 Hz, the observer will perceive a frequency shift of a whole tone, 240 Hz, if the emitter travels at 2 meters per second.

Application

Sirens

The siren on a passing emergency vehicle will start out higher than its stationary pitch, slide down as it passes, and continue lower than its stationary pitch as it recedes from the observer. Astronomer John Dobson explained the effect thus:

> "The reason the siren slides is because it doesn't hit you."

In other words, if the siren approached the observer directly, the pitch would remain constant, at a higher than stationary pitch, until the vehicle hit him, and then immediately jump to a new lower pitch. Because the vehicle passes by the observer, the radial velocity does not remain constant, but instead varies as a function of the angle between his line of sight and the siren's velocity:

$$v_{radial} = v_s \cdot \cos\theta$$

where θ is the angle between the object's forward velocity and the line of sight from the object to the observer.

Astronomy

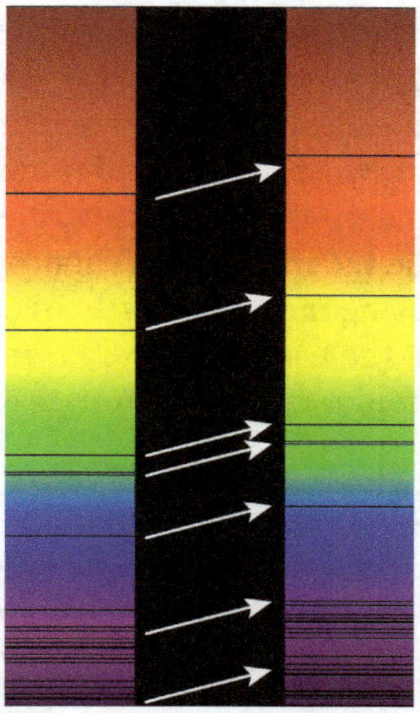

Redshift of spectral lines in the optical spectrum of a supercluster of distant galaxies (right), as compared to that of the Sun (left)

The Doppler effect for electromagnetic waves such as light is of great use in astronomy and results in either a so-called redshift or blueshift. It has been used to measure the speed at which stars and galaxies are approaching or receding from us; that is, their radial velocities. This may be used to detect if an apparently single star is, in reality, a close binary, to measure the rotational speed of stars and galaxies, or to detect exoplanets. This redshift and blueshift happens on a very small scale, if an object is moving toward earth, there would not be a noticeable difference in visible light.

Note that redshift is also used to measure the expansion of space, but that this is not truly a Doppler effect. Rather, redshifting due to the expansion of space is known as cosmological redshift, which can be derived purely from the Robertson-Walker metric under the formalism of General Relativity. Having said this, it also happens that there *are* detectable Doppler effects on cosmological scales, which, if incorrectly interpreted as cosmological in origin, lead to the observation of redshift-space distortions.

The use of the Doppler effect for light in astronomy depends on our knowledge that the spectra of stars are not homogeneous. They exhibit absorption lines at well defined frequencies that are correlated with the energies required to excite electrons in various elements from one level to another. The Doppler effect is recognizable in the fact that the absorption lines are not always at the frequencies that are obtained from the spectrum of a stationary light source. Since blue light has a higher frequency than red light, the spectral lines of an approaching astronomical light source exhibit a blueshift and those of a receding astronomical light source exhibit a redshift.

Among the nearby stars, the largest radial velocities with respect to the Sun are +308 km/s (BD-15°4041, also known as LHS 52, 81.7 light-years away) and −260 km/s (Woolley 9722, also known as Wolf 1106 and LHS 64, 78.2 light-years away). Positive radial velocity means the star is receding from the Sun, negative that it is approaching.

Radar

The Doppler effect is used in some types of radar, to measure the velocity of detected objects. A radar beam is fired at a moving target — e.g. a motor car, as police use radar to detect speeding motorists — as it approaches or recedes from the radar source. Each successive radar wave has to travel farther to reach the car, before being reflected and re-detected near the source. As each wave has to move farther, the gap between each wave increases, increasing the wavelength. In some situations, the radar beam is fired at the moving car as it approaches, in which case each successive wave travels a lesser distance, decreasing the wavelength. In either situation, calculations from the Doppler effect accurately determine the car's velocity. Moreover, the proximity fuze, developed during World War II, relies upon Doppler radar to detonate explosives at the correct time, height, distance, etc.

Because the doppler shift affects the wave incident upon the target as well as the wave reflected back to the radar, the change in frequency observed by a radar due to a target moving at relative velocity Δv is twice that from the same target emitting a wave:

$$\Delta f = \frac{2\Delta v}{c} f_0.$$

Medical

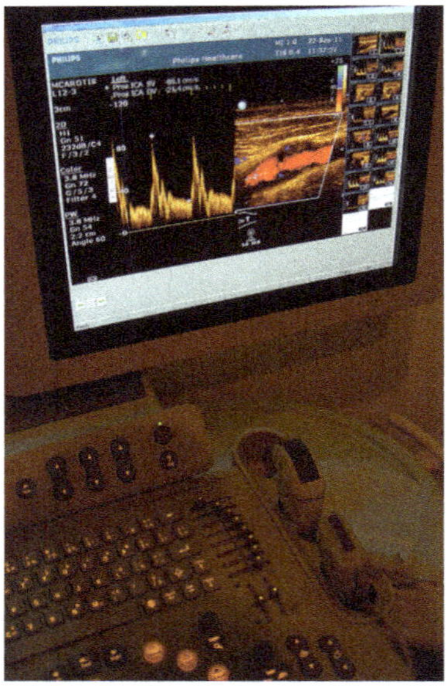

Colour flow ultrasonography (Doppler) of a carotid artery – scanner and screen

An echocardiogram can, within certain limits, produce an accurate assessment of the direction of blood flow and the velocity of blood and cardiac tissue at any arbitrary point using the Doppler effect. One of the limitations is that the ultrasound beam should be as parallel to the blood flow as possible. Velocity measurements allow assessment of cardiac valve areas and function, abnormal communications between the left and right side of the heart, leaking of blood through the valves (valvular regurgitation), and calculation of the cardiac output. Contrast-enhanced ultrasound using gas-filled microbubble contrast media can be used to improve velocity or other flow-related medical measurements.

Although "Doppler" has become synonymous with "velocity measurement" in medical imaging, in many cases it is not the frequency shift (Doppler shift) of the received signal that is measured, but the phase shift (*when* the received signal arrives).

Velocity measurements of blood flow are also used in other fields of medical ultrasonography, such as obstetric ultrasonography and neurology. Velocity measurement of blood flow in arteries and veins based on Doppler effect is an effective tool for diagnosis of vascular problems like stenosis.

Flow Measurement

Instruments such as the laser Doppler velocimeter (LDV), and acoustic Doppler velocimeter (ADV) have been developed to measure velocities in a fluid flow. The LDV emits a light beam and the ADV emits an ultrasonic acoustic burst, and measure the Doppler shift in wavelengths of reflections from particles moving with the flow. The actual flow is computed as a function of the water velocity and phase. This technique allows non-intrusive flow measurements, at high precision and high frequency.

Velocity Profile Measurement

Developed originally for velocity measurements in medical applications (blood flow), Ultrasonic Doppler Velocimetry (UDV) can measure in real time complete velocity profile in almost any liquids containing particles in suspension such as dust, gas bubbles, emulsions. Flows can be pulsating, oscillating, laminar or turbulent, stationary or transient. This technique is fully non-invasive.

Satellite Communication

Fast moving satellites can have a Doppler shift of dozens of kilohertz relative to a ground station. The speed, thus magnitude of Doppler effect, changes due to earth curvature. Dynamic Doppler compensation, where the frequency of a signal is changed multiple times during transmission, is used so the satellite receives a constant frequency signal.

Audio

The Leslie speaker, most commonly associated with and predominantly used with the famous Hammond organ, takes advantage of the Doppler effect by using an electric motor to rotate an acoustic horn around a loudspeaker, sending its sound in a circle. This results at the listener's ear in rapidly fluctuating frequencies of a keyboard note.

Vibration Measurement

A laser Doppler vibrometer (LDV) is a non-contact instrument for measuring vibration. The laser beam from the LDV is directed at the surface of interest, and the vibration amplitude and frequency are extracted from the Doppler shift of the laser beam frequency due to the motion of the surface.

Developmental Biology

During the segmentation of vertebrate embryos, waves of gene expression sweep across the presomitic mesoderm, the tissue from which the precursors of the vertebrae (somites) are formed. A new somite is formed upon arrival of a wave at the anterior end of the presomitic mesoderm. In zebrafish, it has been shown that the shortening of the presomitic mesoderm during segmentation leads to a Doppler effect as the anterior end of the tissue moves into the waves. This Doppler effect contributes to the period of segmentation.

Inverse Doppler effect

Since 1968 scientists such as Victor Veselago have speculated about the possibility of an inverse Doppler effect. The experiment that claimed to have detected this effect was conducted by Nigel Seddon and Trevor Bearpark in Bristol, United Kingdom in 2003.

Resonance

The word resonance has Latin origin, which in English means 'to sound' to produce sound when a body produce sound which vibrates with the same sound wave length of another object. The definition of Resonance in physics is a phenomenon in which an external force or a vibrating system forces another system around it to vibrate with greater amplitude at a specified frequency of operation. The frequency at which the second body starts oscillating or vibrating at higher amplitude is called the resonant frequency of the body. The best examples of resonance can be observed in various musical instruments around us.

Whenever any person hits, strikes, strums, drums or tweaks any musical instrument, the instrument is set into oscillation or vibration at the natural frequency of vibration of the instrument. A unique standing wave pattern defines each frequency of vibration as specific instrument. These natural frequencies of a musical instrument are known widely as the harmonics of the specified instrument. If a second interconnected object or instrument vibrates or oscillates at that specified frequency then the first object can be forced to vibrate at a frequency higher that its natural harmonic frequency. This phenomenon is known as resonance i.e. one object vibrating or oscillating at the natural frequency of another object forces the other object to vibrate at a frequency higher than its natural frequency.

A classic example of resonance is the swinging of a person sitting on a swing. A swing is a very good example of an object in oscillating motion. Initially the motion is slow and the swing doesn't extend to its maximum potential. But once when the swing reaches its natural frequency of oscillation, a gentle push to the swing helps it maintain that amplitude of swing all throughout due to resonance.

In an ideal situation, with no friction at all, even that slight push won't be necessary once the swing reaches its natural frequency for it to sustain the maximum amplitude forever. Also almost all musical instruments, like the flute, guitar etc work on the principle of resonance itself.

Another common physics demonstration that serves as an excellent model of resonance is the famous "singing rod" demonstration. A long hollow aluminum rod is held at its center. Being a trained musician, teacher reaches in a rosin bag to prepare for the event. Then with great enthusiasm, he/she slowly slides her hand across the length of the aluminum rod, causing it to sound out with a loud sound. This is an example of resonance. As the hand slides across the surface of the aluminum rod, slip-stick friction between the hand and the rod produces vibrations of the aluminum. The vibrations of the aluminum force the air column inside of the rod to vibrate at its natural frequency. The match between the vibrations of the air column and one of the natural frequencies of the singing rod causes resonance. The result of resonance is always a big vibration - that is, a loud sound.

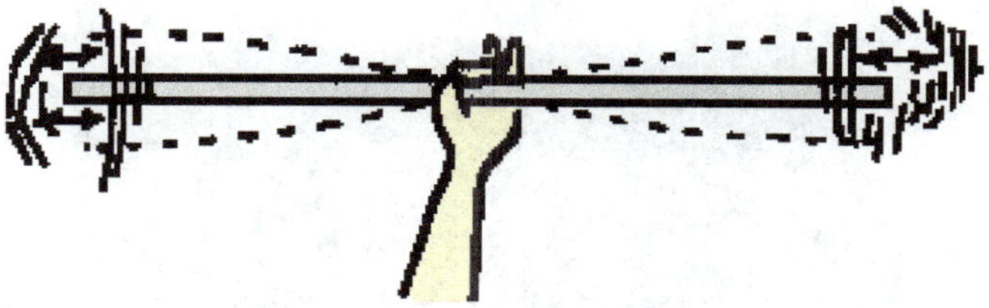

A vibrating metal rod forces the air column inside into vibrations at the same frequency-resonance occurs

The familiar sound of the sea that is heard when a seashell is placed up to your ear is also explained by resonance. Even in an apparently quiet room, there are sound waves with a range of frequencies. These sounds are mostly inaudible due to their low intensity. This so-called background noise fills the seashell, causing vibrations within the seashell. But the seashell has a set of natural frequencies at which it will vibrate. If one of the frequencies in the room forces air within the seashell to vibrate at its natural frequency, a resonance situation is created. And always, the result of resonance is a big vibration - that is, a loud sound. In fact, the sound is loud enough to hear. So the next time you hear the sound of the sea in a seashell, remember that all that you are hearing is the amplification of one of the many background frequencies in the room.

Resonance occurs when a system is able to store and easily transfer energy between two or more different storage modes (such as kinetic energy and potential energy in the case of a simple pendulum). However, there are some losses from cycle to cycle, called damping. When damping is small, the resonant frequency is approximately equal to the natural frequency of the system, which is a frequency of unforced vibrations. Some systems have multiple, distinct, resonant frequencies.

Resonance phenomena occur with all types of vibrations or waves: there is mechanical resonance, acoustic resonance, electromagnetic resonance, nuclear magnetic resonance (NMR), electron spin resonance (ESR) and resonance of quantum wave functions. Resonant systems can be used to generate vibrations of a specific frequency (e.g., musical instruments), or pick out specific frequencies from a complex vibration containing many frequencies (e.g., filters).

The term *resonance* (from Latin *resonantia*, 'echo', from *resonare*, 'resound') originates from the field of acoustics, particularly observed in musical instruments, e.g., when strings started to vibrate and to produce sound without direct excitation by the player.

The name of the musical note "Re" in the solfège scale may come from the word *resonare*, as it appears in a religious anthem for John the Baptist in Latin. This was due to the naming of musical notes by the Italian medieval scholar Guido of Arezzo.

Examples

A familiar example is a playground swing, which acts as a pendulum. Pushing a person in a swing in time with the natural interval of the swing (its resonant frequency) makes the swing go higher and higher (maximum amplitude), while attempts to push the swing at a faster or slower tempo produce smaller arcs. This is because the energy the swing absorbs is maximized when the pushes match the swing's natural oscillations.

Pushing a person in a swing is a common example of resonance. The loaded swing, a pendulum, has a natural frequency of oscillation, its resonant frequency, and resists being pushed at a faster or slower rate

Resonance occurs widely in nature, and is exploited in many manmade devices. It is the mechanism by which virtually all sinusoidal waves and vibrations are generated. Many sounds we hear, such as when hard objects of metal, glass, or wood are struck, are caused by brief resonant vibrations in the object. Light and other short wavelength electromagnetic radiation is produced by resonance on an atomic scale, such as electrons in atoms. Other examples of resonance:

- Timekeeping mechanisms of modern clocks and watches, e.g., the balance wheel in a mechanical watch and the quartz crystal in a quartz watch
- Tidal resonance of the Bay of Fundy
- Acoustic resonances of musical instruments and the human vocal tract
- Shattering of a crystal wineglass when exposed to a musical tone of the right pitch (its resonant frequency)
- Friction idiophones, such as making a glass object (glass, bottle, vase) vibrate by rubbing around its rim with a fingertip
- Electrical resonance of tuned circuits in radios and TVs that allow radio frequencies to be selectively received
- Creation of coherent light by optical resonance in a laser cavity
- Orbital resonance as exemplified by some moons of the solar system's gas giants
- Material resonances in atomic scale are the basis of several spectroscopic techniques that are used in condensed matter physics
    - Electron spin resonance
    - Mössbauer effect
    - Nuclear magnetic resonance

Tacoma Narrows Bridge

The dramatically visible, rhythmic twisting that resulted in the 1940 collapse of "Galloping Gertie", the original Tacoma Narrows Bridge, is misleadingly characterized as an example of resonance phenomenon in certain textbooks. The catastrophic vibrations that destroyed the bridge were not due to simple mechanical resonance, but to a more complicated interaction between the bridge and the winds passing through it—a phenomenon known as aeroelastic flutter, which is a kind of "self-sustaining vibration" as referred to in the nonlinear theory of vibrations. Robert H. Scanlan, father of bridge aerodynamics, has written an article about this misunderstanding.

International Space Station

The rocket engines for the International Space Station (ISS) are controlled by an autopilot. Ordinarily, uploaded parameters for controlling the engine control system for the Zvezda module make the rocket engines boost the International Space Station to a higher orbit. The rocket engines are hinge-mounted, and ordinarily the crew doesn't notice the operation. On January 14, 2009, however, the uploaded parameters made the autopilot swing the rocket engines in larger and larger oscillations, at a frequency of 0.5 Hz. These oscillations were captured on video, and lasted for 142 seconds.

Types of Resonance

Mechanical and Acoustic Resonance

Mechanical resonance is the tendency of a mechanical system to absorb more energy when the frequency of its oscillations matches the system's natural frequency of vibration than it does at other frequencies. It may cause violent swaying motions and even catastrophic failure in improperly constructed structures including bridges, buildings, trains, and aircraft. When designing objects, engineers must ensure the mechanical resonance frequencies of the component parts do not match driving vibrational frequencies of motors or other oscillating parts, a phenomenon known as resonance disaster.

Avoiding resonance disasters is a major concern in every building, tower, and bridge construction project. As a countermeasure, shock mounts can be installed to absorb resonant frequencies and thus dissipate the absorbed energy. The Taipei 101 building relies on a 660-tonne pendulum (730-short-ton)—a tuned mass damper—to cancel resonance. Furthermore, the structure is designed to resonate at a frequency that does not typically occur. Buildings in seismic zones are often constructed to take into account the oscillating frequencies of expected ground motion. In addition, engineers designing objects having engines must ensure that the mechanical resonant frequencies of the component parts do not match driving vibrational frequencies of the motors or other strongly oscillating parts.

Clocks keep time by mechanical resonance in a balance wheel, pendulum, or quartz crystal.

The cadence of runners has been hypothesized to be energetically favorable due to resonance between the elastic energy stored in the lower limb and the mass of the runner.

Acoustic resonance is a branch of mechanical resonance that is concerned with the mechanical vibrations across the frequency range of human hearing, in other words sound. For humans, hearing is normally limited to frequencies between about 20 Hz and 20,000 Hz (20 kHz), Many objects and materials act as resonators with resonant frequencies within this range, and when struck vibrate mechanically, pushing on the surrounding air to create sound waves. This is the source of many percussive sounds we hear.

Acoustic resonance is an important consideration for instrument builders, as most acoustic instruments use resonators, such as the strings and body of a violin, the length of tube in a flute, and the shape of, and tension on, a drum membrane.

Like mechanical resonance, acoustic resonance can result in catastrophic failure of the object at resonance. The classic example of this is breaking a wine glass with sound at the precise resonant frequency of the glass, although this is difficult in practice.

Electrical Resonance

Electrical resonance occurs in an electric circuit at a particular *resonant frequency* when the impedance of the circuit is at a minimum in a series circuit or at maximum in a parallel circuit (or when the transfer function is at a maximum). Resonance in circuits are used for both transmitting and receiving wireless communications such as television, cell phones and radio.

Optical Resonance

An optical cavity, also called an *optical resonator*, is an arrangement of mirrors that forms a standing wave cavity resonator for light waves. Optical cavities are a major component of lasers, surrounding the gain medium and providing feedback of the laser light. They are also used in optical parametric oscillators and some interferometers. Light confined in the cavity reflects multiple times producing standing waves for certain resonant frequencies. The standing wave patterns produced are called "modes". Longitudinal modes differ only in frequency while transverse modes differ for different frequencies and have different intensity patterns across the cross-section of the beam. Ring resonators and whispering galleries are examples of optical resonators that do not form standing waves.

Different resonator types are distinguished by the focal lengths of the two mirrors and the distance between them; flat mirrors are not often used because of the difficulty of aligning them precisely. The geometry (resonator type) must be chosen so the beam remains stable, i.e., the beam size does not continue to grow with each reflection. Resonator types are also designed to meet other criteria such as minimum beam waist or having no focal point (and therefore intense light at that point) inside the cavity.

Optical cavities are designed to have a very large Q factor. A beam reflects a large number of times with little attenuation—therefore the frequency line width of the beam is small compared to the frequency of the laser.

Additional optical resonances are guided-mode resonances and surface plasmon resonance, which result in anomalous reflection and high evanescent fields at resonance. In this case, the resonant modes are guided modes of a waveguide or surface plasmon modes of a dielectric-metallic interface. These modes are usually excited by a subwavelength grating.

Orbital Resonance

In celestial mechanics, an orbital resonance occurs when two orbiting bodies exert a regular, periodic gravitational influence on each other, usually due to their orbital periods being related by a ratio of two small integers. Orbital resonances greatly enhance the mutual gravitational influence of the bodies. In most cases, this results in an *unstable* interaction, in which the bodies exchange momentum and shift orbits until the resonance no longer exists. Under some circumstances, a resonant system can be stable and self-correcting, so that the bodies remain in resonance. Examples are the 1:2:4 resonance of Jupiter's moons Ganymede, Europa, and Io, and the 2:3 resonance between Pluto and Neptune. Unstable resonances with Saturn's inner moons give rise to gaps in the rings of Saturn. The special case of 1:1 resonance (between bodies with similar orbital radii) causes large Solar System bodies to clear the neighborhood around their orbits by ejecting nearly everything else around them; this effect is used in the current definition of a planet.

Atomic, Particle, and Molecular Resonance

NMR Magnet at HWB-NMR, Birmingham, UK. In its strong 21.2-tesla field, the proton resonance is at 900 MHz

Nuclear magnetic resonance (NMR) is the name given to a physical resonance phenomenon involving the observation of specific quantum mechanical magnetic properties of an atomic nucleus in the presence of an applied, external magnetic field. Many scientific techniques exploit NMR phenomena to study molecular physics, crystals, and non-crystalline materials through NMR spectroscopy. NMR is also routinely used in advanced medical imaging techniques, such as in magnetic resonance imaging (MRI).

All nuclei containing odd numbers of nucleons have an intrinsic magnetic moment and angular momentum. A key feature of NMR is that the resonant frequency of a particular substance is directly proportional to the strength of the applied magnetic field. It is this feature that is exploited in imaging techniques; if a sample is placed in a non-uniform magnetic field then the resonant frequencies of the sample's nuclei depend on where in the field they are located. Therefore, the particle can be located quite precisely by its resonant frequency.

Electron paramagnetic resonance, otherwise known as *Electron Spin Resonance* (ESR) is a spectroscopic technique similar to NMR, but uses unpaired electrons instead. Materials for which this can be applied are much more limited since the material needs to both have an unpaired spin and be paramagnetic.

The Mössbauer effect is the resonant and recoil-free emission and absorption of gamma ray photons by atoms bound in a solid form.

Resonance in particle physics appears in similar circumstances to classical physics at the level of quantum mechanics and quantum field theory. However, they can also be thought of as unstable particles, with the formula above valid if Γ is the decay rate and Ω replaced by the particle's mass M. In that case, the formula comes from the particle's propagator, with its mass replaced by the complex number $M + i\Gamma$. The formula is further related to the particle's decay rate by the optical theorem.

Theory

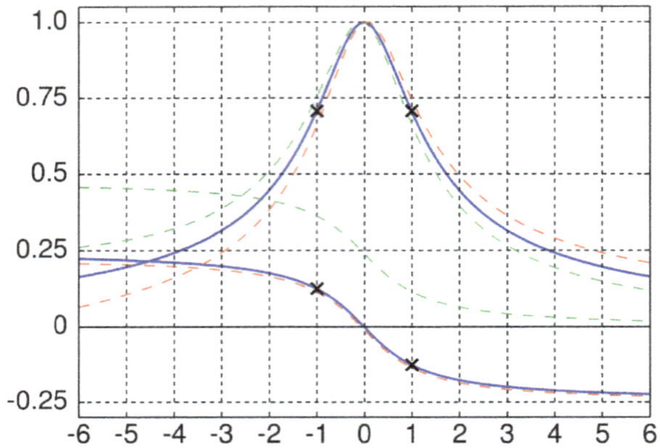

"Universal Resonance Curve", a symmetric approximation to the normalized response of a resonant circuit

Abscissa values are deviation from center frequency, in units of center frequency divided by 2Q; ordinate is relative amplitude, and phase in cycles; dashed curves compare the range of responses of real two-pole circuits for a Q value of 5; for higher Q values, there is less deviation from the universal curve. Crosses mark the edges of the 3 dB bandwidth (gain 0.707, phase shift 45° or 0.125 cycle).

The exact response of a resonance, especially for frequencies far from the resonant frequency, depends on the details of the physical system, and is usually not exactly symmetric about the resonant frequency, as illustrated for the simple harmonic oscillator above. For a lightly damped linear oscillator with a resonance frequency Ω, the *intensity* of oscillations I when the system is driven with a driving frequency ω is typically approximated by a formula that is symmetric about the resonance frequency:

$$I(\omega) \equiv |\chi|^2 \propto \frac{1}{(\omega-\Omega)^2 + \left(\frac{\Gamma}{2}\right)^2}.$$

Where the susceptibility $\chi(\omega)$ links the amplitude of the oscillator to the driving force in frequency space:

$$x(\omega) = \chi(\omega)F(\omega)$$

The intensity is defined as the square of the amplitude of the oscillations. This is a Lorentzian function, or Cauchy distribution, and this response is found in many physical situations involving resonant systems. Γ is a parameter dependent on the damping of the oscillator, and is known as the *linewidth* of the resonance. Heavily damped oscillators tend to have broad linewidths, and respond to a wider range of driving frequencies around the resonant frequency. The linewidth is inversely proportional to the Q factor, which is a measure of the sharpness of the resonance.

In electrical engineering, this approximate symmetric response is known as the *universal resonance curve*, a concept introduced by Frederick E. Terman in 1932 to simplify the approximate analysis of radio circuits with a range of center frequencies and Q values.

Resonators

A physical system can have as many resonant frequencies as it has degrees of freedom; each degree of freedom can vibrate as a harmonic oscillator. Systems with one degree of freedom, such as a mass on a spring, pendulums, balance wheels, and LC tuned circuits have one resonant frequency. Systems with two degrees of freedom, such as coupled pendulums and resonant transformers can have two resonant frequencies. As the number of coupled harmonic oscillators grows, the time it takes to transfer energy from one to the next becomes significant. The vibrations in them begin to travel through the coupled harmonic oscillators in waves, from one oscillator to the next.

Extended objects that can experience resonance due to vibrations inside them are called resonators, such as organ pipes, vibrating strings, quartz crystals, microwave and laser cavities. Since these can be viewed as being made of millions of coupled moving parts (such as atoms), they can have millions of resonant frequencies. The vibrations inside them travel as waves, at an approximately constant velocity, bouncing back and forth between the sides of the resonator. If the distance between the sides is d, the length of a roundtrip is $2d$. To cause resonance, the phase of a sinusoidal wave after a roundtrip must be equal to the initial phase, so the waves reinforce the oscillation. So the condition for resonance in a resonator is that the roundtrip distance, $2d$, be equal to an integer number of wavelengths λ of the wave:

$$2d = N\lambda, \qquad N \in \{1, 2, 3, \ldots\}$$

If the velocity of a wave is v, the frequency is $f = \frac{v}{\lambda}$ so the resonant frequencies are:

$$f = \frac{Nv}{2d} \qquad N \in \{1, 2, 3, \ldots\}$$

So the resonant frequencies of resonators, called normal modes, are equally spaced multiples of a lowest frequency called the fundamental frequency. The multiples are often called overtones. There may be several such series of resonant frequencies, corresponding to different modes of oscillation.

Q Factor

The Q factor or *quality factor* is a dimensionless parameter that describes how under-damped an oscillator or resonator is, or equivalently, characterizes a resonator's bandwidth relative to its center frequency. Higher Q indicates a lower rate of energy loss relative to the stored energy of the oscillator, i.e., the oscillations die out more slowly. A pendulum suspended from a high-quality bearing, oscillating in air, has a high Q, while a pendulum immersed in oil has a low Q. To sustain a system in resonance in constant amplitude by providing power externally, the energy provided in each cycle must be less than the energy stored in the system (i.e., the sum of the potential and kinetic) by a factor of $Q/2\pi$. Oscillators with high-quality factors have low damping, which tends to make them ring longer.

Sinusoidally driven resonators having higher Q factors resonate with greater amplitudes (at the resonant frequency) but have a smaller range of frequencies around the frequency at which they resonate. The range of frequencies at which the oscillator resonates is called the bandwidth. Thus, a high-Q tuned circuit in a radio receiver would be more difficult to tune, but would have greater selectivity, it would do a better job of filtering out signals from other stations that lie nearby on the spectrum. High Q oscillators operate over a smaller range of frequencies and are more stable.

The quality factor of oscillators varies substantially from system to system. Systems for which damping is important (such as dampers keeping a door from slamming shut) have $Q = 1/2$. Clocks, lasers, and other systems that need either strong resonance or high frequency stability need high-quality factors. Tuning forks have quality factors around $Q = 1000$. The quality factor of atomic clocks and some high-Q lasers can reach as high as 10^{11} and higher.

There are many alternate quantities used by physicists and engineers to describe how damped an oscillator is that are closely related to its quality factor. Important examples include: the damping ratio, relative bandwidth, linewidth, and bandwidth measured in octaves.

Natural Frequency

Nearly all objects, when hit or struck or plucked or strummed or somehow disturbed, will vibrate. If you drop a meter stick or pencil on the floor, it will begin to vibrate. If you pluck a guitar string, it will begin to vibrate. If you blow over the top of a pop bottle, the air inside will vibrate. When each of these objects vibrates, they tend to vibrate at a particular frequency or a set of frequencies. The frequency or frequencies at which an object tends to vibrate with when hit, struck, plucked, strummed or somehow disturbed is known as the natural frequency of the object. If the amplitudes of the vibrations are large enough and if natural frequency is within the human frequency range, then the vibrating object will produce sound waves that are audible.

All objects have a natural frequency or set of frequencies at which they vibrate. The quality or timbre of the sound produced by a vibrating object is dependent upon the natural frequencies of the sound waves produced by the objects. Some objects tend to vibrate at a single frequency and they are often said to produce a pure tone. A flute tends to vibrate at a single frequency, producing a very pure tone. Other objects vibrate and produce more complex waves with a set of frequencies that have a whole number mathematical relationship between them; these are said to produce a rich sound. A tuba tends to vibrate at a set of frequencies that are mathematically related by whole number ratios; it produces a rich tone. Still other objects will vibrate at a set of multiple frequencies that have no simple mathematical relationship between them. These objects are not musical at all and the sounds that they create could be described as noise. When a meter stick or pencil is dropped on the floor, it vibrates with a number of frequencies, producing a complex sound wave that is clanky and noisy.

Flute	Tuba	Dropped Pencil
200 Hz	200 Hz	197 Hz
	400 Hz	211 Hz
	600 Hz	217 Hz
	800 Hz	219 Hz
	1000 Hz	287 Hz
		311 Hz
		329 Hz
		399 Hz
		407 Hz

Free vibrations of an elastic body are called *natural vibrations* and occur at a frequency called the natural frequency. Natural vibrations are different from forced vibrations which happen at frequency of applied force (forced frequency). If forced frequency is equal to the natural frequency, the amplitude of vibration increases manyfold. This phenomenon is known as resonance.

In electrical circuits, s_1 is a natural frequency of variable x if the zero-input response of x includes the term $K_1 e^{-s_1 t}$, where $K_1 \neq 0$ is a constant dependent on initial state of the circuit, network topology, and element values. In a network, s_k is a natural frequency of the network if it is a natural frequency of some voltage or current in the network. Natural frequencies depend *only* on network topology and element values but not the input. It can be shown that the set of natural frequencies in a network can be obtained by calculating the poles of all impedance and admittance functions of the network. All poles of the network transfer function are also natural frequencies of the corresponding response variable; however there may exist some natural frequencies that are not a pole of the network function. These frequencies happen at some special initial states.

In LC and RLC circuits, natural frequency of circuit can be calculated as:

$$\omega_0 = \frac{1}{\sqrt{LC}}$$

Equation of Motion : Natural Frequency

Figure shows a simple undamped spring-mass system, which is assumed to move only along the vertical direction. It has one degree of freedom (DOF), because its motion is described by a single coordinate x.

When placed into motion, oscillation will take place at the natural frequency fn which is a property of the system. We now examine some of the basic concepts associated with the free vibration of systems with one degree of freedom.

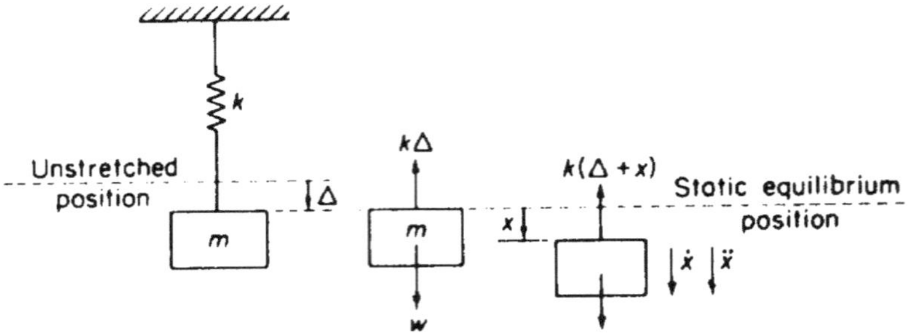

Figure- Spring-Mass System and Free-Body Diagram

Newton's second law is the first basis for examining the motion of the system. As shown in figure the deformation of the spring in the static equilibrium position is D, and the spring force *kD* is equal to the gravitational force *w* acting on mass m

$$k\Delta = w = mg$$

By measuring the displacement *x* from the static equilibrium position, the forces acting on *m* are $k(\Delta + x)$ and *w*. With *x* chosen to be positive in the downward direction, all quantities - force, velocity, and acceleration are also positive in the downward direction.

We now apply Newton's second law of motion to the mass *m* :

$$mx = \sum F = w - k(\Delta + x)$$

and because *kD* = *w*, we obtain :

$$mx = -kx$$

It is evident that the choice of the static equilibrium position as reference for *x* has eliminated *w*, the force due to gravity, and the static spring force *kD* from the equation of motion, and the resultant force on *m* is simply the spring force due to the displacement *x*.

By defining the circular frequency w *n* by the equation

$$\omega_n^2 = \frac{k}{m}$$

Eq. can be written as

$$x = \omega_n^2 x = 0$$

and we conclude that the motion is harmonic. Equation, a homogeneous second order linear differential equation, has the following general solution :

$$x = A\sin\omega_n t + B\cos\omega_n t$$

where A and *B* are the two necessary constants. These constants are evaluated from initial conditions $x(0) and\ x\,(0)$, and Eq. can be shown to reduce to

$$x = \frac{x(0)}{\omega_n} \sin \omega_n t + x(0) \cos \omega_n t$$

The natural period of the oscillation is established from $\omega_n \tau = 2\pi$, or

$$\tau = 2\pi \sqrt{\frac{m}{k}}$$

and the natural frequency is

$$f_n = \frac{1}{\tau} = \frac{1}{2_\pi} \sqrt{\frac{k}{m}}$$

These quantities can be expressed in terms of the static deflection D by observing Eq. $k\Delta = mg$. Thus, Eq. can be expressed in terms of the static deflection D as

$$f_n = \frac{1}{2_\pi} \sqrt{\frac{g}{\Delta}}$$

Note that $\tau, f_n and\ \omega_n$, depend only on the mass and stiffness of the system, which are properties of the system.

Factors Affecting the Natural Frequency

The actual frequency at which an object vibrates at is determined by a variety of factors. Each of these factors will affect either the wavelength or the speed of the object. From frequency = speed / wavelength.

A change of velocity or wavelength will result in a change in the natural frequency. The role of a musician is to control these variables to produce a given frequency from the instrument that is played. Consider as an example a guitar. There are six channels, each having a different linear density (the larger strings are denser on a per meter), a different voltage (which can be controlled by the guitar), and a different length (also controllable by guitarist) . The speed at which they move through the waves of channels depends on the medium properties – in this case, the sealing (pressure) of the chain and the linear density of channels. Variations in these properties would affect the natural frequency of the particular channel. The vibrating part of a particular channel can be shortened by pressing the string against one of the frets on the neck of the guitar. This change in the length of the chain would affect the wavelength of the wave and in turn the natural frequency at which a particular string vibrates. Control of the speed and wavelength in this manner allows a guitarist to control the natural frequency of the vibrating object (a string) and therefore producing musical sounds intended. The same principles can be applied to any string instrument – whether the harp, harpsichord, violin or guitar.

As another example, consider the trombone with its long cylindrical tube which is folded twice on itself and terminates in a flared end. The trombone is an example of a wind instrument. The tube of any wind instrument acts as a container for a vibrating air column. The air inside the tube is vibrated by a vibrating reed or vibration of lips of a musician against a mouthpiece. Although the speed of the sound waves inside the air column is not modifiable by the musician (they can not be modified by changes in ambient temperature), the length of the air column is. For a paper clip, the length is changed by pushing the tube outward away from the mouth of the stretch or pull to shorten it. This causes the length of the column of air that needs to be modified, and subsequently changes the wavelength of the wave that it produces. And of course, a change in wavelength will result in a variation in the frequency. Thus, the natural frequency of a wind instrument such as the trombone depends on the length of the air column of the device. The same principles can be applied to any other similar instrument (tuba, flute, bells, organ pipe, clarinet, or pop bottle) whose sound is produced by vibrations of the air inside a tube.

There were a variety of classroom demonstrations (some are fun and some are corny) that illustrate the idea of natural frequencies and their modification. A bottle of soft drink may be partially filled with water, leaving a volume of inside air and which is capable of vibrating. When a person blows on the top of the bottle, indoor air is set vibrating movement; turbulence above the lip of the bottle creates disturbances inside the bottle. These vibrations cause a sound wave that is audible for students. Of course, the frequency can be changed by changing the volume of the air column (addition or removal of water), which changes the wavelength and the frequency in turn. The principle is similar to the relationship of air columns frequency wave-length; a smaller volume of air inside the bottle to a shorter wavelength and a higher average frequency.

A toilet roll band can be created from different lengths of toilet paper rolls (or wrapping paper rolls). The rollers will vibrate at different frequencies when struck against the head of a student. A set of properly selected rolls will result in the production of sounds which are capable of Tony Award rendition of “Mary Had A Little Lamb”.

Perhaps you are familiar with the cup of water prom popular trick that is often demonstrated in Physics class. Obtain a cup of water and clean your fingers. Then gently slide your finger on the rim of the cup of water. If you are lucky enough, you might be able to fix the cup vibrated by means of slip-stick friction. (It is necessary to use a crystal goblet It is often said that the crystal goblets work best; But the trick is just as easy to achieve with clean fingers and an inexpensive cup). As a violin string being pulled through a violin string, the finger sticks to glass molecules, pulling at one time until the voltage becomes so great. Then slides the finger on the glass, then is another microscopic surface paste; the finger pulls the molecules at the surface, and then slides sticks to another location. This process of stick-slip friction occurring at a high frequency is sufficient to bind the molecules of the glass in vibration at its natural frequency. The result is enough to impress your guests. Try it at home.

Maybe you saw a pendulum vibrant back and forth about its equilibrium position. Although a pendulum does not produce a sound when it oscillates, it illustrates an important principle. A pendulum consisting of a longer string vibrates with a longer period and thus a lower frequency. Once again, there is an inverse relationship between the length of the vibrating object and the natural frequency at which the object vibrates. This relationship carries over all very vibrant instrument – whether it is a guitar string, xylophone, an instrument pop bottle or cup.

To conclude, all objects have a natural frequency or set of frequencies at which they vibrate when struck, plucked, strummed or somehow disturbed. The actual frequency depends on the properties of the object material comprises (this affects the wave velocity) and the material length (which affects the wavelength of the wave). It is the aim of musicians to find instruments that have the ability to vibrate with sets of frequencies that are musically sound (ie, mathematically related by reports of simple integers) and to vary the lengths and (if possible) properties to create the desired sound.

Harmonic

A harmonic is a signal or wave whose frequency is an integral (whole-number) multiple of the frequency of some reference signal or wave. The term can also refer to the ratio of the frequency of such a signal or wave to the frequency of the reference signal or wave.

Let f represent the main, or fundamental, frequency of an alternating current (AC) signal, electromagnetic field , or sound wave. This frequency, usually expressed in hertz , is the frequency at which most of the energy is contained, or at which the signal is defined to occur. If the signal is displayed on an oscilloscope, the waveform will appear to repeat at a rate corresponding to f Hz.

For a signal whose fundamental frequency is f , the second harmonic has a frequency 2 f , the third harmonic has a frequency of 3 f , and so on. Let w represent the wavelength of the signal or wave in a specified medium. The second harmonic has a wavelength of w /2, the third harmonic has a wavelength of w /3, and so on. Signals occurring at frequencies of 2 f , 4 f , 6 f , etc. are called even harmonics; the signals at frequencies of 3 f , 5 f , 7 f , etc. are called odd harmonics. A signal can, in theory, have infinitely many harmonics.

Nearly all signals contain energy at harmonic frequencies, in addition to the energy at the fundamental frequency. If all the energy in a signal is contained at the fundamental frequency, then that signal is a perfect sine wave. If the signal is not a perfect sine wave, then some energy is contained in the harmonics. Some waveforms contain large amounts of energy at harmonic frequencies. Examples are square waves, sawtooth waves, and triangular waves.

In wireless communications and broadcasting, transmitters are designed so they emit a minimum of energy at harmonic frequencies. Normally, a wireless device is intended for use at only one frequency. Signal output at harmonic frequencies can cause interference to other communications or broadcasting. For example, a broadcast signal at 90.5 MHz (in the standard FM band) would have a second harmonic at 181 MHz, a third harmonic at 271.5 MHz, a fourth harmonic at 362 MHz, and so on. Some or all of these harmonic signals could, if strong, disrupt activities in other wireless services.

Characteristics

A whizzing, whistling tonal character, distinguishes all the harmonics both natural and artificial from the firmly stopped intervals; therefore their application in connection with the latter must always be carefully considered.

Most acoustic instruments emit complex tones containing many individual partials (component simple tones or sinusoidal waves), but the untrained human ear typically does not perceive those partials as separate phenomena. Rather, a musical note is perceived as one sound, the quality or timbre of that sound being a result of the relative strengths of the individual partials. Many acoustic oscillators, such as the human voice or a bowed violin string, produce complex tones that are more or less periodic, and thus are composed of partials that are near matches to integer multiples of the fundamental frequency and therefore resemble the ideal harmonics and are called "harmonic partials" or simply "harmonics" for convenience (although it's not strictly accurate to call a partial a harmonic, the first being real and the second being ideal).

Oscillators that produce harmonic partials behave somewhat like one-dimensional resonators, and are often long and thin, such as a guitar string or a column of air open at both ends (as with the modern orchestral transverse flute). Wind instruments whose air column is open at only one end, such as trumpets and clarinets, also produce partials resembling harmonics. However they only produce partials matching the odd harmonics, at least in theory. The reality of acoustic instruments is such that none of them behaves as perfectly as the somewhat simplified theoretical models would predict.

Partials whose frequencies are not integer multiples of the fundamental are referred to as inharmonic partials. Some acoustic instruments emit a mix of harmonic and inharmonic partials but still produce an effect on the ear of having a definite fundamental pitch, such as pianos, strings plucked pizzicato, vibraphones, marimbas, and certain pure-sounding bells or chimes. Antique singing bowls are known for producing multiple harmonic partials or multiphonics. Other oscillators, such as cymbals, drum heads, and other percussion instruments, naturally produce an abundance of inharmonic partials and do not imply any particular pitch, and therefore cannot be used melodically or harmonically in the same way other instruments can.

Partials, Overtones, and Harmonics

An overtone is any partial higher than the lowest partial in a compound tone. The relative strengths and frequency relationships of the component partials determine the timbre of an instrument. The similarity between the terms overtone and partial sometimes leads to their being loosely used interchangeably in a musical context, but they are counted differently, leading to some possible confusion. In the special case of instrumental timbres whose component partials closely match a harmonic series (such as with most strings and winds) rather than being inharmonic partials (such as with most pitched percussion instruments), it is also convenient to call the component partials "harmonics" but not strictly correct (because harmonics are numbered the same even when missing, while partials and overtones are only counted when present).

In many musical instruments, it is possible to play the upper harmonics without the fundamental note being present. In a simple case (e.g., recorder) this has the effect of making the note go up in pitch by an octave, but in more complex cases many other pitch variations are obtained. In some cases it also changes the timbre of the note. This is part of the normal method of obtaining higher notes in wind instruments, where it is called overblowing. The extended technique of playing multiphonics also produces harmonics. On string instruments it is possible to produce very pure sounding notes, called harmonics or flageolets by string players, which have an eerie quality, as well as being high in pitch. Harmonics may be used to check at a unison the tuning of strings that are not tuned to the unison. For example, lightly fingering the node found halfway down the highest string of a cello produces the same pitch as lightly fingering the node ⅓ of the way down the second highest string.

While it is true that electronically produced periodic tones (e.g. square waves or other non-sinusoidal waves) have "harmonics" that are whole number multiples of the fundamental frequency, practical instruments do not all have this characteristic. For example, higher "harmonics"' of piano notes are not true harmonics but are "overtones" and can be very sharp, i.e. a higher frequency than given by a pure harmonic series. This is especially true of instruments other than stringed or brass/woodwind ones, e.g., xylophone, drums, bells etc., where not all the overtones have a simple whole number ratio with the fundamental frequency. The fundamental frequency is the reciprocal of the period of the periodic phenomenon.

On Stringed Instruments

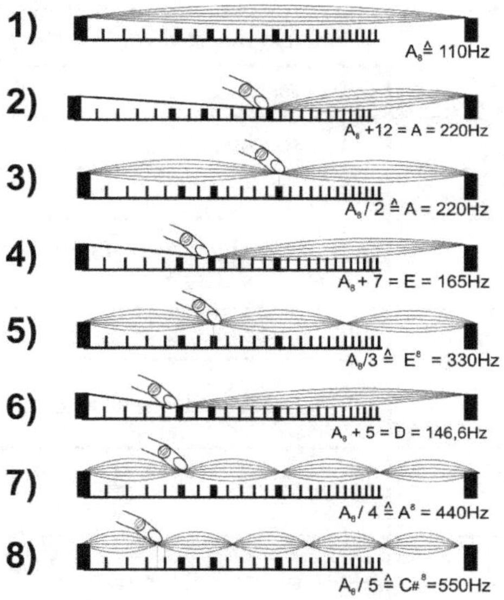

Playing a harmonic on a string

Harmonics may be singly produced [on stringed instruments] (1) by varying the point of contact with the bow, or (2) by slightly pressing the string at the nodes, or divisions of its aliquot parts ($\frac{1}{2}, \frac{1}{3}, \frac{1}{4}$ etc.). (1) In the first case, advancing the bow from the usual place where the fundamental note is produced, towards the bridge, the whole scale of harmonics may be produced in succession, on an old and highly resonant instrument. The employment of this means produces the effect called 'sul ponticello.' (2) The production of harmonics by the slight pressure of the finger on the open string is more useful. When produced by pressing slightly on the various nodes of the open strings they are called 'Natural harmonics.' Violinists are well aware that the longer the string in proportion to its thickness, the

—Grove's Dictionary of Music and Musicians (1879)

The following table displays the stop points on a stringed instrument, such as the guitar (guitar harmonics), at which gentle touching of a string will force it into a harmonic mode when vibrated. String harmonics (flageolet tones) are described as having a "flutelike, silvery quality" that can be highly effective as a special color or tone color (timbre) when used and heard in orchestration. It is unusual to encounter natural harmonics higher than the fifth partial on any stringed instrument except the double bass, on account of its much longer strings. Harmonics are widely used in plucked string instruments, such as acoustic guitar, electric guitar and electric bass. On an electric guitar played loudly through a guitar amplifier with distortion, harmonics are more sustained and can be used in guitar solos. In the heavy metal music lead guitar style known as shred guitar, harmonics, both natural and artificial, are widely used.

Harmonic	Stop note	Sounded note relative to open string	Cents above open string	Cents reduced to one octave
2	octave	octave (P8)	1,200.0	0.0
3	just perfect fifth	P8 + just perfect fifth (P5)	1,902.0	702.0
4	just perfect fourth	2P8	2,400.0	0.0
5	just major third	2P8 + just major third (M3)	2,786.3	386.3
6	just minor third	2P8 + P5	3,102.0	702.0
7	septimal minor third	2P8 + septimal minor seventh (m7)	3,368.8	968.8
8	septimal major second	3P8	3,600.0	0.0
9	Pythagorean major second	3P8 + Pythagorean major second (M2)	3,803.9	203.9
10	just minor whole tone	3P8 + just M3	3,986.3	386.3
11	greater undecimal neutral second	3P8 + lesser undecimal tritone	4,151.3	551.3
12	lesser undecimal neutral second	3P8 + P5	4,302.0	702.0
13	tridecimal 2/3-tone	3P8 + tridecimal neutral sixth (n6)	4,440.5	840.5
14	2/3-tone	3P8 + P5 + septimal minor third (m3)	4,568.8	968.8
15	septimal (or major) diatonic semitone	3P8 + just major seventh (M7)	4,688.3	1,088.3
16	just (or minor) diatonic semitone	4P8	4,800.0	0.0

Table

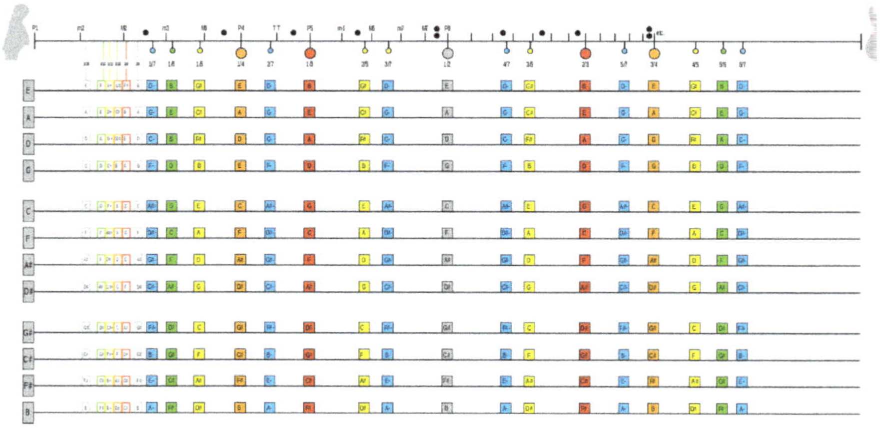

Table of harmonics of a stringed instrument with colored dots indicating which positions can be lightly fingered to generate just intervals up to the 7th harmonic

Artificial Harmonics

Although harmonics are most often used on open strings (natural harmonics), occasionally a score will call for an artificial harmonic, produced by playing an overtone on an already stopped string. As a performance technique, it is accomplished by using two fingers on the fingerboard, the first to shorten the string to the desired fundamental, with the second touching the node corresponding to the appropriate harmonic. On fretted instruments, such as an electric guitar, the performer can look at the frets to determine where to stop the string and where to touch the node. On unfretted instruments, such as the violin and related instruments, playing artificial harmonics is an advanced technique, as it requires the performer to find two precise locations on the same string.

Other Information

Harmonics may be either used or considered as the basis of just intonation systems. Composer Arnold Dreyblatt is able to bring out different harmonics on the single string of his modified double bass by slightly altering his unique bowing technique halfway between hitting and bowing the strings. Composer Lawrence Ball uses harmonics to generate music electronically.

Harmonic Wave Equation

For the rest of the course we will focus on infinite repeating waves of a specific type: harmonic waves. Mechanical harmonic waves can be expressed mathematically as

$$y(x,t) - y_0 = A\sin\left(2\pi\frac{t}{T} \pm 2\pi\frac{x}{\lambda} + \phi\right)$$

The displacement of a piece of the wave at equilibrium position x and time t is given by the whole left hand side $\left(y(x,t) - y_0\right)$. y_0 is the position of the medium without any wave, and $y(x,t)$ is its actual position. Earlier we carelessly used $y(x,t)$ to describe the displacement, but to be precise we must describe the displacement using the entire left side.

On the right hand side, we're familiar with A the amplitude T the period, and λ the wavelength. The fixed phase constant ϕ is new; it describes what the wave looks like at

$(x = 0, t = 0)$ The symbol $\pm$ asks us to choose $+$ or $-$ which describes the direction that the wave travels.

Before going too much further, it is worthwhile noting the difference between *variables* and *parameters.*

- The parameters $(A, T, \lambda, \phi, y_0$ and the choice of $+$ or $-$ are defined for any given harmonic wave. They describe the wave and its behavior.

- The wave exists in all space and at any time (for any x and any t) In the formula above x and t are chosen by us to answer a question about the displacement of a specific piece in the medium at a specific time. This distinction makes x and t variables.

  In other words, we can ask about different locations and times by changing variables, but parameters for a wave are *fixed* values.

Extending the Harmonic Wave Model

While we have framed this discussion in terms of material waves because it is the easiest to visualize, we should be aware that the harmonic wave is a much more general concept. It can apply to the variation in pressure (for sound waves):

$$P(x,t) - P_0 = A\sin\left(2\pi\frac{t}{T} \pm 2\pi\frac{x}{\lambda} + \phi\right)$$

or the variation in electric field due to a light wave.

$$E(x,t) - E_0 = A\sin\left(2\pi\frac{t}{T} \pm 2\pi\frac{x}{\lambda} + \phi\right)$$

In general we can let $y(x,t)$ stand for any of these physical quantities, not just position. We shall refer to $y(x,t)$ in this general form as the wave function. Sometimes harmonic waves are also called sinusoidal waves as the wave function represents a sine or cosine function.

While waves in the real world do not go on forever, and do not exist for all time, we can still use harmonic waves of this form as a good approximation. They offer a considerable simplification.

Total Phase Φ

If we wanted to, we could define a new quantity Φ as a function of x and t so that

$$\Phi(x,t) = 2\pi\frac{t}{T} \pm 2\pi\frac{x}{\lambda} + \phi$$

Using this, we can rewrite our general harmonic wave formula as

$$y(x,t) - y_0 = A\sin\phi(x,t)$$

Because the sine function is periodic with period 2π changing Φ by $2\pi, 4\pi, \ldots$ does not change $y(x,t)$. This ambiguity exists partially because the wave repeats so that many places on the wave look exactly the same. Let us try and make our example more concrete: because $\sin\frac{\pi}{2} = 1$ is the maximum of the sine function, $\Phi = \frac{\pi}{2}$ labels a peak in our wave. Note that $\Phi = \frac{5\pi}{2}$ and $\Phi = \frac{3\pi}{2}$ also label peaks, but they label different peaks. When we imagine ourselves riding the wave, or when we watch a wave peak travel, we are really following a point of constant total phase. The next example should make this more clear.

We are going to look at the wave described by this equation

$$y(x,t) = (25\text{ cm})\sin\left(2\pi\frac{t}{4s} + 2\pi\frac{x}{4\text{ cm}} + \frac{\pi}{2}\right)$$

a. One of the peaks of the wave has a total phase $\Phi = \pi/2$.What is the location of this peak when t $t = 0, t = 1s, t = 2s, t = 3s,$ and $t = 4s$?

b. Is the wave travelling to the left or right?

Solution

a. We only need information about the total phase to solve this. We are given

$$\phi(x,t) = 2\pi\frac{t}{4\ s} + 2\pi\frac{x}{4\ \text{cm}} + \frac{\pi}{2}$$

We are asked to find where (for which x) $\Phi = \pi/2$ is at t=0. We can solve this rather simply:

$$\phi(x,t) = \frac{\pi}{2} = 2\pi\frac{0}{4\ \text{s}} + 2\pi\frac{x}{4\ \text{cm}} + \frac{\pi}{2}$$

Which is only satisfied for $x = 0$ Therefore the $\Phi = \pi/2$ peak is at $x = 0$ when *t*=0. Substituting the remaining values for t we find that for $t = 1\ s,\ 2\ s,\ 3\ s,$ and $4\,\text{s}$ we have the peak is located at $x = 0\ \text{cm}, -1\ \text{cm}, -2\ \text{cm}, -3\ \text{cm}, \text{and} -4\ \text{cm}$ respectively.

b. We see that a particular peak goes from 0 cm to -4 cm. The peak moves in the direction of negative x, which is to the left by our convention.

We can see that the +sign in front of the spatial term is responsible for this. Remember that by riding the wave we are actually looking at a piece of constant total phase Φ. Going back to our equation, to ensure the left side of our equation remains constant as t increases, another term must decrease. Φ the phase constant of our wave does not change with time. Therefore our x term must decrease, showing again that our wave travels to the left.

Fixed Phase Constant ϕ

Note that the phase expression is very similar to the mathematical description we developed for the motion of a particle vibrating in simple harmonic motion. The first term in the argument of the sine function, $2\pi t/T$, like in harmonic motion, gives information about the phase for different values of t. The fixed phase constant ϕ gives the proper value of y at $t = 0$ and $x = 0$ The new term, (2 \pi x/ \lambda\), resembles the time-dependent term very closely. This term involving and λ gives the change in phase as we look along different values of x.

The total phase Φ goes through a complete cycle of 2π radians each time x increases or decreases by an amount equal to the wavelength λ Likewise the total phase goes through a complete cycle of 2π radians each time t increases by one period T. This is a reminder that λ ontrols repetition in space, while T controls repetition in time. Note that if $y = 0$ and $t = 0$ we have $\Phi = \phi$ the total phase is given by the fixed phase constant.

Relationship between v_{wave}, λ, and f

We have already learned that the speed of the wave depends on the properties of the medium. When dealing with repeating waves we must consider three additional parameters: the wavelength λ the period T and the frequency f. Recall that frequency and period are related by $f = 1/T$ so only two of the three parameters are independent. Below, we go through two different arguments to show that these parameters are related to wave speed v_{wave} by

$$v_{wave} = \lambda f.$$

One Argument, Distance Over Time

Our definition of speed (more specifically, velocity) from Physics 7B can be written as

$$\text{speed} = \frac{\text{Distance travelled}}{\text{Time spent}}$$

Let us look at the wave at a particular time, and focus on a particular peak indicated by the solid dot.

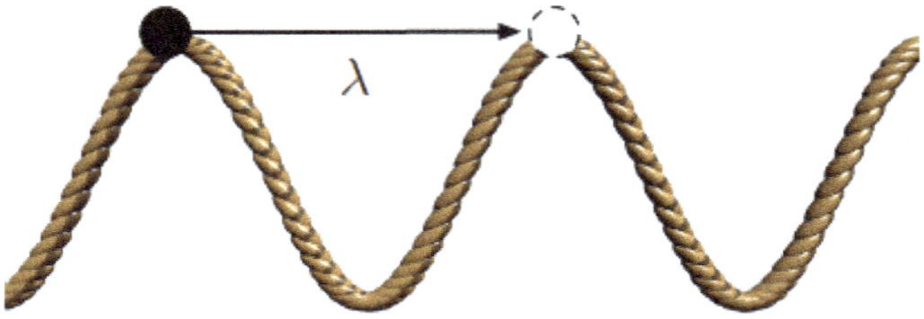

Recall that one period is the shortest amount of time before the wave looks *exactly* the same. For the wave to look exactly the same, the peak indicated by the solid black dot must have moved to the location of the dashed circle, which means it moved a distance of one wavelength. We can now calculate the speed of the peak (which is the same as the speed of the entire wave).

$$\text{Distance peak travels} = \lambda, \text{Time spent} = 1 \text{ period} = T$$

$$v_{wave} = \frac{Distance}{Time} = \frac{\lambda}{T}$$

Using the fact that $1/T$ is another way of writing the frequency f we can write this formula in a more familiar form

$$v_{wave} = \lambda f$$

Note that this method works for a wave traveling in the opposite direction as well. We would still see a peak travel one wavelength in one period, so we would still obtain $v_{wave} = \lambda f$

Another Argument, Following the Total Phase

A superficially different way of finding the wave speed is to follow a piece of the wave, that is look at a piece of the wave with a constant total phase Φ. Let us pick a phase and look at it at two different times t_1 (where it is at x_1) and (where that piece of the wave is at x_2) This gives us the relationships

$$\Phi = 2\pi\frac{t_1}{T} \pm 2\pi\frac{x_1}{\lambda} + \phi$$

$$\Phi = 2\pi\frac{t_2}{T} \pm 2\pi\frac{x_2}{\lambda} + \phi$$

Now we can subtract these equations from one another and rewrite them as follows:

$$2\pi\frac{t_2 - t_1}{T} \pm 2\pi\frac{x_1 - x_2}{\lambda} = 0$$

$$2\pi\frac{\Delta t}{T} \pm 2\pi\frac{\Delta x}{\lambda} = 0$$

where we use Δ to mean "final minus initial." Dividing the whole equation by 2π and rearranging, we get

$$\frac{\Delta x}{\Delta t} = \mp\frac{\lambda}{T}$$

But this expression tells us how far (Δx) the "disturbance" with phase Φ moved in a time Δt. This is exactly what we mean by velocity. Taking the absolute value of this gives us the wave speed:

$$v_{wave} = \left|\frac{\text{distance traveled}}{\text{time taken}}\right| = \left|\frac{\Delta x}{\Delta t}\right| = \frac{\lambda}{T} = \lambda f$$

References

- Lev A. Ostrovsky & Alexander I. Potapov (2001). Modulated waves: theory and application. Johns Hopkins University Press. ISBN 0-8018-7325-8
- Haifeng Wang, Luping Shi, Boris Luk›yanchuk, Colin Sheppardand Chong Tow Chong, "Creation of a needle of longitudinally polarized light in vacuum using binary optics," Nature Photonics, Vol.2, pp 501-505, 2008, doi:10.1038/nphoton.2008.127
- Ling, R. T.; Scholler, J. D.; Ufimtsev, P. Ya. (1998). "Propagation and excitation of surface waves in an absorbing layer" (PDF). Northrop Grumman Corporation. Progress in Electromagnetics Research (PIER). 19: 49–91. Retrieved 2018-05-10
- Davies, MJ; Newton, JD (2 July 2017). "Non-invasive imaging in cardiology for the generalist". British journal of hospital medicine (London, England : 2005). 78 (7): 392–398. doi:10.12968/hmed.2017.78.7.392. PMID 2 8692375
- Michael A. Slawinski (2003). "Wave equations". Seismic waves and rays in elastic media. Elsevier. pp. 131 ff. ISBN 0-08-043930-6
- Hill, D. A., and J. R. Wait (1978), Excitation of the Zenneck surface wave by a vertical aperture, Radio Sci., 13(6), 969–977, doi:10.1029/RS013i006p00969

- Milsom, J. (2003). Field Geophysics. The geological field guide series. 25. John Wiley and Sons. p. 232. ISBN 978-0-470-84347-5. Retrieved 2010-02-25
- Petrescu, Florian Ion T (2015). “Improving Medical Imaging and Blood Flow Measurement by using a New Doppler Effect Relationship”. American Journal of Engineering and Applied Sciences. 8.4: 582–588 – via Pro-quest
- Karl F Graaf (1991). Wave motion in elastic solids (Reprint of Oxford 1975 ed.). Dover. pp. 13–14. ISBN 978-0-486-66745-4
- S. Zeng; Baillargeat, Dominique; Ho, Ho-Pui; Yong, Ken-Tye (2014). “Nanomaterials enhanced surface plasmon resonance for biological and chemical sensing applications”. Chemical Society Reviews. 43 (10): 3426–3452. doi:10.1039/C3CS60479A. PMID 24549396
- “Photo Glossary of Earthquakes”. U.S. Geological Survey”. Archived from the original on February 27, 2009. Retrieved March 8,2009
- Wait, James, “Excitation of Surface Waves on Conducting, Stratified, Dielectric-Clad, and Corrugated Surfaces,” Journal of Research of the National Bureau of Standards Vol. 59, No.6, December 1957
- Barbosa Aguiar, Ana C. (2010). “A laboratory model of Saturn›s North Polar Hexagon”. Icarus. 206: 755–763. Bibcode:2010Icar..206..755B. doi:10.1016/j.icarus.2009.10.022
- Henderson, Tom (2017). “The Doppler Effect – Lesson 3, Waves”. Physics tutorial. The Physics Classroom. Retrieved September 4, 2017
- For an example derivation, see the steps leading up to eq. (17) in Francis Redfern. “Kinematic Derivation of the Wave Equation”. Physics Journal
- Mark Lundstrom (2000). Fundamentals of carrier transport. Cambridge University Press. p. 33. ISBN 0-521-63134-3.
- GR Helffrich & BJ Wood (2002). “The Earth›s Mantle” (PDF). Nature. Macmillan Magazines. 412 (2 August): 501; Figure 1. doi:10.1038/35087500
- Scott Russell, John (1848). “On certain effects produced on sound by the rapid motion of the observer”. Report of the Eighteenth Meeting of the British Association for the Advancement of Science. John Murray, London in 1849. 18 (7): 37–38. Retrieved 2008-07-08
- Chin-Lin Chen (2006). “§13.7.3 Pulse envelope in nondispersive media”. Foundations for guided-wave optics. Wiley. p. 363. ISBN 0-471-75687-3

5

Sound

Any vibration, which propagates as an audible pressure wave through a solid, liquid or gas, is known as sound. The human ear can only perceive sound between the frequencies of 20 Hz and 20 kHz. This chapter elaborates the different characteristics of sound such as speed, timbre, pitch and loudness.

All sounds are vibrations traveling through the air as sound waves. Sound waves are caused by the vibrations of objects and radiate outward from their source in all directions. A vibrating object compresses the surrounding air molecules (squeezing them closer together) and then rarefies them (pulling them farther apart). Although the fluctuations in air pressure travel outward from the object, the air molecules themselves stay in the same average position. As sound travels, it reflects off objects in its path, creating further disturbances in the surrounding air. When these changes in air pressure vibrate your eardrum, nerve signals are sent to your brain and are interpreted as sound.

Fundamentals of a Sound Wave

The simplest kind of sound wave is a sine wave. Pure sine waves rarely exist in the natural world, but they are a useful place to start because all other sounds can be broken down into combinations of sine waves. A sine wave clearly demonstrates the three fundamental characteristics of a sound wave: frequency, amplitude, and phase.

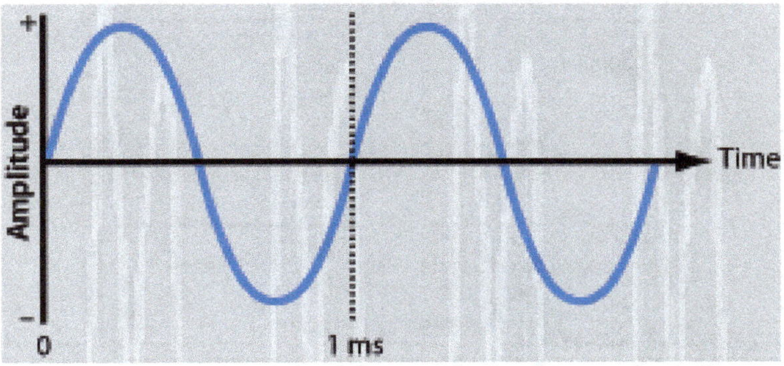

Frequency

Frequency is the rate, or number of times per second, that a sound wave cycles from positive to negative to positive again. Frequency is measured in cycles per second, or hertz (Hz). Humans have a range of hearing from 20 Hz (low) to 20,000 Hz (high). Frequencies beyond this range exist, but they are inaudible to humans.

Amplitude

Amplitude (or intensity) refers to the strength of a sound wave, which the human ear interprets as volume or loudness. People can detect a very wide range of volumes, from the sound of a pin dropping in a quiet room to a loud rock concert. Because the range of human hearing is so large, audio meters use a logarithmic scale (decibels) to make the units of measurement more manageable.

Phase

Phase compares the timing between two similar sound waves. If two periodic sound waves of the same frequency begin at the same time, the two waves are said to be in phase. Phase is measured in degrees from 0 to 360, where 0 degrees means both sounds are exactly in sync (in phase) and 180 degrees means both sounds are exactly opposite (out of phase). When two sounds that are in phase are added together, the combination makes an even stronger result. When two sounds that are out of phase are added together, the opposing air pressures cancel each other out, resulting in little or no sound. This is known as *phase cancelation.*

Phase cancelation can be a problem when mixing similar audio signals together, or when original and reflected sound waves interact in a reflective room. For example, when the left and right channels of a stereo mix are combined to create a mono mix, the signals may suffer from phase cancelation.

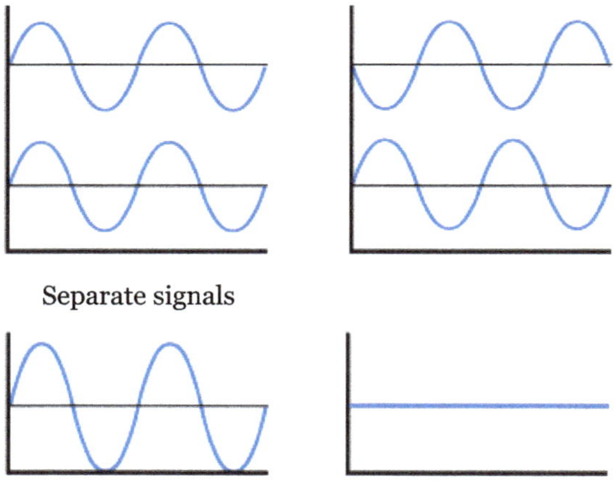

In phase Mixed signals Out of phase

Frequency Spectrum of Sounds

With the exception of pure sine waves, sounds are made up of many different frequency components vibrating at the same time. The particular characteristics of a sound are the result of the unique combination of frequencies it contains.

Sounds contain energy in different frequency ranges, or *bands*. If a sound has a lot of low-frequency energy, it has a lot of bass. The 250–4000 Hz frequency band, where humans hear best, is described as *midrange*. High-frequency energy beyond the midrange is called treble, and this adds crispness or brilliance to a sound. The graph below shows how the sounds of different musical instruments fall within particular frequency bands.

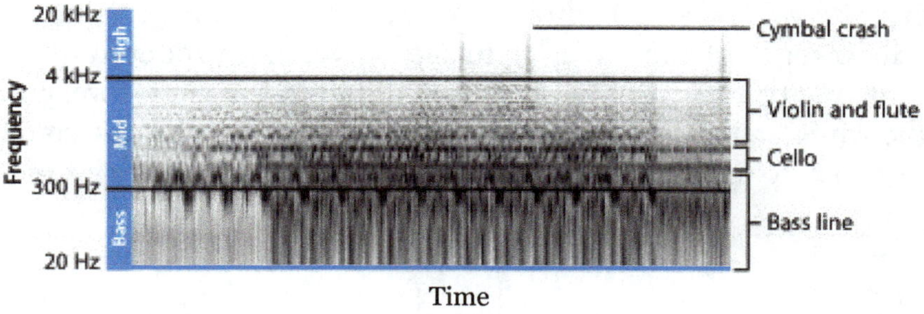

Note: Different manufacturers and mixing engineers define the ranges of these frequency bands differently, so the numbers described above are approximate.

The human voice produces sounds that are mostly in the 250–4000 Hz range, which likely explains why people's ears are also the most sensitive to this range. If the dialogue in your movie is harder to hear when you add music and sound effects, try reducing the midrange frequencies of the nondialogue tracks using an equalizer filter. Reducing the midrange creates a "sonic space" in which the dialogue can be heard more easily.

Musical sounds typically have a regular frequency, which the human ear hears as the sound's pitch. Pitch is expressed using musical notes, such as C, E flat, and F sharp. The pitch is usually only the lowest, strongest part of the sound wave, called the fundamental frequency. Every musical sound also has higher, softer parts called overtones or harmonics, which occur at regular multiples of the fundamental frequency. The human ear doesn't hear the harmonics as distinct pitches, but rather as the tone color (also called the timbre) of the sound, which allows the ear to distinguish one instrument or voice from another, even when both are playing the same pitch.

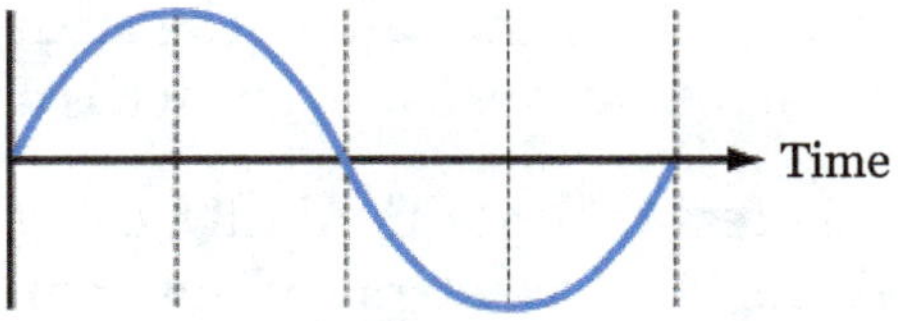

Fundamental

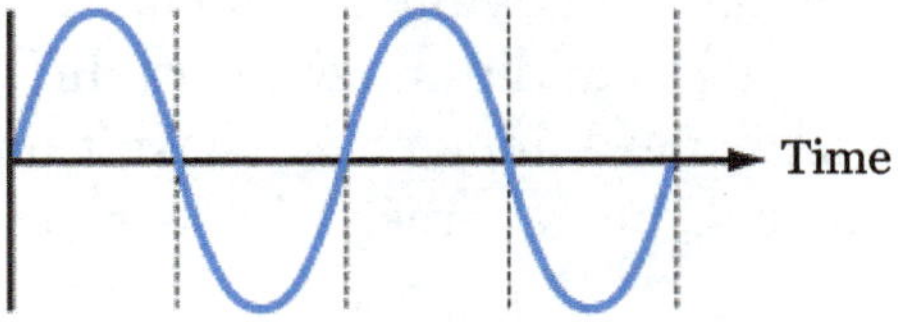

First harmonic

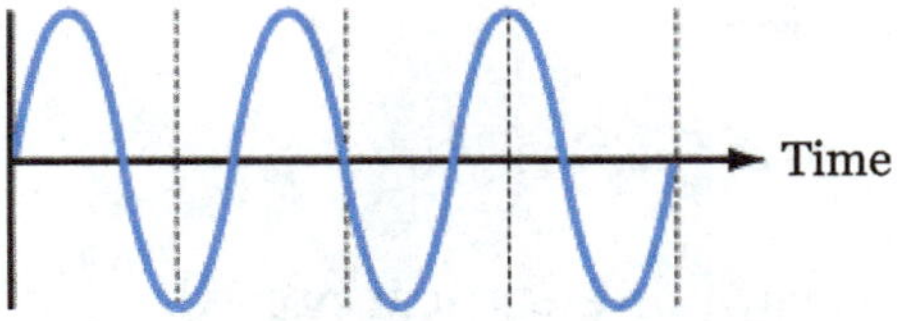

Second harmonic

Musical sounds also typically have a *volume envelope*. Every note played on a musical instrument has a distinct curve of rising and falling volume over time. Sounds produced by some instruments, particularly drums and other percussion instruments, start at a high volume level but quickly decrease to a much lower level and die away to silence. Sounds produced by other instruments, for example, a violin or a trumpet, can be sustained at the same volume level and can be raised or lowered in volume while being sustained. This volume curve is called the sound's envelope and acts like a signature to help the ear recognize what instrument is producing the sound.

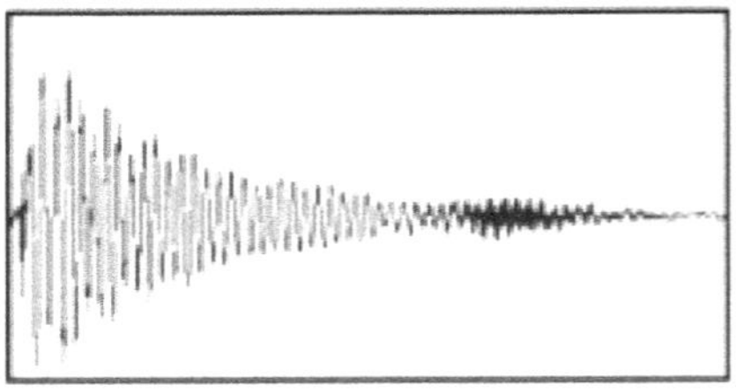

Percussive volume envelope

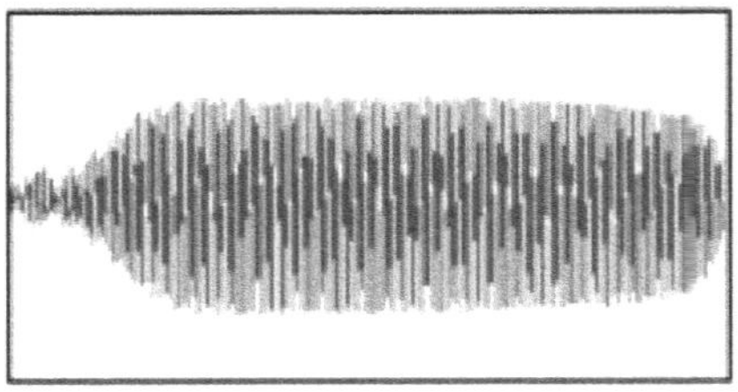

Sustained volume envelope

Measuring Sound Intensity

Human ears are remarkably sensitive to vibrations in the air. The threshold of human hearing is around 20 microPascals (μP), which is an extremely small amount of atmospheric pressure. At the other extreme, the loudest sound a person can withstand without pain or ear damage is about 200,000,000 μP: for example, a loud rock concert or a nearby jet airplane taking off.

Because the human ear can handle such a large range of intensities, measuring sound pressure levels on a linear scale is inconvenient. For example, if the range of human hearing were measured on a ruler, the scale would go from 1 foot (quietest) to over 3000 miles (loudest)! To make this huge range of numbers easier to work with, a logarithmic unit—the decibel—is used. Logarithms map exponential values to a linear scale. For example, by taking the base-ten logarithm of 10 (101) and 1,000,000,000 (109), this large range of numbers can be written as 1–9, which is a much more convenient scale.

Because the ear responds to sound pressure logarithmically, using a logarithmic scale corresponds to the way humans perceive loudness. Audio meters and sound measurement equipment are specifically designed to show audio levels in decibels. Small changes at the bottom of an audio meter may represent large changes in signal level, while small changes toward the top may represent small changes in signal level. This makes audio meters very different from linear measuring devices like rulers, thermometers, and speedometers. Each unit on an audio meter represents an exponential increase in sound pressure, but a perceived linear increase in loudness.

Important: When you mix audio, you don't need to worry about the mathematics behind logarithms and decibels. Just be aware that to hear incremental increases in sound volume, exponentially more sound pressure is required.

Decibel

The decibel measures sound pressure or electrical pressure (voltage) levels. It is a logarithmic unit that describes a ratio of two intensities, such as two different sound pressures, two different voltages, and so on. A bel (named after Alexander Graham Bell) is a base-ten logarithm of the ratio between two signals. This means that for every additional bel on the scale, the signal represented is ten times stronger. For example, the sound pressure level of a loud sound can be billions of times stronger than a quiet sound. Written logarithmically, one billion (1,000,000,000 or 109) is simply 9. Decibels make the numbers much easier to work with.

In practice, a bel is a bit too large to use for measuring sound, so a one-tenth unit called the decibel is used instead. The reason for using decibels instead of bels is no different from the reason for measuring shoe size in, say, centimeters instead of meters; it is a more practical unit.

Number of decibels	Relative increase in power
0	1
1	1.26
3	2
10	10
20	100
30	1000
50	100,000
100	10,000,000,000

Decibel Units

Audio meters are labeled with decibels. Several reference levels have been used in audio meters over the years, starting with the invention of the telephone and evolving to present day systems. Some of these units are only applicable to older equipment. Today, most professional equipment uses dBu, and most consumer equipment uses dBV. Digital meters use dBFS.

- *dBm:* The *m* stands for *milliwatt* (*m*W), which is a unit for measuring electrical power. (Power is different from electrical voltage and current, though it is related to both.) This was the standard used in the early days of telephone technology and remained the professional audio standard for years.
- *dBu:* This reference level measures voltage instead of power, using a reference level of 0.775 volts. dBu has mostly replaced dBm on professional audio equipment. The *u* stands for *unloaded*, because the electrical load in an audio circuit is no longer as relevant as it was in the early days of audio equipment.
- *dBV:* This also uses a reference voltage like dBu, but in this case the reference level is 1 volt, which is more convenient than 0.775 volts in dBu. dBV is often used on consumer and semiprofessional devices.
- *dBFS:* This scale is very different from the others because it is used for measuring digital audio levels. *FS* stands for *full-scale*, which is used because, unlike analog audio signals that have an optimum signal voltage, the entire range of digital values is equally acceptable when using digital audio. 0 dBFS is the highest-possible digital audio signal you can record without distortion. Unlike analog audio scales like dBV and dBu, there is no headroom past 0 dBFS.

Signal-to-Noise Ratio

Every electrical system produces a certain amount of low-level electrical activity called *noise*. The noise floor is the level of noise inherent in a system. It is nearly impossible to eliminate all the noise in an electrical system, but you don't have to worry about the noise if you record your signals significantly higher than the noise floor. If you record audio too low, you raise the volume to hear it, which also raises the volume of the noise floor, causing a noticeable hiss.

The more a signal is amplified, the louder the noise becomes. Therefore, it is important to record most audio around the nominal (ideal) level of the device, which is labeled 0 dB on an analog audio meter.

The *signal-to-noise ratio*, typically measured in dB, is the difference between the nominal recording level and the noise floor of the device. For example, the signal-to-noise ratio of an analog tape deck may be 60 dB, which means the inherent noise in the system is 60 dB lower than the ideal recording level.

Headroom and Distortion

If an audio signal is too strong, it will overdrive the audio circuit, causing the shape of the signal to distort. In analog equipment, distortion increases gradually the more the audio signal overdrives the circuit. For some audio recordings, this kind of distortion can add a unique "warmth" to the recording that is difficult to achieve with digital equipment. However, for audio post-production, the goal is to keep the signal clean and undistorted.

0 dB on an analog meter refers to the ideal recording level, but there is some allowance for stronger signals before distortion occurs. This safety margin is known as headroom, meaning that the signal can occasionally go higher than the ideal recording level without distorting. Having headroom is critical when recording, especially when the audio level is very dynamic and unpredictable. Even though you can adjust the recording level while you record, you can't always anticipate quick, loud sounds. The extra headroom above 0 dB on the meter is there in case the audio abruptly becomes loud.

Dynamic Range and Compression

Dynamic range is the difference between the quietest and loudest sound in your mix. A mix that contains quiet whispers and loud screams has a large dynamic range. A recording of a constant drone such as an air conditioner or steady freeway traffic has very little amplitude variation, so it has a small dynamic range.

You can actually see the dynamic range of an audio clip by looking at its waveform. For example, two waveforms are shown below. The top one is a section from a well-known piece of classical music. The bottom one is from a piece of electronic music. From the widely varied shape of the waveform, you can tell that the classical piece has the greater dynamic range.

Waveform from a well-known cllasical piece

Waveform from an excerpt of electronic music

Notice that the loud and soft parts of the classical piece vary more frequently, as compared to the fairly consistent levels of the electronic music. The long, drawn-out part of the waveform at the left end of the top piece is not silence—it's actually a long, low section of the music.

Dynamic sound has drastic volume changes. Sound can be made less dynamic by reducing, or compressing, the loudest parts of the signal to be closer to the quiet parts. Compression is a useful technique because it makes the sounds in your mix more equal. For example, a train pulling into the station, a man talking, and the quiet sounds of a cricket-filled evening are, in absolute terms, very different volumes. Because televisions and film theaters must compete with ambient noise in the real world, it is important that the quiet sounds are not lost.

The goal is to make the quiet sounds (in this case, the crickets) louder so they can compete with the ambient noise in the listening environment. One approach to making the crickets louder is to simply raise the level of the entire soundtrack, but when you increase the level of the quiet sounds, the loud sounds (such as the train) get too loud and distort. Instead of raising the entire volume of your mix, you can compress the loud sounds so they are closer to the quiet sounds. Once the loud sounds are quieter (and the quiet sounds remain the same level), you can raise the overall level of the mix, bringing up the quiet sounds without distorting the loud sounds.

When used sparingly, compression can help you bring up the overall level of your mix to compete with noise in the listening environment. However, if you compress a signal too far, it sounds very unnatural. For example, reducing the sound of an airplane jet engine to the sound of a quiet forest at night and then raising the volume to maximum would cause the noise in the forest to be amplified immensely.

Different media and genres use different levels of compression. Radio and television commercials use compression to achieve a consistent wall of sound. If the radio or television becomes too quiet, the audience may change the channel—a risk advertisers and broadcasters don't want to take. Films in theaters have a slightly wider dynamic range because the ambient noise level of the theater is lower, so quiet sounds can remain quiet.

Stereo Audio

The human ear hears sounds in stereo, and the brain uses the subtle differences in sounds entering the left and right ears to locate sounds in the environment. To re-create this sonic experience, stereo recordings require two audio channels throughout the recording and playback process. The microphones must be properly positioned to accurately capture a stereo image, and speakers must also be spaced properly to re-create a stereo image accurately.

If any part of the audio reproduction pathway eliminates one of the audio channels, the stereo image will most likely be compromised. For example, if your playback system has a CD player (two audio channels) connected to only one speaker, you will not hear the intended stereo image.

Important: All stereo recordings require two channels, but two-channel recordings are not necessarily stereo. For example, if you use a single-capsule microphone to record the same signal on two tracks, you are not making a stereo recording.

Identifying Two-Channel Mono Recordings

When you are working with two-channel audio, it is important to be able to distinguish between true stereo recordings and two tracks used to record two independent mono channels. These are called dual mono recordings.

Examples of dual mono recordings include:

- Two independent microphones used to record two independent sounds, such as two different actors speaking. These microphones independently follow each actor's voice and are never positioned in a stereo left-right configuration. In this case, the intent is not a stereo recording but two discrete mono channels of synchronized sound.
- Two channels with exactly the same signal. This is no different than a mono recording, because both channels contain exactly the same information. Production audio is sometimes recorded this way, with slightly different gain settings on each channel. This way, if one channel distorts, you have a safety channel recorded at a lower level.
- Two completely unrelated sounds, such as dialogue on track 1 and a timecode audio signal on track 2, or music on channel 1 and sound effects on channel 2. Conceptually, this is not much different than recording two discrete dialogue tracks in the example above.

The important point to remember is that if you have a two-track recording system, each track can be used to record anything you want. If you use the two tracks to record properly positioned left and right microphones, you can make a stereo recording. Otherwise, you are simply making a two-channel mono recording.

Identifying Stereo Recordings

When you are trying to decide how to work with an audio clip, you need to know whether a two-channel recording was intended to be stereo or not. Usually, the person recording production sound will have labeled the tapes or audio files to indicate whether they were recorded as stereo recordings or dual-channel mono recordings. However, things don't always go as planned, and tapes aren't always labeled as thoroughly as they should be. As an editor, it's important to learn how to differentiate between the two.

Here are some tips for distinguishing stereo from dual mono recordings:

- Stereo recordings must have two independent tracks. If you have a tape with only one track of audio, or a one-channel audio file, your audio is mono, not stereo.

  Note: It is possible that a one-channel audio file is one half of a stereo pair. These are known as split stereo files, because the left and right channels are contained in independent files. Usually, these files are labeled accordingly: AudioFile.L and AudioFile.R are two audio files that make up the left and right channels of a stereo sound.
- Almost all music, especially commercially available music, is mixed in stereo.
- Listen to a clip using two (stereo) speakers. If each side sounds subtly different, it is probably stereo. If each side sounds absolutely the same, it may be a mono recording. If each side is completely unrelated, it is a dual mono recording.

Interleaved Versus Split Stereo Audio Files

Digital audio can send a stereo signal within a single stream by interleaving the digital samples during transmission and deinterleaving them on playback. The way the signal is stored is unimportant as long as the samples are properly split to left and right channels during playback. With analog technology, the signal is not nearly as flexible.

Split stereo files are two independent audio files that work together, one for the left channel (AudioFile.L) and one for the right channel (AudioFile.R). This mirrors the traditional analog method of one track per channel (or in this case, one file per channel).

Acoustics

Acoustics is the interdisciplinary science that deals with the study of mechanical waves in gases, liquids, and solids including vibration, sound, ultrasound, and infrasound. A scientist who works in the field of acoustics is an acoustician, while someone working in the field of acoustical engineering may be called an acoustical engineer. An audio engineer, on the other hand, is concerned with the recording, manipulation, mixing, and reproduction of sound.

Applications of acoustics are found in almost all aspects of modern society, subdisciplines include aeroacoustics, audio signal processing, architectural acoustics, bioacoustics, electro-acoustics, environmental noise, musical acoustics, noise control, psychoacoustics, speech, ultrasound, underwater acoustics, and vibration.

Definition

Sound is defined as "(a) Oscillation in pressure, stress, particle displacement, particle velocity, etc., propagated in a medium with internal forces (e.g., elastic or viscous), or the superposition of such propagated oscillation. (b) Auditory sensation evoked by the oscillation described in (a). " Sound can be viewed as a wave motion in air or other elastic media. In this case, sound is a stimulus. Sound can also be viewed as an excitation of the hearing mechanism that results in the perception of sound. In this case, sound is a sensation .

Physics of Sound

Sound can propagate through a medium such as air, water and solids as longitudinal waves and also as a transverse wave in solids. The sound waves are generated by a sound source, such as the vibrating diaphragm of a stereo speaker. The sound source creates vibrations in the surrounding medium. As the source continues to vibrate the medium, the vibrations propagate away from the source at the speed of sound, thus forming the sound wave. At a fixed distance from the source, the pressure, velocity, and displacement of the medium vary in time. At an instant in time, the pressure, velocity, and displacement vary in space. Note that the particles of the medium do not travel with the sound wave. This is intuitively obvious for a solid, and the same is true for liquids and gases (that is, the vibrations of particles in the gas or liquid transport the vibrations, while the average position of the particles over time does not change). During propagation, waves can be reflected, refracted, or attenuated by the medium.

The behavior of sound propagation is generally affected by three things:

- A complex relationship between the density and pressure of the medium. This relationship, affected by temperature, determines the speed of sound within the medium.
- Motion of the medium itself. If the medium is moving, this movement may increase or decrease the absolute speed of the sound wave depending on the direction of the movement. For example, sound moving through wind will have its speed of propagation increased by the speed of the wind if the sound and wind are moving in the same direction. If the sound and wind are moving in opposite directions, the speed of the sound wave will be decreased by the speed of the wind.
- The viscosity of the medium. Medium viscosity determines the rate at which sound is attenuated. For many media, such as air or water, attenuation due to viscosity is negligible.

When sound is moving through a medium that does not have constant physical properties, it may be refracted (either dispersed or focused).

The mechanical vibrations that can be interpreted as sound can travel through all forms of matter: gases, liquids, solids, and plasmas. The matter that supports the sound is called the medium. Sound cannot travel through a vacuum.

Longitudinal and Transverse Waves

Sound is transmitted through gases, plasma, and liquids as longitudinal waves, also called compression waves. It requires a medium to propagate. Through solids, however, it can be transmitted as both longitudinal waves and transverse waves. Longitudinal sound waves are waves of alternating pressure deviations from the equilibrium pressure, causing local regions of compression and rarefaction, while transverse waves (in solids) are waves of alternating shear stress at right angle to the direction of propagation.

Sound waves may be "viewed" using parabolic mirrors and objects that produce sound.

The energy carried by an oscillating sound wave converts back and forth between the potential energy of the extra compression (in case of longitudinal waves) or lateral displacement strain (in case of transverse waves) of the matter, and the kinetic energy of the displacement velocity of particles of the medium.

Sound Wave Properties and Characteristics

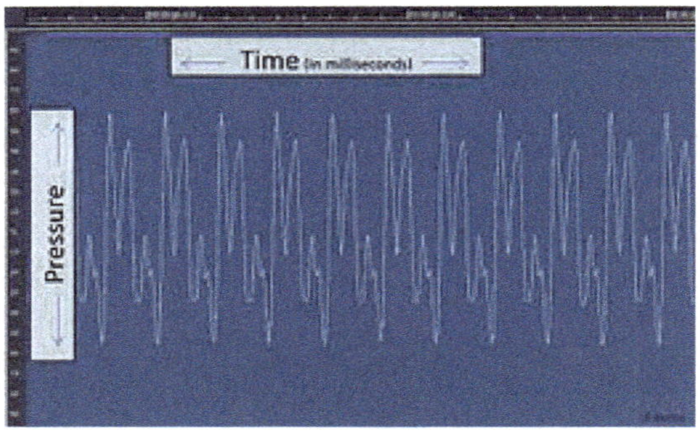

A 'pressure over time' graph of a 20 ms recording of a clarinet tone demonstrates the two fundamental elements of sound: Pressure and Time

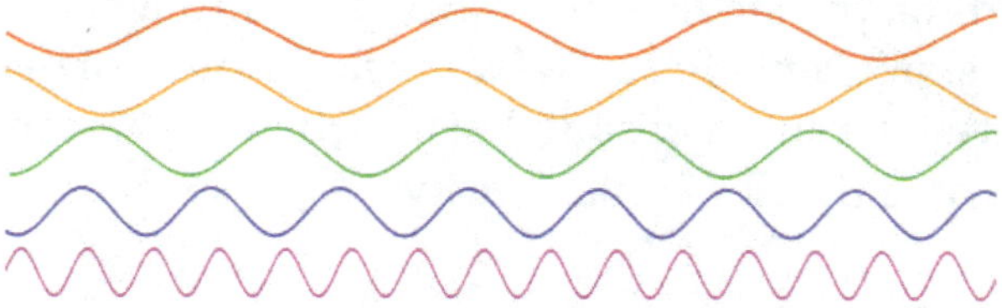

Sounds can be represented as a mixture of their component Sinusoidal waves of different frequencies. The bottom waves have higher frequencies than those above. The horizontal axis represents time.

Although there are many complexities relating to the transmission of sounds, at the point of reception (i.e. the ears), sound is readily dividable into two simple elements: pressure and time. These fundamental elements form the basis of all sound waves. They can be used to describe, in absolute terms, every sound we hear.

In order to understand the sound more fully, a complex wave such as the one shown in a blue background on the right of this text, is usually separated into its component parts, which are a combination of various sound wave frequencies (and noise).

Sound waves are often simplified to a description in terms of sinusoidal plane waves, which are characterized by these generic properties:

- Frequency, or its inverse, wavelength
- Amplitude, sound pressure or Intensity
- Speed of sound
- Direction

Sound that is perceptible by humans has frequencies from about 20 Hz to 20,000 Hz. In air at standard temperature and pressure, the corresponding wavelengths of sound waves range from 17 m to 17 mm. Sometimes speed and direction are combined as a velocity vector; wave number and direction are combined as a wave vector.

Transverse waves, also known as shear waves, have the additional property, polarization, and are not a characteristic of sound waves.

Speed of Sound

U.S. Navy F/A-18 approaching the speed of sound. The white halo is formed by condensed water droplets thought to result from a drop in air pressure around the aircraft

The speed of sound depends on the medium the waves pass through, and is a fundamental property of the material. The first significant effort towards measurement of the speed of sound was made by Isaac Newton. He believed the speed of sound in a particular substance was equal to the square root of the pressure acting on it divided by its density:

$$c = \sqrt{\frac{p}{\rho}}$$

This was later proven wrong when found to incorrectly derive the speed. The French mathematician Laplace corrected the formula by deducing that the phenomenon of sound travelling is not isothermal, as believed by Newton, but adiabatic. He added another factor to the equation—*gamma*—and multiplied $\sqrt{\gamma}$ by $\sqrt{\frac{p}{\rho}}$, thus coming up with the equation $c = \sqrt{\gamma \cdot \frac{p}{\rho}}$. Since $K = \gamma \cdot p$, the final equation came up to be $c = \sqrt{\frac{K}{\rho}}$, which is also known as the Newton-Laplace equation.

In this equation, K = elastic bulk modulus, c = velocity of sound, and ρ = density. Thus, the speed of sound is proportional to the square root of the ratio of the bulk modulus of the medium to its density.

Those physical properties and the speed of sound change with ambient conditions. For example, the speed of sound in gases depends on temperature. In 20 °C (68 °F) air at sea level, the speed of sound is approximately 343 m/s (1,230 km/h; 767 mph) using the formula "v = (331 + 0.6 T) m/s". In fresh water, also at 20 °C, the speed of sound is approximately 1,482 m/s (5,335 km/h; 3,315 mph). In steel, the speed of sound is about 5,960 m/s (21,460 km/h; 13,330 mph). The speed of sound is also slightly sensitive, being subject to a second-order anharmonic effect, to the sound amplitude, which means there are non-linear propagation effects, such as the production of harmonics and mixed tones not present in the original sound.

Perception of Sound

A distinct use of the term sound from its use in physics is that in physiology and psychology, where the term refers to the subject of perception by the brain. The field of psychoacoustics is dedicated to such studies. Historically the word "sound" referred exclusively to an effect in the mind. Webster's 1947 dictionary defined sound as: "that which is heard; the effect which is produced by the vibration of a body affecting the ear." This meant (at least in 1947) the correct response to the question: "if a tree falls in the forest with no one to hear it fall, does it make a sound?" was "no". However, owing to contemporary usage, definitions of sound as a physical effect are prevalent in most dictionaries. Consequently, the answer to the same question is now "yes, a tree falling in the forest with no one to hear it fall does make a sound".

The physical reception of sound in any hearing organism is limited to a range of frequencies. Humans normally hear sound frequencies between approximately 20 Hz and 20,000 Hz (20 kHz). The upper limit decreases with age. Sometimes sound refers to only those vibrations with frequencies that are within the hearing range for humans or sometimes it relates to a particular animal. Other species have different ranges of hearing. For example, dogs can perceive vibrations higher than 20 kHz.

As a signal perceived by one of the major senses, sound is used by many species for detecting danger, navigation, predation, and communication. Earth's atmosphere, water, and virtually any physical phenomenon, such as fire, rain, wind, surf, or earthquake, produces (and is characterized by) its unique sounds. Many species, such as frogs, birds, marine and terrestrial mammals, have also developed special organs to produce sound. In some species, these produce song and speech. Furthermore, humans have developed culture and technology (such as music, telephone and radio) that allows them to generate, record, transmit, and broadcast sound.

Noise is a term often used to refer to an unwanted sound. In science and engineering, noise is an undesirable component that obscures a wanted signal. However, in sound perception it can often be used to identify the source of a sound and is an important component of timbre perception.

Soundscape is the component of the acoustic environment that can be perceived by humans. The acoustic environment is the combination of all sounds (whether audible to humans or not) within a given area as modified by the environment and understood by people, in context of the surrounding environment.

There are, historically, six experimentally separable ways in which sound waves are analysed. They are: pitch, duration, loudness, timbre, sonic texture and spatial location. Some of these terms have a standardised definition (for instance in the ANSI Acoustical Terminology ANSI/ASA S1.1-2013). More recent approaches have also considered temporal envelope (ENV) and temporal fine structure (TFS) as perceptually relevant analyses.

Pitch

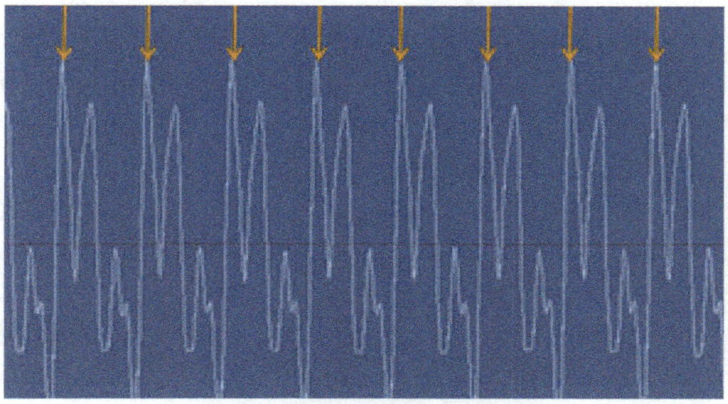

Figure- Pitch perception

Pitch is perceived as how "low" or "high" a sound is and represents the cyclic, repetitive nature of the vibrations that make up sound. For simple sounds, pitch relates to the frequency of the slowest vibration in the sound (called the fundamental harmonic). In the case of complex sounds, pitch perception can vary. Sometimes individuals identify different pitches for the same sound, based on their personal experience of particular sound patterns. Selection of a particular pitch is determined by pre-conscious examination of vibrations, including their frequencies and the balance between them. Specific attention is given to recognising potential harmonics. Every sound is placed on a pitch continuum from low to high. For example: white noise (random noise spread evenly across all frequencies) sounds higher in pitch than pink noise (random noise spread evenly across octaves) as white noise has more high frequency content. Figure shows an example of pitch recognition.

During the listening process, each sound is analysed for a repeating pattern and the results forwarded to the auditory cortex as a single pitch of a certain height (octave) and chroma (note name).

Duration

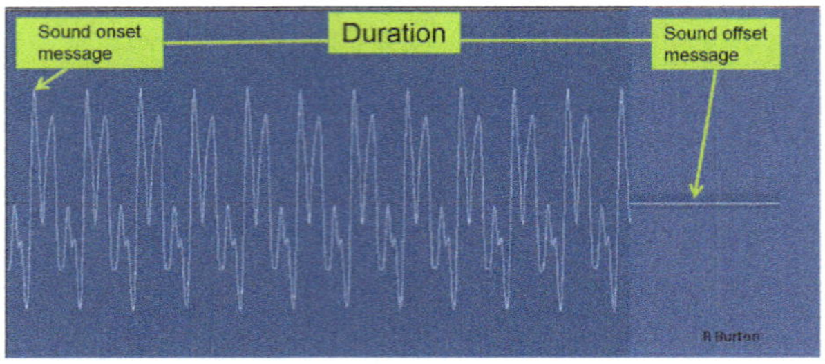

Figure- Duration perception

Duration is perceived as how "long" or "short" a sound is and relates to onset and offset signals created by nerve responses to sounds. The duration of a sound usually lasts from the time the sound is first noticed until the sound is identified as having changed or ceased. Sometimes this is not directly related to the physical duration of a sound. For example; in a noisy environment, gapped sounds (sounds that stop and start) can sound as if they are continuous because the offset messages are missed owing to disruptions from noises in the same general bandwidth. This can be of great benefit in understanding distorted messages such as radio signals that suffer from interference, as (owing to this effect) the message is heard as if it was continuous. Figure gives an example of duration identification. When a new sound is noticed, a sound onset message is sent to the auditory cortex. When the repeating pattern is missed, a sound offset messages is sent.

Loudness

Loudness is perceived as how "loud" or "soft" a sound is and relates to the totalled number of auditory nerve stimulations over short cyclic time periods, most likely over the duration of theta wave cycles. This means that at short durations, a very short sound can sound softer than a longer sound even though they are presented at the same intensity level. Past around 200 ms this is no longer the case and the duration of the sound no longer affects the apparent loudness of the sound. Louder signals create a greater 'push' on the Basilar membrane and thus stimulate more nerves, creating a stronger loudness signal. A more complex signal also creates more nerve firings and so sounds louder (for the same wave amplitude) than a simpler sound, such as a sine wave.

Timbre

Timbre is perceived as the quality of different sounds (e.g. the thud of a fallen rock, the whir of a drill, the tone of a musical instrument or the quality of a voice) and represents the pre-conscious allocation of a sonic identity to a sound (e.g. "it's an oboe!"). This identity is based on information gained from frequency transients, noisiness, unsteadiness, perceived pitch and the spread and intensity of overtones in the sound over an extended time frame. The way a sound changes over time provides most of the information for timbre identification. Even though a small section of the wave form from each instrument looks very similar, differences in changes over time between the clarinet and the piano are evident in both loudness and harmonic content. Less noticeable are the different noises heard, such as air hisses for the clarinet and hammer strikes for the piano.

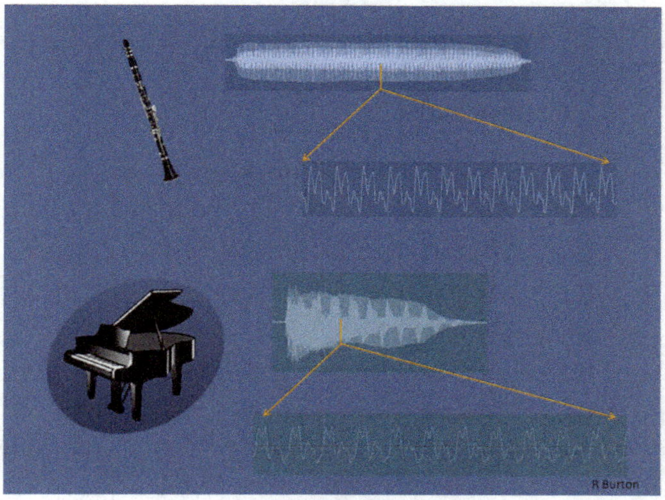

Timbre perception

Sonic Texture

Sonic texture relates to the number of sound sources and the interaction between them. The word 'texture', in this context, relates to the cognitive separation of auditory objects. In music, texture is often referred to as the difference between unison, polyphony and homophony, but it can also relate (for example) to a busy cafe; a sound which might be referred to as 'cacophony'. However texture refers to more than this. The texture of an orchestral piece is very different to the texture of a brass quintet because of the different numbers of players. The texture of a market place is very different to a school hall because of the differences in the various sound sources.

Spatial Location

Spatial location represents the cognitive placement of a sound in an environmental context; including the placement of a sound on both the horizontal and vertical plane, the distance from the sound source and the characteristics of the sonic environment. In a thick texture, it is possible to identify multiple sound sources using a combination of spatial location and timbre identification. It is the main reason why we can pick the sound of an oboe in an orchestra and the words of a single person at a cocktail party.

Sound Pressure Level

Sound measurements	
Characteristic	Symbols
Sound pressure	p, SPL, L_{PA}
Particle velocity	v, SVL
Particle displacement	δ
Sound intensity	I, SIL
Sound power	P, SWL, LWA
Sound energy	W

Sound energy density	w
Sound exposure	E, SEL
Acoustic impedance	Z
Speed of sound	c
Audio frequency	AF
Transmission loss	TL

Sound pressure is the difference, in a given medium, between average local pressure and the pressure in the sound wave. A square of this difference (i.e., a square of the deviation from the equilibrium pressure) is usually averaged over time and/or space, and a square root of this average provides a root mean square (RMS) value. For example, 1 Pa RMS sound pressure (94 dBSPL) in atmospheric air implies that the actual pressure in the sound wave oscillates between (1 atm $-\sqrt{2}$ Pa) and (1 atm $+\sqrt{2}$ Pa), that is between 101323.6 and 101326.4 Pa. As the human ear can detect sounds with a wide range of amplitudes, sound pressure is often measured as a level on a logarithmic decibel scale. The sound pressure level (SPL) or L_p is defined as

$$L_p = 10\log_{10}\left(\frac{p^2}{p_{\text{ref}}^2}\right) = 20\log_{10}\left(\frac{p}{p_{\text{ref}}}\right)\text{ dB}$$

where p is the root-mean-square sound pressure and p_{ref} is a reference sound pressure. Commonly used reference sound pressures, defined in the standard ANSI S1.1-1994, are 20 μPa in air and 1 μPa in water. Without a specified reference sound pressure, a value expressed in decibels cannot represent a sound pressure level.

Since the human ear does not have a flat spectral response, sound pressures are often frequency weighted so that the measured level matches perceived levels more closely. The International Electrotechnical Commission (IEC) has defined several weighting schemes. A-weighting attempts to match the response of the human ear to noise and A-weighted sound pressure levels are labeled dBA. C-weighting is used to measure peak levels.

Ultrasound

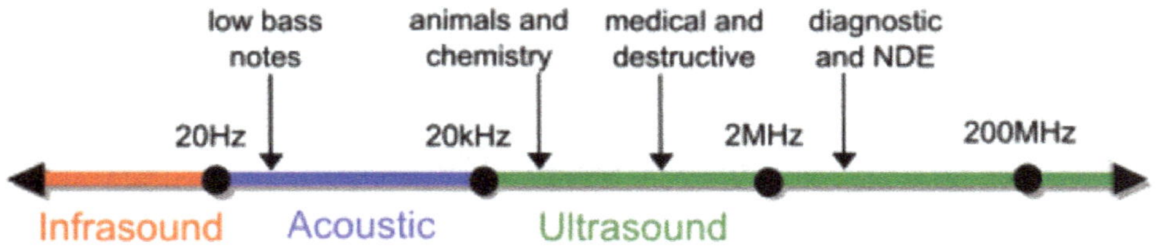

Approximate frequency ranges corresponding to ultrasound, with rough guide of some applications

Ultrasound is sound waves with frequencies higher than the upper audible limit of human hearing. Ultrasound is no different from 'normal' (audible) sound in its physical properties, except in that humans cannot hear it. Ultrasound devices operate with frequencies from 20 kHz up to several gigahertz.

Ultrasound is commonly used for medical diagnostics such as sonograms.

Infrasound

Infrasound is sound waves with frequencies lower than 20 Hz. Although sounds of such low frequency are too low for humans to hear, whales, elephants and other animals can detect infrasound and use it to communicate. It can be used to detect volcanic eruptions and is used in some types of music.

The Wave Equation

Having introduced the necessary notation, we can now derive the wave equation for sound waves. The derivation consists of four main steps, so let's go through them systematically. Our strategy will be to find the net force on a given volume of air, and then write down the F = ma equation for that volume.

1. How the volume changes: First, we need to determine how the volume of a gas changes when the pressure is changed. Qualitatively, if we increase the pressure on a given volume, the volume decreases. By how much? The decrease should certainly be proportional to the volume, because if we put two copies of a given volume next to each other, we will obtain twice the decrease. And the decrease should also be proportional to the pressure increase, provided that the increase is small. This is a reasonable claim, but not terribly obvious. We'll derive it in step 2 below. Assuming that it is true, we can write (recalling that ψ_p is defined to be the increase in pressure relative to equilibrium) the change in volume from equilibrium as $\Delta V \; \alpha - V\psi_p$. This correctly incorporates the two proportionality facts above. The minus sign is due to the fact that an increase in pressure causes a decrease in volume. This equation is valid as long as ΔV is small compared with V, as we'll see below.

   Let's define k to be the constant of proportionality in $\Delta V \propto -V\psi_p \cdot k$ is known as the *compressibility*. The larger k is, the more the volume is compressed (or expanded), for a given increase (or decrease) in pressure, ψ_p . In terms of k, we have

   $$\Delta V = -kV\psi_p.$$

   But from the first picture in figure, we see that the volume of gas is $V = A\Delta x$, where A is the cross-sectional area. And the change in the volume between the two pictures shown is $\Delta V = A\Delta\psi$.[1] Eq. therefore becomes

   $$\frac{\Delta V}{V} = -k\psi_p \Rightarrow \frac{AV\psi}{AVx} = -k\psi_p \Rightarrow \frac{\partial\psi}{\partial x} = -k\psi_p$$

   where we have taken the infinitesimal limit and changed the Δ's to differentials (partial ones, since ψ is a function of t also). The quantity $\partial\psi / \partial x$ indicates how the displacement from equilibrium grows as a function of x. Equivalently, $\partial\psi / \partial x$ is the "stretching fraction." If the displacement ψ grows by, say, 1 mm over the course of a distance of 10 cm, then the length (and hence the volume) of the region has increased by 1/100, and this equals $\partial\psi / \partial x$. Eq. says that the stretching fraction is proportional to the change in pressure, which is quite reasonable.

2. Calculating the compressibility, k: We'll now be rigorous about the above statement that the decrease in volume should be proportional to the pressure increase, provided that the increase is small. In the course of doing this, we'll find the value of k.

Let's first give a derivation that isn't quite correct. From the ideal gas law (which we'll accept here; we have to start somewhere), we have $pV = nRT$. If the temperature T is constant, then any changes in p and V must satisfy $(p + dp)(V + dV) = nRT$. Subtracting the original $pV = nRT$ equation from this one, and ignoring the second- order $dp\, dV$ term (this is where the assumption of small changes comes in), we obtain $p\, dV + dp\, V = 0 \Rightarrow dV = -(1/p)V\, dp$. But dp is what we've been calling ψ_p above, so if we compare this result with Eq. we see that the compressibility k equals $1/p$. Note that what we did here was basically take the differential of the $pV = C$ equation (where C is a constant) to obtain $p\, dV + dp\, V = 0$.

However, this $k = 1/p$ result isn't correct, because it is based on the assumption that T is constant in the relation $pV = nRT$. But T isn't constant in a sound wave (in either space or time). The compressions are actually adiabatic, meaning that heat can't flow quickly enough to redistribute itself and even things out. Basically, the time scale of the heat flow is large compared with the time scale of the wave oscillations. So a region that heats up due to high pressure stays hot, until the pressure decreases. The correct relation (which we'll just accept here) for adiabatic processes turns out to be $pV^{\gamma} = C$, where γ happens to be about 7/5 for air (γ is 7/5 for a diatomic gas, and air is 99% N_2 and O_2). Taking the differential of our new $pV^{\gamma} = C$ equation gives

$$p \cdot \gamma V^{\gamma-1}\, dV + dp V^{\gamma} = 0 \Rightarrow dV = -\left(\frac{1}{\gamma p}\right) V\, dp.$$

Therefore, the correct value of the compressibility k is

$$k = \frac{1}{\gamma p0} \approx \frac{5}{7p0}.$$

The above incorrect result wasn't so bad; it was off by only a factor of 7/5.

3. Calculating the difference in pressure, $\Delta\psi_P$: Eq. involves the quantity ψ_P. However, what we're actually concerned with is not ψ_P but the *change* in ψ_P from one end of the volume to the other, because in writing down the $F = ma$ equation for the volume, we're concerned with the *net* force on it, and this involves the difference in the pressures at the ends. So let's see how ψ_P changes with x. Differentiating Eq. with respect to x gives

$$\frac{\partial \psi_P}{\partial x} = -\frac{1}{k}\frac{\partial^2 \psi}{\partial x^2} \Rightarrow \Delta\psi_P = -\frac{1}{k}\frac{\partial^2 \psi}{\partial x^2}\Delta x,$$

where we have multiplied both sides by ∂x and switched back to the Δ notation. Or equivalently, we have multiplied both sides by Δx, and then used the relation $\Delta\psi_P = (\partial\psi_P / \partial x)\Delta x$. Note that the second derivative $\partial^2\psi / \partial x^2$ appears here. From Eq. a *constant* value of $\partial\psi / \partial x$ corresponds to a constant excess pressure $\psi_P(x)$; the tube of air just stretches uniformly if the pressure is the same everywhere. So to obtain a varying value of ψ_P, we need a varying value of $\partial^2\psi / \partial x^2$. That is, we need a nonzero value of $\partial^2\psi / \partial x^2$.

4. The $F = ma$ equation: We can now write down the $F = ma$ equation for a given volume of air. The net rightward force on the tube of air in the second picture in figure is the

cross-sectional area times the difference in the pressure at the ends. So we have, recalling the definition of $\Delta\psi_p$,

$$\mathrm{F}_{\text{net}} = \mathrm{A}\left(P(x) - p(x+\Delta x)\right) = A\left(-\Delta\psi_p\right).$$

Using the expression for $\Delta\psi_p$ we found in Eq. the *F*net = *ma* equation on a given tube of air is (with p being the mass density).

$$-A\Delta\psi_p = (\rho V)\frac{\partial^2\psi}{\partial t^2}$$

$$\Rightarrow -A\left(-\frac{1}{k}\frac{\partial^2\psi}{\partial x^2}\Delta x\right) = \rho(A\Delta x)\frac{\partial^2\psi}{\partial t^2}.$$

Sound Waves

Canceling the common factor of $A\Delta x$ and using $k = 1/\gamma p0$ yields

$$\frac{\partial^2\psi}{\partial t^2} = \frac{\gamma p_0}{\rho}\cdot\frac{\partial^2\psi}{\partial x^2}$$

This is the desired wave equation for sound waves in air. We see that $\gamma p_0 / p$ is the coefficient that replaces the E/μ coefficient for the longitudinal spring/mass system, or the T/μ coefficient for the transverse string.

The solutions to the wave equation are the usual exponentials,

$$\psi(x,t) = Ae^{i(\pm kx \pm \omega t)},$$

where ω and k satisfy $\omega/k = \sqrt{\gamma p_0/\rho}$. And since ω/k is the speed c of the wave, we have $c = \sqrt{\gamma p_0/\rho}$. As always, the speed is simply the square root of the factor on the righthand side. As a double check on the units, we have

$$\frac{\gamma p_0}{\rho} = \frac{(\text{unitless})\cdot(\text{force/ area})}{\text{mass/ volume}} = \frac{\left(\text{kg m/ s}^2\right)/\text{m}^2}{\text{kg/ m}^3} = \frac{\text{m}^2}{\text{s}^2},$$

which has the correct units of velocity squared. What is the numerical value of the speed of sound waves? For air at sea level, the pressure is $p_0 \approx\sim 10^5$ kg/ms² and the density is $p_0 \approx 1.3$ kg/m³ . So we have

$$c = \sqrt{\frac{\gamma p_0}{\rho}} \approx \sqrt{\frac{(7/5)\left(10^5\text{ kg/ ms}^2\right)}{1.3\text{kg/ m}^3}} \approx 330\text{ m/ s}.$$

This decreases with p, which makes sense because the larger the density, the more inertia the air has, so the harder it is to accelerate it. The speed increases with p_0. This follows from the fact that if po is large, then the compressibility k is small (meaning the gas is not easily compressed). So for a given value of $\partial^2\psi/\partial x^2$, the force on the left side of Eq. is large, which implies large accelerations.

The two partial derivatives in Eq. come about in the usual way. The second time derivative comes from the "a" in $F = ma$, and the second space derivative comes from the fact that it is the difference in the first derivatives that gives the net force. Eq. tells us that the first space derivative of the displacement gives a measure of the force at a given location (just as with the spring/mass system, the first space derivative told us how much the springs were stretched, which in turn gave the force). The difference in the force at the two ends is therefore proportional to the second space derivative (again as it was with the spring/mass system).

As with the other wave equations we have encountered thus far in this book, the speed of sound waves is independent of w and k. So all frequencies travel at the same speed. This is fortunate, because if it weren't true, then a music concert would sound like a complete mess.

Speed of Sound

The speed of "sound" is actually the speed of transmission of a small disturbance through a medium. Sound itself is a sensation created in the human brain in response to sensory inputs from the inner ear.

Disturbances are transmitted through a gas as a result of collisions between the randomly moving molecules in the gas. The transmission of a small disturbance through a gas is an isentropic process. The conditions in the gas are the same before and after the disturbance passes through. Because the speed of transmission depends on molecular collisions, the speed of sound depends on the state of the gas. The speed of sound is a constant within a given gas and the value of the constant depends on the type of gas (air, pure oxygen, carbon dioxide, etc.) and the temperature of the gas. An analysis based on conservation of mass and momentum shows that the speed of sound a is equal to the square root of the ratio of specific heats g times the gas constant R times the temperature T.

a = sqrt [g * R * T]

Notice that the temperature must be specified on an absolute scale (Kelvin or Rankine). The dependence on the type of gas is included in the gas constant R. which equals the universal gas constant divided by the molecular weight of the gas, and the ratio of specific heats.

The speed of sound in air depends on the type of gas and the temperature of the gas. On Earth, the atmosphere is composed of mostly diatomic nitrogen and oxygen, and the temperature depends on the altitude in a rather complex way. Scientists and engineers have created a mathematical model of the atmosphere to help them account for the changing effects of temperature with altitude. Mars also has an atmosphere composed of mostly carbon dioxide. There is a similar mathematical model of the Martian atmosphere.

Measuring the Speed of Sound

Once it was recognized that sound is in fact a wave, measurement of the speed of sound became a serious goal. The French scientist and philosopher Pierre Gassendi made the earliest known attempt at measuring the speed of sound in air. Assuming correctly that the speed of light is effectively infinite compared with the speed of sound, Gassendi measured the time difference between spotting the flash of a gun and hearing its report over a long distance on a still day. Although the value he obtained was too high—about 478.4 metres per second (1,569.6 feet per second)—he correctly concluded that the speed of sound is independent of frequency. In the 1650s, Italian physicists Giovanni Alfonso Borelli and Vincenzo Viviani obtained the much better value of 350 metres per second using the same technique. Their compatriot G.L. Bianconi demonstrated in 1740 that the speed of sound in air increases with temperature. The earliest precise experimental value for the speed of sound, obtained at the Academy of Sciences in Paris in 1738, was 332 metres per second—incredibly close to the presently accepted value, considering the rudimentary nature of the measuring tools of the day. A more recent value for the speed of sound, 331.45 metres per second (1,087.4 feet per second), was obtained in 1942; it was amended in 1986 to 331.29 metres per second at 0° C (1,086.9 feet per second at 32° F).

The speed of sound in water was first measured by Daniel Colladon, a Swiss physicist, in 1826. Strangely enough, his primary interest was not in measuring the speed of sound in water but in calculating water's compressibility—a theoretical relationship between the speed of sound in a material and the material's compressibility having been established previously. Colladon came up with a speed of 1,435 metres per second at 8° C; the presently accepted value interpolated at that temperature is about 1,439 metres per second.

Two approaches were employed to determine the velocity of sound in solids. In 1808 Jean-Baptiste Biot, a French physicist, conducted direct measurements of the speed of sound in 1,000 metres of iron pipe by comparing it with the speed of sound in air. A better measurement had earlier been carried out by a German, Ernst Florenz Friedrich Chladni, using analysis of the nodal pattern in standing-wave vibrations in long rods.

Basic Concepts

The transmission of sound can be illustrated by using a model consisting of an array of spherical objects interconnected by springs.

In real material terms, the spheres represent the material's molecules and the springs represent the bonds between them. Sound passes through the system by compressing and expanding the springs, transmitting the acoustic energy to neighboring spheres. This helps transmit the energy in-turn to the neighboring sphere's springs (bonds), and so on.

The speed of sound through the model depends on the stiffness/rigidity of the springs, and the mass of the spheres. As long as the spacing of the spheres remains constant, stiffer springs/bonds transmit energy quicker, while larger spheres transmit the energy slower.

In a real material, the stiffness of the springs is known as the "elastic modulus", and the mass corresponds to the material density. Given that all other things being equal (ceteris paribus), sound will travel slower in spongy materials, and faster in stiffer ones. Effects like dispersion and reflection can also be understood using this model.

For instance, sound will travel 1.59 times faster in nickel than in bronze, due to the greater stiffness of nickel at about the same density. Similarly, sound travels about 1.41 times faster in light hydrogen (protium) gas than in heavy hydrogen (deuterium) gas, since deuterium has similar properties but twice the density. At the same time, "compression-type" sound will travel faster in solids than in liquids, and faster in liquids than in gases, because the solids are more difficult to compress than liquids, while liquids in turn are more difficult to compress than gases.

Some textbooks mistakenly state that the speed of sound increases with increasing density. This is usually illustrated by presenting data for three materials, such as air, water and steel, which also have vastly different levels compressibilities which more than make up for the density differences. An illustrative example of the two effects is that sound travels only 4.3 times faster in water than air, despite enormous differences in compressibility of the two media. The reason is that the larger density of water, which works to *slow* sound in water relative to air, nearly makes up for the compressibility differences in the two media.

A practical example can be observed in Edinburgh, when the "One o' Clock Gun" is fired at the eastern end of Edinburgh Castle. Standing at the base of the western end of the Castle Rock, the sound of the Gun can be heard through the rock, slightly before it arrives by the air route, partly delayed by the slightly longer route. It is particularly effective if a multi-gun salute such as for "The Queen's Birthday" is being fired.

Compression and Shear Waves

In a gas or liquid, sound consists of compression waves. In solids, waves propagate as two different types. A longitudinal wave is associated with compression and decompression in the direction of travel, and is the same process in gases and liquids, with an analogous compression-type wave in solids. Only compression waves are supported in gases and liquids. An additional type of wave, the transverse wave, also called a shear wave, occurs only in solids because only solids support elastic deformations. It is due to elastic deformation of the medium perpendicular to the direction of wave travel; the direction of shear-deformation is called the "polarization" of this type of wave. In general, transverse waves occur as a pair of orthogonal polarizations.

These different waves (compression waves and the different polarizations of shear waves) may have different speeds at the same frequency. Therefore, they arrive at an observer at different times, an extreme example being an earthquake, where sharp compression waves arrive first, and rocking transverse waves seconds later.

The speed of a compression wave in fluid is determined by the medium's compressibility and density. In solids, the compression waves are analogous to those in fluids, depending on compressibility and density, but with the additional factor of shear modulus which affects compression waves due to off-axis elastic energies which are able to influence effective tension and relaxation in a compression. The speed of shear waves, which can occur only in solids, is determined simply by the solid material's shear modulus and density.

Equations

The speed of sound in mathematical notation is conventionally represented by c, from the Latin *celeritas* meaning "velocity".

For fluids in general, the speed of sound c is given by the Newton–Laplace equation:

$$c = \sqrt{\frac{K_s}{\rho}},$$

where

- K_s is a coefficient of stiffness, the isentropic bulk modulus (or the modulus of bulk elasticity for gases);
- ρ is the density.

Thus the speed of sound increases with the stiffness (the resistance of an elastic body to deformation by an applied force) of the material, and decreases with increase in density. For ideal gases the bulk modulus *K* is simply the gas pressure multiplied by the dimensionless adiabatic index, which is about 1.4 for air under normal conditions of pressure and temperature.

For general equations of state, if classical mechanics is used, the speed of sound c is given by

$$c = \sqrt{\left(\frac{\partial p}{\partial \rho}\right)_s},$$

where

- *p* is the pressure;
- ρ is the density and the derivative is taken isentropically, that is, at constant entropy *s*.

If relativistic effects are important, the speed of sound is calculated from the relativistic Euler equations.

In a non-dispersive medium, the speed of sound is independent of sound frequency, so the speeds of energy transport and sound propagation are the same for all frequencies. Air, a mixture of oxygen and nitrogen, constitutes a non-dispersive medium. However, air does contain a small amount of CO_2 which *is* a dispersive medium, and causes dispersion to air at ultrasonic frequencies (> 28 kHz).

In a dispersive medium, the speed of sound is a function of sound frequency, through the dispersion relation. Each frequency component propagates at its own speed, called the phase velocity, while the energy of the disturbance propagates at the group velocity. The same phenomenon occurs with light waves.

Dependence on the Properties of the Medium

The speed of sound is variable and depends on the properties of the substance through which the wave is travelling. In solids, the speed of transverse (or shear) waves depends on the shear deformation under shear stress (called the shear modulus), and the density of the medium. Longitudinal (or compression) waves in solids depend on the same two factors with the addition of a dependence on compressibility.

In fluids, only the medium's compressibility and density are the important factors, since fluids do not transmit shear stresses. In heterogeneous fluids, such as a liquid filled with gas bubbles, the density of the liquid and the compressibility of the gas affect the speed of sound in an additive manner, as demonstrated in the hot chocolate effect.

In gases, adiabatic compressibility is directly related to pressure through the heat capacity ratio (adiabatic index), while pressure and density are inversely related to the temperature and molecular weight, thus making only the completely independent properties of *temperature and molecular structure* important (heat capacity ratio may be determined by temperature and molecular structure, but simple molecular weight is not sufficient to determine it).

In low molecular weight gases such as helium, sound propagates faster as compared to heavier gases such as xenon. For monatomic gases, the speed of sound is about 75% of the mean speed that the atoms move in that gas.

For a given ideal gas the molecular composition is fixed, and thus the speed of sound depends only on its temperature. At a constant temperature, the gas pressure has no effect on the speed of sound, since the density will increase, and since pressure and density (also proportional to pressure) have equal but opposite effects on the speed of sound, and the two contributions cancel out exactly. In a similar way, compression waves in solids depend both on compressibility and density—just as in liquids—but in gases the density contributes to the compressibility in such a way that some part of each attribute factors out, leaving only a dependence on temperature, molecular weight, and heat capacity ratio which can be independently derived from temperature and molecular composition. Thus, for a single given gas (assuming the molecular weight does not change) and over a small temperature range (for which the heat capacity is relatively constant), the speed of sound becomes dependent on only the temperature of the gas.

In non-ideal gas behavior regimen, for which the van der Waals gas equation would be used, the proportionality is not exact, and there is a slight dependence of sound velocity on the gas pressure.

Humidity has a small but measurable effect on the speed of sound (causing it to increase by about 0.1%–0.6%), because oxygen and nitrogen molecules of the air are replaced by lighter molecules of water. This is a simple mixing effect.

Altitude Variation and Implications for Atmospheric Acoustics

In the Earth's atmosphere, the chief factor affecting the speed of sound is the temperature. For a given ideal gas with constant heat capacity and composition, the speed of sound is dependent *solely* upon temperature. In such an ideal case, the effects of decreased density and decreased pressure of altitude cancel each other out, save for the residual effect of temperature.

Since temperature (and thus the speed of sound) decreases with increasing altitude up to 11 km, sound is refracted upward, away from listeners on the ground, creating an acoustic shadow at some distance from the source. The decrease of the speed of sound with height is referred to as a negative sound speed gradient.

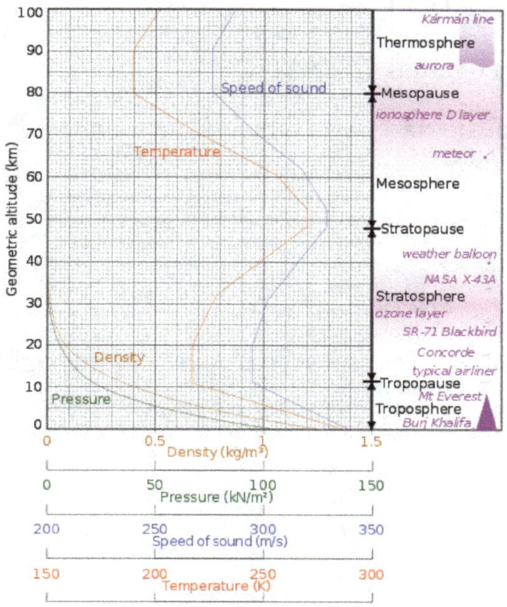

Density and pressure decrease smoothly with altitude, but temperature (red) does not. The speed of sound (blue) depends only on the complicated temperature variation at altitude and can be calculated from it

However, there are variations in this trend above 11 km. In particular, in the stratosphere above about 20 km, the speed of sound increases with height, due to an increase in temperature from heating within the ozone layer. This produces a positive speed of sound gradient in this region. Still another region of positive gradient occurs at very high altitudes, in the aptly-named thermosphere above 90 km.

Practical Formula for Dry Air

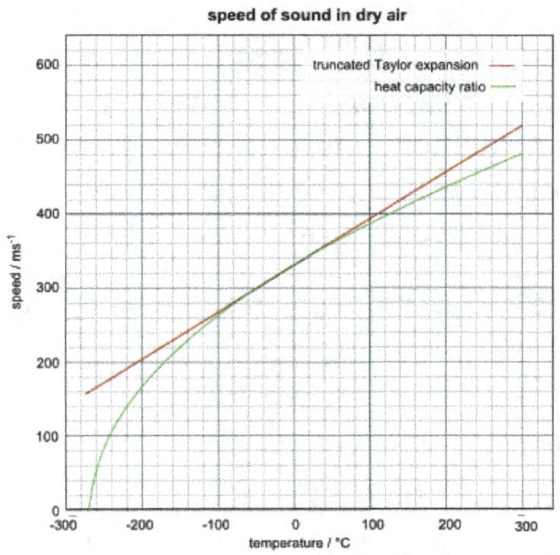

Approximation of the speed of sound in dry air based on the heat capacity ratio (in green) against the truncated Taylor expansion (in red)

The approximate speed of sound in dry (0% humidity) air, in meters per second, at temperatures near 0 °C, can be calculated from

$$c_{\text{air}} = (331.3 + 0.606 \cdot \vartheta) \quad \text{m/s},$$

where ϑ is the temperature in degrees Celsius (°C).

This equation is derived from the first two terms of the Taylor expansion of the following more accurate equation:

$$c_{air} = 331.3\sqrt{1 + \frac{\vartheta}{273.15}} \quad \text{m/s}.$$

Dividing the first part, and multiplying the second part, on the right hand side, by $\sqrt{273.15}$ gives the exactly equivalent form

$$c_{air} = 20.05\sqrt{\vartheta + 273.15} \quad \text{m/s}.$$

which can also be written as

$$c_{air} = 20.05\sqrt{T/K} \quad \text{m/s}.$$

where T denotes the thermodynamic temperature.

The value of 331.3 m/s, which represents the speed at 0 °C (or 273.15 K), is based on theoretical (and some measured) values of the heat capacity ratio, γ, as well as on the fact that at 1 atm real air is very well described by the ideal gas approximation. Commonly found values for the speed of sound at 0 °C may vary from 331.2 to 331.6 due to the assumptions made when it is calculated. If ideal gas γ is assumed to be $7/5 = 1.4$ exactly, the 0 °C speed is calculated to be 331.3 m/s, the coefficient used above.

This equation is correct to a much wider temperature range, but still depends on the approximation of heat capacity ratio being independent of temperature, and for this reason will fail, particularly at higher temperatures. It gives good predictions in relatively dry, cold, low pressure conditions, such as the Earth's stratosphere. The equation fails at extremely low pressures and short wavelengths, due to dependence on the assumption that the wavelength of the sound in the gas is much longer than the average mean free path between gas molecule collisions.

A graph comparing results of the two equations is at right, using the slightly different value of 331.5 m/s for the speed of sound at 0 °C.

Details

Speed of Sound in Ideal Gases and Air

For an ideal gas, K (the bulk modulus in equations above, equivalent to C, the coefficient of stiffness in solids) is given by

$$K = \gamma \cdot p,$$

thus, from the Newton–Laplace equation above, the speed of sound in an ideal gas is given by

$$c = \sqrt{\gamma \cdot \frac{p}{\rho}},$$

where

- γ is the adiabatic index also known as the *isentropic expansion factor*. It is the ratio of specific heats of a gas at a constant-pressure to a gas at a constant-volume(C_p / C_v), and arises because a classical sound wave induces an adiabatic compression, in which the heat of the compression does not have enough time to escape the pressure pulse, and thus contributes to the pressure induced by the compression;
- p is the pressure;
- ρ is the density.

Using the ideal gas law to replace p with nRT/V, and replacing ρ with nM/V, the equation for an ideal gas becomes

$$c_{ideal} = \sqrt{\gamma \cdot \frac{p}{\rho}} = \sqrt{\frac{\gamma \cdot R \cdot T}{M}} = \sqrt{\frac{\gamma \cdot k \cdot T}{m}},$$

where

- c_{ideal} is the speed of sound in an ideal gas;
- R (approximately 8.314,5 J · mol^{-1} · K^{-1}) is the molar gas constant(universal gas constant);
- k is the Boltzmann constant;
- γ (gamma) is the adiabatic index. At room temperature, where thermal energy is fully partitioned into rotation (rotations are fully excited) but quantum effects prevent excitation of vibrational modes, the value is 7/5 = 1.400 for diatomic molecules, according to kinetic theory. Gamma is actually experimentally measured over a range from 1.399,1 to 1.403 at 0 °C, for air. Gamma is exactly 5/3 = 1.6667 for monatomic gases such as noble gases;
- T is the absolute temperature;
- M is the molar mass of the gas. The mean molar mass for dry air is about 0.028,964,5 kg/mol;
- n is the number of moles;
- m is the mass of a single molecule.

This equation applies only when the sound wave is a small perturbation on the ambient condition, and the certain other noted conditions are fulfilled, as noted below. Calculated values for c_{air} have been found to vary slightly from experimentally determined values.

Newton famously considered the speed of sound before most of the development of thermodynamics and so incorrectly used isothermal calculations instead of adiabatic. His result was missing the factor of γ but was otherwise correct.

Numerical substitution of the above values gives the ideal gas approximation of sound velocity for gases, which is accurate at relatively low gas pressures and densities (for air, this includes standard Earth sea-level conditions). Also, for diatomic gases the use of γ = 1.4000 requires that the gas exists in a temperature range high enough that rotational heat capacity is fully excited (i.e., molecular rotation is fully used as a heat energy "partition" or reservoir); but at the same time the temperature must be low enough that molecular vibrational modes contribute no heat capacity (i.e., insignificant heat goes into vibration, as all vibrational quantum modes above the minimum-energy-mode, have energies too high to be populated by a significant number of molecules at this temperature). For air, these conditions are fulfilled at room temperature, and also temperatures considerably below room temperature.

For air, we introduce the shorthand

$$R_* = R / M_{air}.$$

In addition, we switch to the Celsius temperature $\vartheta = T - 273.15$, which is useful to calculate air speed in the region near 0°C (about 273 kelvin). Then, for dry air,

$$c_{ideal} = \sqrt{\gamma \cdot R_* \cdot T} = \sqrt{\gamma \cdot R_* \cdot (\vartheta + 273.15)},$$

$$c_{ideal} = \sqrt{\gamma \cdot R_* \cdot 273.15} \cdot \sqrt{1 + \frac{\vartheta}{273.15}},$$

where ϑ (theta) is the temperature in degrees Celsius(°C).

Substituting numerical values

$$R = 8.314510\, J/(mol \cdot K)$$

for the molar gas constant in J/mole/Kelvin, and

$$M_{air} = 0.0289645\, kg/mol$$

for the mean molar mass of air, in kg; and using the ideal diatomic gas value of γ = 1.4000, we have

$$c_{air} = 331.3 \sqrt{1 + \frac{\vartheta}{273.15}} \;\; m/s.$$

Finally, Taylor expansion of the remaining square root in ϑ yields

$$c_{air} = 331.3\, (1 + \frac{\vartheta}{2 \cdot 273.15}) \;\; m/s,$$

$$c_{air} = (331.3 + 0.606 \cdot \vartheta) \;\; m/s.$$

The above derivation includes the first two equations given in the "Practical formula for dry air" section above.

Effects due to Wind Shear

The speed of sound varies with temperature. Since temperature and sound velocity normally decrease with increasing altitude, sound is refracted upward, away from listeners on the ground, creating an acoustic shadow at some distance from the source. Wind shear of 4 m/(s · km) can produce refraction equal to a typical temperature lapse rate of 7.5 °C/km. Higher values of wind gradient will refract sound downward toward the surface in the downwind direction, eliminating the acoustic shadow on the downwind side. This will increase the audibility of sounds downwind. This downwind refraction effect occurs because there is a wind gradient; the sound is not being carried along by the wind.

For sound propagation, the exponential variation of wind speed with height can be defined as follows:

$$U(h) = U(0)h^{\zeta},$$

$$\frac{U}{\mathrm{d}H}(h) = \zeta\frac{U(h)}{h},$$

where

- $U(h)$ is the speed of the wind at height h;
- ζ is the exponential coefficient based on ground surface roughness, typically between 0.08 and 0.52;
- $\mathrm{d}U/\mathrm{d}H(h)$ is the expected wind gradient at height h.

Tables

In the standard atmosphere:

- T_0 is 273.15 K (= 0 °C = 32 °F), giving a theoretical value of 331.3 m/s (= 1086.9 ft/s = 1193 km/h = 741.1 mph = 644.0 kn). Values ranging from 331.3 to 331.6 m/s may be found in reference literature, however;
- T_{20} is 293.15 K (= 20 °C = 68 °F), giving a value of 343.2 m/s (= 1126.0 ft/s = 1236 km/h = 767.8 mph = 667.2 kn);
- T_{25} is 298.15 K (= 25 °C = 77 °F), giving a value of 346.1 m/s (= 1135.6 ft/s = 1246 km/h = 774.3 mph = 672.8 kn).

In fact, assuming an ideal gas, the speed of sound c depends on temperature only, not on the pressure or density (since these change in lockstep for a given temperature and cancel out). Air is almost an ideal gas. The temperature of the air varies with altitude, giving the following variations in the speed of sound using the standard atmosphere—*actual conditions may vary.*

Effect of temperature on properties of air			
Temperature T (°C)	**Speed of sound c (m/s)**	**Density of air ρ (kg/m³)**	**Characteristic specific acoustic impedance z_0 (Pa·s/m)**
35	351.88	1.1455	403.2
30	349.02	1.1644	406.5
25	346.13	1.1839	409.4
20	343.21	1.2041	413.3
15	340.27	1.2250	416.9
10	337.31	1.2466	420.5
5	334.32	1.2690	424.3
0	331.30	1.2922	428.0
−5	328.25	1.3163	432.1
−10	325.18	1.3413	436.1
−15	322.07	1.3673	440.3
−20	318.94	1.3943	444.6
−25	315.77	1.4224	449.1

Given normal atmospheric conditions, the temperature, and thus speed of sound, varies with altitude:

Altitude	Temperature	m/s	km/h	mph	kn
Sea level	15 °C (59 °F)	340	1,225	761	661
11,000 m–20,000 m (Cruising altitude of commercial jets, and first supersonic flight)	−57 °C (−70 °F)	295	1,062	660	573
29,000 m (Flight of X-43A)	−48 °C (−53 °F)	301	1,083	673	585

Effect of Frequency and Gas Composition

General Physical Considerations

The medium in which a sound wave is travelling does not always respond adiabatically, and as a result the speed of sound can vary with frequency.

The limitations of the concept of speed of sound due to extreme attenuation are also of concern. The attenuation which exists at sea level for high frequencies applies to successively lower frequencies as atmospheric pressure decreases, or as the mean free path increases. For this reason, the concept of speed of sound (except for frequencies approaching zero) progressively loses its range of applicability at high altitudes. The standard equations for the speed of sound apply with reasonable accuracy only to situations in which the wavelength of the soundwave is considerably longer than the mean free path of molecules in a gas.

The molecular composition of the gas contributes both as the mass (M) of the molecules, and their heat capacities, and so both have an influence on speed of sound. In general, at the same molecular mass, monatomic gases have slightly higher speed of sound (over 9% higher) because they have a higher γ (5/3 = 1.66...) than diatomics do (7/5 = 1.4). Thus, at the same molecular mass, the speed of sound of a monatomic gas goes up by a factor of

$$\frac{c_{gas,monatomic}}{c_{gas,diatomic}} = \sqrt{\frac{5/3}{7/5}} = \sqrt{\frac{25}{21}} = 1.091\ldots$$

This gives the 9% difference, and would be a typical ratio for speeds of sound at room temperature in helium vs. deuterium, each with a molecular weight of 4. Sound travels faster in helium than deuterium because adiabatic compression heats helium more, since the helium molecules can store heat energy from compression only in translation, but not rotation. Thus helium molecules (monatomic molecules) travel faster in a sound wave and transmit sound faster. (Sound travels at about 70% of the mean molecular speed in gases; the figure is 75% in monatomic gases and 68% in diatomic gases).

Note that in this example we have assumed that temperature is low enough that heat capacities are not influenced by molecular vibration. However, vibrational modes simply cause gammas which decrease toward 1, since vibration modes in a polyatomic gas gives the gas additional ways to store heat which do not affect temperature, and thus do not affect molecular velocity and sound velocity. Thus, the effect of higher temperatures and vibrational heat capacity acts to increase the difference between the speed of sound in monatomic vs. polyatomic molecules, with the speed remaining greater in monatomics.

Practical Application to Air

By far the most important factor influencing the speed of sound in air is temperature. The speed is proportional to the square root of the absolute temperature, giving an increase of about 0.6 m/s per degree Celsius. For this reason, the pitch of a musical wind instrument increases as its temperature increases.

The speed of sound is raised by humidity but decreased by carbon dioxide. The difference between 0% and 100% humidity is about 1.5 m/s at standard pressure and temperature, but the size of the humidity effect increases dramatically with temperature. The carbon dioxide content of air is not fixed, due to both carbon pollution and human breath (e.g., in the air blown through wind instruments).

The dependence on frequency and pressure are normally insignificant in practical applications. In dry air, the speed of sound increases by about 0.1 m/s as the frequency rises from 10 Hz to 100 Hz. For audible frequencies above 100 Hz it is relatively constant. Standard values of the speed of sound are quoted in the limit of low frequencies, where the wavelength is large compared to the mean free path.

Mach Number

Mach number, a useful quantity in aerodynamics, is the ratio of air speed to the local speed of sound. At altitude, for reasons explained, Mach number is a function of temperature. Aircraft flight instruments, however, operate using pressure differential to compute Mach number, not temperature. The assumption is that a particular pressure represents a particular altitude and, therefore, a standard temperature. Aircraft flight instruments need to operate this way because the stagnation pressure sensed by a Pitot tube is dependent on altitude as well as speed.

Experimental Methods

A range of different methods exist for the measurement of sound in air.

The earliest reasonably accurate estimate of the speed of sound in air was made by William Derham, and acknowledged by Isaac Newton. Derham had a telescope at the top of the tower of the Church of St Laurence in Upminster, England. On a calm day, a synchronized pocket watch would be given to an assistant who would fire a shotgun at a pre-determined time from a conspicuous point some miles away, across the countryside. This could be confirmed by telescope. He then measured the interval between seeing gunsmoke and arrival of the sound using a half-second pendulum. The distance from where the gun was fired was found by triangulation, and simple division (distance/time) provided velocity. Lastly, by making many observations, using a range of different distances, the inaccuracy of the half-second pendulum could be averaged out, giving his final estimate of the speed of sound. Modern stopwatches enable this method to be used today over distances as short as 200–400 meters, and not needing something as loud as a shotgun.

Single-shot Timing Methods

The simplest concept is the measurement made using two microphones and a fast recording device such as a digital storage scope. This method uses the following idea.

If a sound source and two microphones are arranged in a straight line, with the sound source at one end, then the following can be measured:

1. The distance between the microphones (x), called microphone basis.
2. The time of arrival between the signals (delay) reaching the different microphones (t).

Then $v = x/t$.

Other Methods

In these methods the time measurement has been replaced by a measurement of the inverse of time (frequency).

Kundt's tube is an example of an experiment which can be used to measure the speed of sound in a small volume. It has the advantage of being able to measure the speed of sound in any gas. This method uses a powder to make the nodes and antinodes visible to the human eye. This is an example of a compact experimental setup.

A tuning fork can be held near the mouth of a long pipe which is dipping into a barrel of water. In this system it is the case that the pipe can be brought to resonance if the length of the air column in the pipe is equal to $(1 + 2n)\lambda/4$ where n is an integer. As the antinodal point for the pipe at the open end is slightly outside the mouth of the pipe it is best to find two or more points of resonance and then measure half a wavelength between these.

Here it is the case that $v = f\lambda$.

High-precision Measurements in Air

The effect from impurities can be significant when making high-precision measurements. Chemical desiccants can be used to dry the air, but will in turn contaminate the sample. The air can be dried cryogenically, but this has the effect of removing the carbon dioxide as well; therefore many high-precision measurements are performed with air free of carbon dioxide rather than with natural air. A 2002 review found that a 1963 measurement by Smith and Harlow using a cylindrical resonator gave "the most probable value of the standard speed of sound to date." The experiment was done with air from which the carbon dioxide had been removed, but the result was then corrected for this effect so as to be applicable to real air. The experiments were done at 30 °C but corrected for temperature in order to report them at 0 °C. The result was 331.45 ± 0.01 m/s for dry air at STP, for frequencies from 93 Hz to 1,500 Hz.

Non-gaseous Media

Speed of sound in Solids

Three-dimensional Solids

In a solid, there is a non-zero stiffness both for volumetric deformations and shear deformations. Hence, it is possible to generate sound waves with different velocities dependent on the deformation mode. Sound waves generating volumetric deformations (compression) and shear deformations (shearing) are called pressure waves (longitudinal waves) and shear waves (transverse waves), respectively. In earthquakes, the corresponding seismic waves are called P-waves (primary waves) and S-waves (secondary waves), respectively. The sound velocities of these two types of waves propagating in a homogeneous 3-dimensional solid are respectively given by

$$c_{solid,p} = \sqrt{\frac{K + \frac{4}{3}G}{\rho}} = \sqrt{\frac{E(1-\nu)}{\rho(1+\nu)(1-2\nu)}},$$

$$c_{solid,s} = \sqrt{\frac{G}{\rho}},$$

where

- K is the bulk modulus of the elastic materials;
- G is the shear modulus of the elastic materials;
- E is the Young's modulus;
- ρ is the density;
- ν is Poisson's ratio.

The last quantity is not an independent one, as $E = 3K(1 - 2\nu)$. Note that the speed of pressure waves depends both on the pressure and shear resistance properties of the material, while the speed of shear waves depends on the shear properties only.

Typically, pressure waves travel faster in materials than do shear waves, and in earthquakes this is the reason that the onset of an earthquake is often preceded by a quick upward-downward shock, before arrival of waves that produce a side-to-side motion. For example, for a typical steel alloy, $K = 170$ GPa, $G = 80$ GPa and $\rho = 7{,}700$ kg/m³, yielding a compressional speed $c_{solic,p}$ of 6,000 m/s. This is in reasonable agreement with $c_{solid,p}$ measured experimentally at 5,930 m/s for a (possibly different) type of steel. The shear speed $c_{solid,s}$ is estimated at 3,200 m/s using the same numbers.

One-dimensional Solids

The speed of sound for pressure waves in stiff materials such as metals is sometimes given for "long rods" of the material in question, in which the speed is easier to measure. In rods where their diameter is shorter than a wavelength, the speed of pure pressure waves may be simplified and is given by:

$$c_{solid,s} = \sqrt{\frac{E}{\rho}},$$

where E is Young's modulus. This is similar to the expression for shear waves, save that Young's modulus replaces the shear modulus. This speed of sound for pressure waves in long rods will always be slightly less than the same speed in homogeneous 3-dimensional solids, and the ratio of the speeds in the two different types of objects depends on Poisson's ratio for the material.

Speed of Sound in Liquids

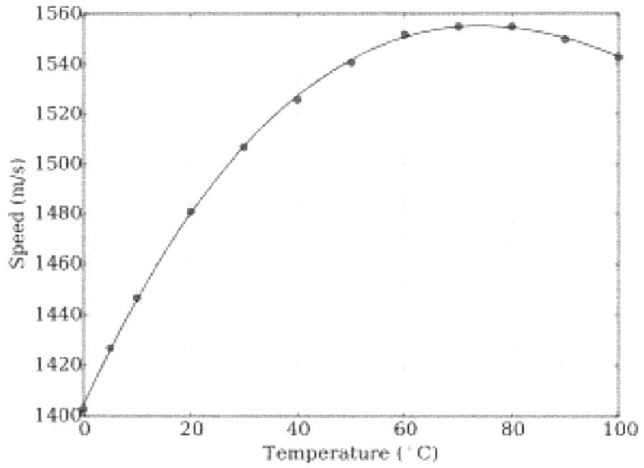

Speed of sound in water vs temperature

In a fluid the only non-zero stiffness is to volumetric deformation (a fluid does not sustain shear forces).

Hence the speed of sound in a fluid is given by

$$c_{fluid} = \sqrt{\frac{K}{\rho}},$$

where K is the bulk modulus of the fluid.

Water

In fresh water, sound travels at about 1481 m/s at 20 °C. Applications of underwater sound can be found in sonar, acoustic communication and acoustical oceanography.

Seawater

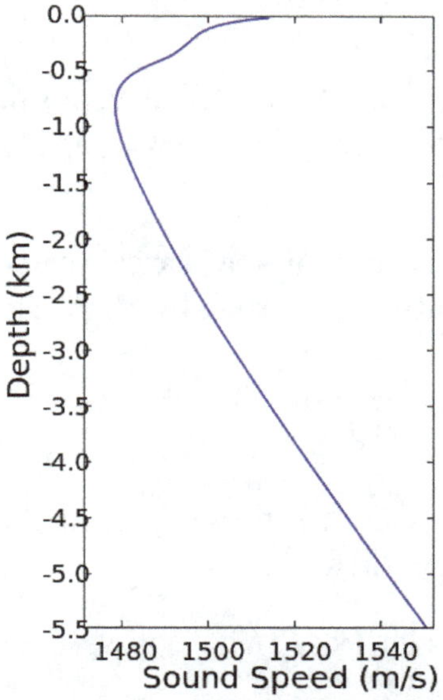

Speed of sound as a function of depth at a position north of Hawaii in the Pacific Ocean derived from the 2005 World Ocean Atlas. The SOFAR channel is centred on the minimum in the speed of sound at ca. 750-m depth

In salt water that is free of air bubbles or suspended sediment, sound travels at about 1500 m/s (1500.235 m/s at 1000 kilopascals, 10 °C and 3% salinity by one method). The speed of sound in seawater depends on pressure (hence depth), temperature (a change of 1 °C ~ 4 m/s), and salinity (a change of 1‰ ~ 1 m/s), and empirical equations have been derived to accurately calculate the speed of sound from these variables. Other factors affecting the speed of sound are minor. Since temperature decreases with depth while pressure and generally salinity increase, the profile of the speed of sound with depth generally shows a characteristic curve which decreases to a minimum at a depth of several hundred meters, then increases again with increasing depth (right).

A simple empirical equation for the speed of sound in sea water with reasonable accuracy for the world's oceans is due to Mackenzie:

$$c(T,S,z) = a_1 + a_2T + a_3T^2 + a_4T^3 + a_5(S-35) + a_6z + a_7z^2 + a_8T(S-35) + a_9Tz^3,$$

where

- T is the temperature in degrees Celsius;
- S is the salinity in parts per thousand;
- z is the depth in meters.

The constants $a_1, a_2, \ldots, a_9$ are

$$a_1 = 1{,}448.96, \quad a_2 = 4.591, \quad a_3 = -5.304\times10{-2},$$
$$a_4 = 2.374\times10^{-4}, \quad a_5 = 1.340, \quad a_6 = 1.630\times10{-2},$$
$$a_7 = 1.675\times10^{-7}, \quad a_8 = -1.025\times10{-2}, \quad a_9 = -7.139\times10{-13},$$

with check value 1550.744 m/s for T = 25 °C, S = 35 parts per thousand, z = 1,000 m. This equation has a standard error of 0.070 m/s for salinity between 25 and 40 ppt. Speed of Sound in Sea-Water for an online calculator.

Other equations for the speed of sound in sea water are accurate over a wide range of conditions, but are far more complicated, e.g., that by V. A. Del Grosso and the Chen-Millero-Li Equation.

Speed of Sound in Plasma

The speed of sound in a plasma for the common case that the electrons are hotter than the ions (but not too much hotter) is given by the formula

$$c_s = (\gamma Z k T_e / m_i)^{1/2} = 9.79\times10^3 (\gamma Z T_e / \mu)^{1/2}\ m/s,$$

where

- m_i is the ion mass;
- μ is the ratio of ion mass to proton mass $\mu = m_i/m_p$;
- T_e is the electron temperature;
- Z is the charge state;
- k is Boltzmann constant;
- γ is the adiabatic index.

In contrast to a gas, the pressure and the density are provided by separate species, the pressure by the electrons and the density by the ions. The two are coupled through a fluctuating electric field.

Gradients

When sound spreads out evenly in all directions in three dimensions, the intensity drops in proportion to the inverse square of the distance. However, in the ocean there is a layer called the 'deep sound channel' or SOFAR channel which can confine sound waves at a particular depth.

In the SOFAR channel, the speed of sound is lower than that in the layers above and below. Just as light waves will refract towards a region of higher index, sound waves will refract towards a region where their speed is reduced. The result is that sound gets confined in the layer, much the way light can be confined in a sheet of glass or optical fiber. Thus, the sound is confined in essentially two dimensions. In two dimensions the intensity drops in proportion to only the inverse of the distance. This allows waves to travel much further before being undetectably faint.

A similar effect occurs in the atmosphere. Project Mogul successfully used this effect to detect a nuclear explosion at a considerable distance.

Timbre

Timbre, also called timber, quality of auditory sensations produced by the tone of a sound wave.

The timbre of a sound depends on its wave form, which varies with the number of overtones, or harmonics, that are present, their frequencies, and their relative intensities. The illustration shows the wave form that results when pure tones of frequencies 100, 300, and 500 hertz (cycles per second) and relative amplitudes of 10, 5, and 2.5 are synthesized into a complex tone. At the right is the resultant of the three sine curves when their ordinates are added point by point along the time scale. In equation form, the amplitude y of the wave form at any time t would be represented by $y = 10 \sin (2\pi\ 100t) + 5 \sin (2\pi\ 300t) + 2.5 \sin (2\pi\ 500t)$. The timbre of this form would be recognizable and different from others having a fundamental tone of 100 hertz but a different harmonic amplitude.

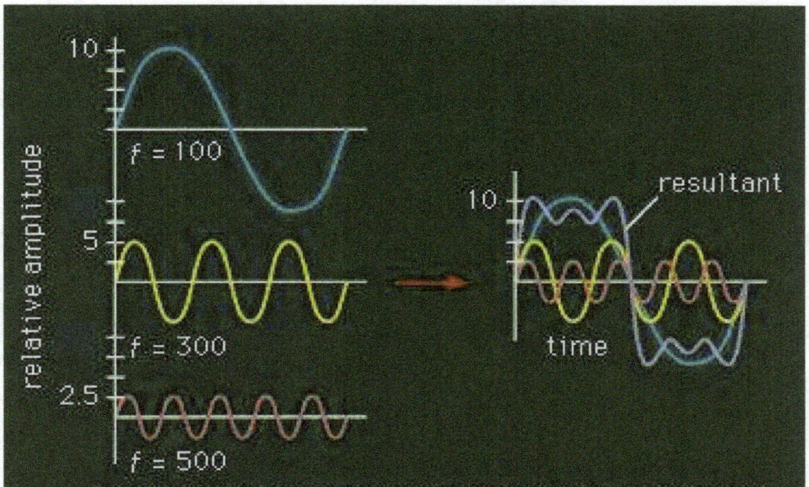

Synthesis of complex tone from three pure tones.

In music timbre is the characteristic tone colour of an instrument or voice, arising from reinforcement by individual singers or instruments of different harmonics, or overtones (q.v.), of a fundamental pitch. Extremely nasal timbre thus stresses different overtones than mellow timbre. The timbre of the tuning fork and of the stopped diapason organ pipe is clear and pure because the sound they produce is almost without overtones. Timbre is determined by an instrument's shape (e.g., the conical or cylindrical pipe of a wind instrument), by the frequency range within which the instrument can produce overtones, and by the envelope of the instrument's sound. The timbre of spoken vowels or of a singing voice is modified by constricting or opening various parts of the vocal tract, such as the lips, tongue, or throat.

Synonyms

Tone quality and *tone color* are synonyms for *timbre*, as well as the "*texture* attributed to a single instrument". However, the word texture can also refer to the type of music, such as multiple, interweaving melody lines versus a singable melody accompanied by subordinate chords. Hermann von Helmholtz used the German *Klangfarbe* (*tone color*), and John Tyndall proposed an English translation, *clangtint*, but both terms were disapproved of by Alexander Ellis, who also discredits *register* and *color* for their pre-existing English meanings. The sound of a musical instrument may be described with words such as *bright*, *dark*, *warm*, *harsh*, and other terms. There are also colors of noise, such as pink and white. In visual representations of sound, timbre corresponds to the shape of the image, while loudness corresponds to brightness; pitch corresponds to the y-shift of the spectrogram.

ASA Definition

The Acoustical Society of America (ASA) Acoustical Terminology definition 12.09 of timbre describes it as "that attribute of auditory sensation which enables a listener to judge that two non-identical sounds, similarly presented and having the same loudness and pitch, are dissimilar", adding, "Timbre depends primarily upon the frequency spectrum, although it also depends upon the sound pressure and the temporal characteristics of the sound".

Attributes

Timbre has been called "the psychoacoustician's multidimensional waste-basket category for everything that cannot be labeled pitch or loudness."

Many commentators have attempted to decompose timbre into component attributes. For example, J. F. Schouten (1968, 42) describes the "elusive attributes of timbre" as "determined by at least five major acoustic parameters", which Robert Erickson (1975, 5) finds, "scaled to the concerns of much contemporary music":

1. Range between tonal and noiselike character
2. Spectral envelope
3. Time envelope in terms of rise, duration, and decay (ADSR, which stands for "attack, decay, sustain, release")
4. Changes both of spectral envelope (formant-glide) and fundamental frequency (micro-intonation)
5. Prefix, or onset of a sound, quite dissimilar to the ensuing lasting vibration

An example of a tonal sound is a musical sound that has a definite pitch, such as pressing a key on a piano; a sound with a noiselike character would be white noise, the sound similar to that produced when a radio is not tuned to a station.

Erickson (1975, 6) gives a table of subjective experiences and related physical phenomena based on Schouten's five attributes:

Subjective	***Objective***
Tonal character, usually pitched	Periodic sound
Noisy, with or without some tonal character, including rustle noise	Noise, including random pulses characterized by the rustle time (the mean interval between pulses)
Coloration	Spectral envelope
Beginning/ending	Physical rise and decay time
Coloration glide or formant glide	Change of spectral envelope
Microintonation	Small change (one up and down) in frequency
Vibrato	Frequency modulation
Tremolo	Amplitude modulation
Attack	Prefix
Final sound	Suffix

Harmonics

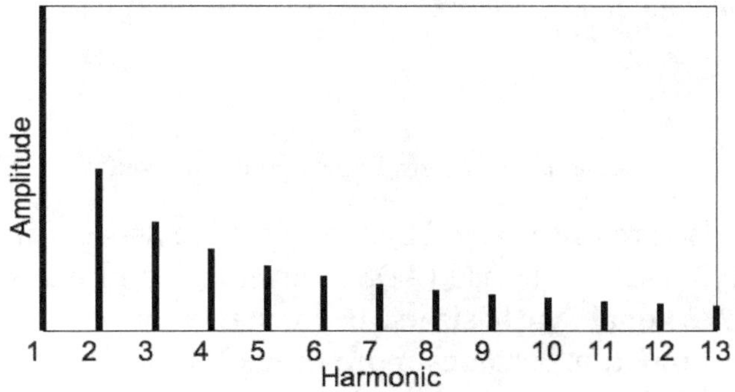

Harmonic spectra

The richness of a sound or note a musical instrument produces is sometimes described in terms of a sum of a number of distinct frequencies. The lowest frequency is called the *fundamental frequency*, and the pitch it produces is used to name the note, but the fundamental frequency is not always the dominant frequency. The dominant frequency is the frequency that is most heard, and it is always a multiple of the fundamental frequency. For example, the dominant frequency for the transverse flute is double the fundamental frequency. Other significant frequencies are called overtones of the fundamental frequency, which may include harmonics and partials. Harmonics are whole number multiples of the fundamental frequency, such as ×2, ×3, ×4, etc. Partials are other overtones. There are also sometimes subharmonics at whole number *divisions* of the fundamental frequency. Most instruments produce harmonic sounds, but many instruments produce partials and inharmonic tones, such as cymbals and other indefinite-pitched instruments.

When the tuning note in an orchestra or concert band is played, the sound is a combination of 440 Hz, 880 Hz, 1320 Hz, 1760 Hz and so on. Each instrument in the orchestra or concert band produces a different combination of these frequencies, as well as harmonics and overtones. The sound waves of the different frequencies overlap and combine, and the balance of these amplitudes is a major factor in the characteristic sound of each instrument.

William Sethares wrote that just intonation and the western equal tempered scale are related to the harmonic spectra/timbre of many western instruments in an analogous way that the inharmonic timbre of the Thai renat (a xylophone-like instrument) is related to the seven-tone near-equal tempered pelog scale in which they are tuned. Similarly, the inharmonic spectra of Balinese metallophones combined with harmonic instruments such as the stringed rebab or the voice, are related to the five-note near-equal tempered slendro scale commonly found in Indonesian gamelan music (Sethares 1998, 6, 211, 318).

Envelope

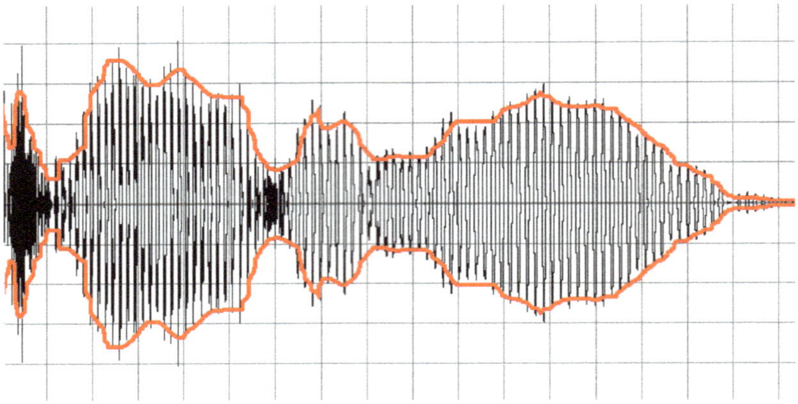

A signal and its envelope marked with red

The timbre of a sound is also greatly affected by the following aspects of its *envelope*: attack time and characteristics, decay, sustain, release (ADSR envelope) and transients. Thus these are all common controls on professional synthesizers. For instance, if one takes away the attack from the sound of a piano or trumpet, it becomes more difficult to identify the sound correctly, since the sound of the hammer hitting the strings or the first blast of the player's lips on the trumpet mouthpiece are highly characteristic of those instruments. The envelope is the overall amplitude structure of a sound, so called because the sound just "fits" inside its envelope: what this means should be clear from a time-domain display of almost any interesting sound, zoomed out enough that the entire waveform is visible.

In Music History

Instrumental timbre played an increasing role in the practice of orchestration during the eighteenth and nineteenth centuries. Berlioz (Macdonald 1969, 51) and Wagner (Latham) made significant contributions to its development during the nineteenth century. Debussy, who composed during the last decades of the nineteenth and the first decades of the twentieth centuries, has been credited with elevating further the role of timbre: "To a marked degree the music of Debussy elevates timbre to an unprecedented structural status; already in *Prélude à l'après-midi d'un faune* the *color* of flute and harp functions referentially".

Mahler's approach to orchestration illustrates the increasing role of differentiated timbres in music of the early twentieth century. Norman Del Mar (1980, 48) describes the following passage from the Scherzo movement of his Sixth Symphony, as "a seven-bar link to the trio consisting of an extension in diminuendo of the repeated As though now rising in a succession of piled octaves which moreover leap-frog with Cs added to the As. The lower octaves then drop away and only the Cs remain so as to dovetail with the first oboe phrase of the trio." During these bars, Mahler passes the repeated notes through a gamut of instrumental colours, mixed and single: starting with horns and pizzicato strings, progressing through trumpet, clarinet, flute, piccolo and finally, oboe:

Mahler, Symphony No. 6, Scherzo, Figure bars 5–12

In rock music from the late 1960s to the 2000s, the timbre of specific sounds is important to a song. For example, in heavy metal music, the sonic impact of the heavily amplified, heavily distorted power chord played on electric guitar through very loud guitar amplifiers and rows of speaker cabinets is an essential part of the style's musical identity. You could not take the heavily amplified electric guitar part and substitute it with the same notes played on a grand piano or pipe organ.

Psychoacoustic Evidence

Often, listeners can identify an instrument, even at different pitches and loudness, in different environments, and with different players. In the case of the clarinet, acoustic analysis shows waveforms irregular enough to suggest three instruments rather than one. David Luce (1963, 16) suggests that this implies that "Certain strong regularities in the acoustic waveform of the above instruments must exist which are invariant with respect to the above variables." However, Robert Erickson argues that there are few regularities and they do not explain our "powers of recognition and identification." He suggests borrowing the concept of subjective constancy from studies of vision and visual perception.

Psychoacoustic experiments from the 1960s onwards tried to elucidate the nature of timbre. One method involves playing pairs of sounds to listeners, then using a multidimensional scaling algorithm to aggregate their dissimilarity judgments into a timbre space. The most consistent outcomes from such experiments are that brightness or spectral energy distribution (Grey 1977), and the *bite*, or rate and synchronicity (Wessel 1979) and rise time (Lakatos 2000), of the attack are important factors.

Tristimulus Timbre Model

The concept of tristimulus originates in the world of color, describing the way three primary colors can be mixed together to create a given color. By analogy, the musical tristimulus measures the mixture of harmonics in a given sound, grouped into three sections. The first tristimulus measures the relative weight of the first harmonic; the second tristimulus measures the relative weight of the second, third, and fourth harmonics taken together; and the third tristimulus measures the relative weight of all the remaining harmonics:

$$T_1 = \frac{a_1}{\sum_{h=1}^{H} a_h} \qquad T_2 = \frac{a_2 + a_3 + a_4}{\sum_{h=1}^{H} a_h} \qquad T_3 = \frac{\sum_{h=5}^{H} a_h}{\sum_{h=1}^{H} a_h}$$

Pitch

A sound wave, like any other wave, is introduced into a medium by a vibrating object. The vibrating object is the source of the disturbance that moves through the medium. The vibrating object that creates the disturbance could be the vocal cords of a person, the vibrating string and sound board of a guitar or violin, the vibrating tines of a tuning fork, or the vibrating diaphragm of a radio speaker. Regardless of what vibrating object is creating the sound wave, the particles of the medium through which the sound moves is vibrating in a back and forth motion at a given frequency. The frequency of a wave refers to how often the particles of the medium vibrate when a wave passes through the medium. The frequency of a wave is measured as the number of complete back-and-forth vibrations of a particle of the medium per unit of time. If a particle of air undergoes 1000 longitudinal vibrations in 2 seconds, then the frequency of the wave would be 500 vibrations per second. A commonly used unit for frequency is the Hertz (abbreviated Hz), where

1 Hertz = 1 vibration/second

As a sound wave moves through a medium, each particle of the medium vibrates at the same frequency. This is sensible since each particle vibrates due to the motion of its nearest neighbor. The first particle of the medium begins vibrating, at say 500 Hz, and begins to set the second particle into vibrational motion at the same frequency of 500 Hz. The second particle begins vibrating at 500 Hz and thus sets the third particle of the medium into vibrational motion at 500 Hz. The process continues throughout the medium; each particle vibrates at the same frequency. And of course the frequency at which each particle vibrates is the same as the frequency of the original source of the sound wave. Subsequently, a guitar string vibrating at 500 Hz will set the air particles in the room vibrating at the same frequency of 500 Hz, which carries a sound signal to the ear of a listener, which is detected as a 500 Hz sound wave.

The back-and-forth vibrational motion of the particles of the medium would not be the only observable phenomenon occurring at a given frequency. Since a sound wave is a pressure wave, a detector could be used to detect oscillations in pressure from a high pressure to a low pressure and back to a high pressure. As the compressions (high pressure) and rarefactions (low pressure) move through the medium, they would reach the detector at a given frequency. For example, a compression would reach the detector 500 times per second if the frequency of the wave were 500 Hz. Similarly, a rarefaction would reach the detector 500 times per second if the frequency of the wave were 500 Hz. The frequency of a sound wave not only refers to the number of back-and-forth vibrations of the particles per unit of time, but also refers to the number of compressions or rarefactions that pass a given point per unit of time. A detector could be used to detect the frequency of these pressure oscillations over a given period of time. The typical output provided by such a detector is a pressure-time plot as shown below.

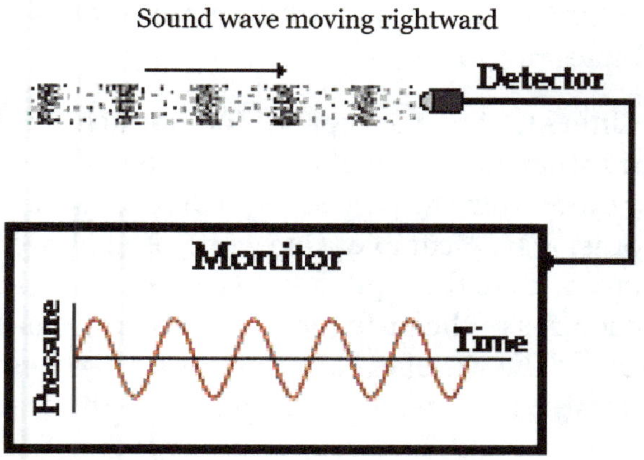

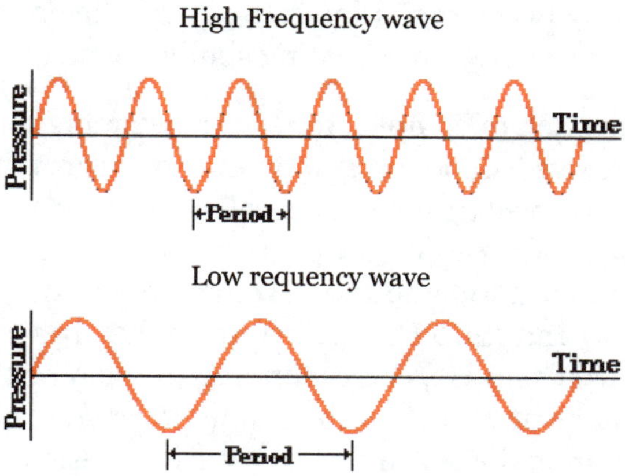

Since a pressure-time plot shows the fluctuations in pressure over time, the period of the sound wave can be found by measuring the time between successive high pressure points (corresponding to the compressions) or the time between successive low pressure points (corresponding to the rarefactions). As discussed in an earlier unit, the frequency is simply the reciprocal of the period. For this reason, a sound wave with a high frequency would correspond to a pressure time plot with a small period - that is, a plot corresponding to a small amount of time between successive high pressure points. Conversely, a sound wave with a low frequency would correspond to a pressure time plot with a large period - that is, a plot corresponding to a large amount of time between successive high pressure points.

Pitch is an auditory sensation in which a listener assigns musical tones to relative positions on a musical scale based primarily on their perception of the frequency of vibration. Pitch is closely related to frequency, but the two are not equivalent. Frequency is an objective, scientific attribute that can be measured. Pitch is each person's *subjective perception* of a sound wave, which cannot be directly measured. However, this does not necessarily mean that most people won't agree on which notes are higher and lower.

Sound waves themselves do not have pitch, but their oscillations can often be characterized in terms of frequency. Pitches are usually associated with, and thus quantified as frequencies in cycles per second, or hertz, by comparing sounds with pure tones, which have periodic, sinusoidal waveforms. Complex and aperiodic sound waves can often be assigned a pitch by this method.

According to the American National Standards Institute, pitch is the auditory attribute of sound according to which sounds can be ordered on a scale from low to high. Since pitch is such a close proxy for frequency, it is almost entirely determined by how quickly the sound wave is making the air vibrate and has almost nothing to do with the intensity, or amplitude, of the wave. That is, "high" pitch means very rapid oscillation, and "low" pitch corresponds to slower oscillation. Despite that, the idiom relating vertical height to sound pitch is shared by most languages. At least in English, it is just one of many deep conceptual metaphors that involve up/down. The exact etymological history of the musical sense of high and low pitch is still unclear. There is evidence that humans do actually perceive that the source of a sound is slightly higher or lower in vertical space when the sound frequency is increased or reduced.

In most cases, the pitch of complex sounds such as speech and musical notes corresponds very nearly to the repetition rate of periodic or nearly-periodic sounds, or to the reciprocal of the time interval between repeating similar events in the sound waveform.

The pitch of complex tones can be ambiguous, meaning that two or more different pitches can be perceived, depending upon the observer. When the actual fundamental frequency can be precisely determined through physical measurement, it may differ from the perceived pitch because of overtones, also known as upper partials, harmonic or otherwise. A complex tone composed of two sine waves of 1000 and 1200 Hz may sometimes be heard as up to three pitches: two spectral pitches at 1000 and 1200 Hz, derived from the physical frequencies of the pure tones, and the combination tone at 200 Hz, corresponding to the repetition rate of the waveform. In a situation like this, the percept at 200 Hz is commonly referred to as the missing fundamental, which is often the greatest common divisor of the frequencies present.

Pitch depends to a lesser degree on the sound pressure level (loudness, volume) of the tone, especially at frequencies below 1,000 Hz and above 2,000 Hz. The pitch of lower tones gets lower as sound pressure increases. For instance, a tone of 200 Hz that is very loud seems one semitone lower in pitch than if it is just barely audible. Above 2,000 Hz, the pitch gets higher as the sound gets louder. These results were obtained in the pioneering works by S.Stevens and W.Snow. Later investigations, i.e. by A.Cohen, had shown that in most cases the apparent pitch shifts were not significantly different from pitch-matching errors. When averaged, the remaining shifts followed the directions of Stevens' curves but were small (2% or less by frequency, i.e. not more than a semitone)

Theories of Pitch Perception

Theories of pitch perception try to explain how the physical sound and specific physiology of the auditory system work together to yield the experience of pitch. In general, pitch perception theories can be divided into place coding and temporal coding. Place theory holds that the perception of pitch is determined by the place of maximum excitation on the basilar membrane.

A place code, taking advantage of the tonotopy in the auditory system, must be in effect for the perception of high frequencies, since neurons have an upper limit on how fast they can phase-lock their action potentials. However, a purely place-based theory cannot account for the accuracy of pitch perception in the low and middle frequency ranges.

Temporal theories offer an alternative that appeals to the temporal structure of action potentials, mostly the phase-locking and mode-locking of action potentials to frequencies in a stimulus. The precise way this temporal structure helps code for pitch at higher levels is still debated, but the processing seems to be based on an autocorrelation of action potentials in the auditory nerve. However, it has long been noted that a neural mechanism that may accomplish a delay—a necessary operation of a true autocorrelation—has not been found. At least one model shows that a temporal delay is unnecessary to produce an autocorrelation model of pitch perception, appealing to phase shifts between cochlear filters; however, earlier work has shown that certain sounds with a prominent peak in their autocorrelation function do not elicit a corresponding pitch percept, and that certain sounds without a peak in their autocorrelation function nevertheless elicit a pitch. To be a more complete model, autocorrelation must therefore apply to signals that represent the output of the cochlea, as via auditory-nerve interspike-interval histograms. Some theories of pitch perception hold that pitch has inherent octave ambiguities, and therefore is best decomposed into a pitch *chroma*, a periodic value around the octave, like the note names in western music—and a pitch *height*, which may be ambiguous, that indicates the octave the pitch is in.

Just-noticeable Difference

The *just-noticeable difference (jnd)* (the threshold at which a change is perceived) depends on the tone's frequency content. Below 500 Hz, the jnd is about 3 Hz for sine waves, and 1 Hz for complex tones; above 1000 Hz, the jnd for sine waves is about 0.6% (about 10 cents). The jnd is typically tested by playing two tones in quick succession with the listener asked if there was a difference in their pitches. The jnd becomes smaller if the two tones are played simultaneously as the listener is then able to discern beat frequencies. The total number of perceptible pitch steps in the range of human hearing is about 1,400; the total number of notes in the equal-tempered scale, from 16 to 16,000 Hz, is 120.

Aural Illusions

The relative perception of pitch can be fooled, resulting in *aural illusions*. There are several of these, such as the tritone paradox, but most notably the Shepard scale, where a continuous or discrete sequence of specially formed tones can be made to sound as if the sequence continues ascending or descending forever.

Definite and Indefinite Pitch

Not all musical instruments make notes with a clear pitch. The unpitched percussion instrument (a class of percussion instrument) does not produce particular pitches. A sound or note of definite pitch is one where a listener can possibly (or relatively easily) discern the pitch. Sounds with definite pitch have harmonic frequency spectra or close to harmonic spectra.

A sound generated on any instrument produces many modes of vibration that occur simultaneously. A listener hears numerous frequencies at once. The vibration with the lowest frequency is called the *fundamental frequency*; the other frequencies are *overtones. Harmonics* are an important class of overtones with frequencies that are integer multiples of the fundamental. Whether or not the higher frequencies are integer multiples, they are collectively called the partials, referring to the different parts that make up the total spectrum.

A sound or note of indefinite pitch is one that a listener finds impossible or relatively difficult to identify as to pitch. Sounds with indefinite pitch do not have harmonic spectra or have altered harmonic spectra—a characteristic known as inharmonicity.

It is still possible for two sounds of indefinite pitch to clearly be higher or lower than one another. For instance, a snare drum sounds higher pitched than a bass drum though both have indefinite pitch, because its sound contains higher frequencies. In other words, it is possible and often easy to roughly discern the relative pitches of two sounds of indefinite pitch, but sounds of indefinite pitch do not neatly correspond to any specific pitch. A special type of pitch often occurs in free nature when sound reaches the ear of an observer directly from the source, and also after reflecting off a sound-reflecting surface. This phenomenon is called *repetition pitch*, because the addition of a true repetition of the original sound to itself is the basic prerequisite.

Pitch Standards and Standard Pitch

A pitch standard (also Concert pitch) is the conventional pitch reference a group of musical instruments are tuned to for a performance. Concert pitch may vary from ensemble to ensemble, and has varied widely over musical history.

Standard pitch is a more widely accepted convention. The A above middle C is usually set at 440 Hz (often written as "A = 440 Hz" or sometimes "A440"), although other frequencies, such as 442 Hz, are also often used as variants. Another standard pitch, the so-called *Baroque pitch*, has been set in the 20th century as A = 415 Hz—approximately an equal-tempered semitone lower than A440 to facilitate transposition.

Transposing instruments have their origin in the variety of pitch standards. In modern times, they conventionally have their parts transposed into different keys from voices and other instruments (and even from each other). As a result, musicians need a way to refer to a particular pitch in an unambiguous manner when talking to each other.

For example, the most common type of clarinet or trumpet, when playing a note written in their part as C, sounds a pitch that is called B♭ on a non-transposing instrument like a violin (which indicates that at one time these wind instruments played at a standard pitch a tone lower than violin pitch). To refer to that pitch unambiguously, a musician calls it *concert B♭*, meaning, "the pitch that someone playing a non-transposing instrument like a violin calls B♭."

Labeling Pitches

Pitches are labeled using:

- Letters, as in Helmholtz pitch notation
- A combination of letters and numbers—as in scientific pitch notation, where notes are labelled upwards from C_0, the 16 Hz C
- Numbers that represent the frequency in hertz (Hz), the number of cycles per second

For example, one might refer to the A above middle C as *a′*, A_4, or *440 Hz*. In standard Western equal temperament, the notion of pitch is insensitive to "spelling": the description "G_4 double sharp" refers to the same pitch as A_4; in other temperaments, these may be distinct pitches. Human perception of musical intervals is approximately logarithmic with respect to fundamental frequency: the perceived interval between the pitches "A220" and "A440" is the same as the perceived interval between the pitches *A440* and *A880*. Motivated by this logarithmic perception, music theorists sometimes represent pitches using a numerical scale based on the logarithm of fundamental frequency. For example, one can adopt the widely used MIDI standard to map fundamental frequency, f, to a real number, p, as follows

$$p = 69 + 12 \times \log_2 \left(\frac{f}{440\ Hz} \right)$$

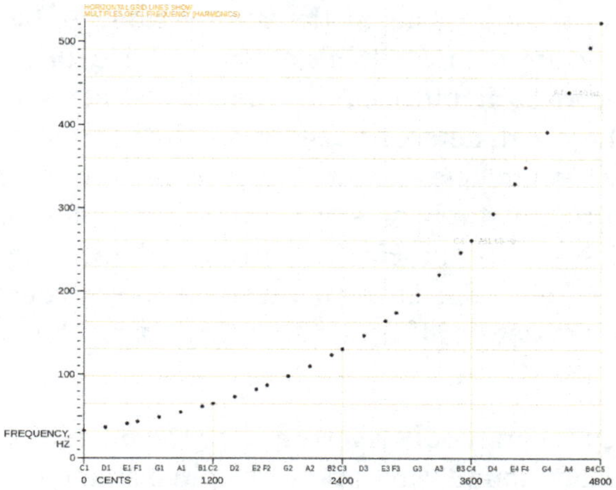

Note frequencies, four-octave C major diatonic scale, starting with C_1.

This creates a linear pitch space in which octaves have size 12, semitones (the distance between adjacent keys on the piano keyboard) have size 1, and A440 is assigned the number 69. Distance in this space corresponds to musical intervals as understood by musicians. An equal-tempered semitone is subdivided into 100 cents. The system is flexible enough to include "microtones" not found on standard piano keyboards. For example, the pitch halfway between C (60) and C# (61) can be labeled 60.5.

The following table shows frequencies in Hz for notes in various octaves, named according to the "German method" of octave nomenclature:

Note	Great	Small	One-lined	Two-lined	Three-lined	Four-lined
A	55.00	110.00	220.00	440.00	880.00	1760.00
A♯/B♭	58.27	116.54	233.08	466.16	932.33	1864.66
B/C♭	61.74	123.47	246.94	493.88	987.77	1975.53
B♯/C	65.41	130.81	261.63	523.25	1046.50	2093.00
C♯/D♭	69.30	138.59	277.18	554.37	1108.73	2217.46
D	73.42	146.83	293.66	587.33	1174.66	2349.32
D♯/E♭	77.78	155.56	311.13	622.25	1244.51	2489.02
E/F♭	82.41	164.81	329.63	659.26	1318.51	2637.02
E♯/F	87.31	174.61	349.23	698.46	1396.91	2793.83
F♯/G♭	92.50	185.00	369.99	739.99	1479.98	2959.96
G	98.00	196.00	392.00	783.99	1567.99	3135.96
G♯/A♭	103.83	207.65	415.30	830.61	1661.22	3322.44

Frequency, Pitch and Human Perception

The ears of a human (and other animals) are sensitive detectors capable of detecting the fluctuations in air pressure that impinge upon the eardrum. The mechanics of the ear's detection ability will be discussed later in this lesson. For now, it is sufficient to say that the human ear is capable of detecting sound waves with a wide range of frequencies, ranging between approximately 20 Hz to 20 000 Hz. Any sound with a frequency below the audible range of hearing (i.e., less than 20 Hz) is known as an infrasound and any sound with a frequency above the audible range of hearing (i.e., more than 20 000 Hz) is known as an ultrasound. Humans are not alone in their ability to detect a wide range of frequencies. Dogs can detect frequencies as low as approximately 50 Hz and as high as 45 000 Hz. Cats can detect frequencies as low as approximately 45 Hz and as high as 85 000 Hz. Bats, being nocturnal creature, must rely on sound echolocation for navigation and hunting. Bats can detect frequencies as high as 120 000 Hz. Dolphins can detect frequencies as high as 200 000 Hz. While dogs, cats, bats, and dolphins have an unusual ability to detect ultrasound, an elephant possesses the unusual ability to detect infrasound, having an audible range from approximately 5 Hz to approximately 10 000 Hz.

The sensation of a frequency is commonly referred to as the pitch of a sound. A high pitch sound corresponds to a high frequency sound wave and a low pitch sound corresponds to a low frequency sound wave. Amazingly, many people, especially those who have been musically trained, are capable of detecting a difference in frequency between two separate sounds that is as little as 2 Hz. When two sounds with a frequency difference of greater than 7 Hz are played simultaneously, most people are capable of detecting the presence of a complex wave pattern resulting from the interference and superposition of the two sound waves. Certain sound waves when played (and heard) simultaneously will produce a particularly pleasant sensation when heard, are said to be consonant. Such sound waves form the basis of intervals in music. For example, any two sounds whose frequencies make a 2:1 ratio are said to be separated by an octave and result in a particularly pleasing sensation when heard. That is, two sound waves sound good when played together if one sound has twice the frequency of the other. Similarly two sounds with a frequency ratio of 5:4 are said to be separated by an interval of a third; such sound waves also sound good when played together. Examples of other sound wave intervals and their respective frequency ratios are listed in the table below.

Interval	Frequency Ratio	Examples
Octave	2:1	512 Hz and 256 Hz
Third	5:4	320 Hz and 256 Hz
Fourth	4:3	342 Hz and 256 Hz
Fifth	3:2	384 Hz and 256 Hz

The ability of humans to perceive pitch is associated with the frequency of the sound wave that impinges upon the ear. Because sound waves traveling through air are longitudinal waves that produce high- and low-pressure disturbances of the particles of the air at a given frequency, the ear has an ability to detect such frequencies and associate them with the pitch of the sound. But pitch is not the only property of a sound wave detectable by the human ear. In the next part, we will investigate the ability of the ear to perceive the intensity of a sound wave.

Loudness

Generally the loudness of a sound is related to the amplitude of the sound wave; a wave with bigger variations in pressure generally sounds louder. For any type of wave the energy carried by the wave is proportional to amplitude squared. This means doubling the amplitude increases the power by a factor of four (two squared). But the amount of energy reaching your ear also depends on the frequency since a wave with more oscillations per second (higher frequency) will mean the same amplitude hits your eardrum more often. Sound intensity is defined to be the energy per second (power in Watts) reaching a given area (measured in square meters). Normal conversation has an intensity of about 10^{-6} W/m^2.

A sound from a small sound source spreads out in all directions in an expanding spherical shape. Because the energy is spread over a larger and larger area as time goes on, the intensity decreases as you move further from the source. The area of a sphere is given by $4\pi r^2$ so the decrease in intensity is proportional to r^2 where r is the distance from the source. This is known as an inverse square law and many other laws in physics follow this same law. For example the gravitational field of the earth (or any other object) decreases as you move away from it proportional to the inverse square of the distance. So does the electric field around an electron or proton. For practical purposes, what this means is doubling the distance decreases the strength by $1/2^2 = 1/4$. If you move three times as far away the strength is $1/3^2$ or $1/9^{th}$ as much. Shortening the distance by half means the intensity will be four times as much.

Source	**Pressure** rms *(Pa)*	**Sound intensity level** SIL *(dB)*	**intensity** *(W/m²)*
Jet engine at 10 m		150	10^3
Jet engine	200	140	100
Jack hammer	60	130	10
Car horn	20	120 (pain threshold)	1
Rock band	6	110	0.1
Machine shop	2	100	0.01
Train	0.6	90	10^{-3}
Vacuum cleaner	0.2	80	10^{-4}
TV	0.06	70	10^{-5}
Conversation	0.02	60	10^{-6}
Office	0.006	50	10^{-7}
Library	0.002	40	10^{-8}
Hospital	0.0006	30	10^{-9}
Broadcast studio	0.0002	20	10^{-10}
Rustle of leaves	0.00006	10	10^{-11}
Threshold of hearing	0.00002	0	10^{-12}

The human ear is an amazing instrument that can detect intensities as low as 10^{-12} W/m^2 and can hear intensities as high as 10^3 W/m^2 (although this is loud enough to cause damage to the ear). To make this huge range easier to write down, a second scale of loudness was created called the sound intensity level, measured in decibels. The relationship between sound intensity, *I* measured in

Watts per meter squared and sound intensity level (*SIL*) measured in decibels (dB), is given by *SIL* $= 10 \log (I/I_o)$. Here *log* is the logarithm and $I_o = 10^{-12}$ W/m² is a reference sound intensity at about the threshold of human hearing.

Here are a few examples and rules of thumb for converting intensity (W/m²) into intensity levels (in dB):

- A 10 fold increase in intensity equals an addition of 10dB. So going from a car horn to a jackhammer *multiplies* the intensity by 10 (1 W/m² to 10 W/m²) but *adds* 10 dB to the intensity level (120 dB to 130 dB).
- A two fold increase in intensity (twice as loud in W/m²) equals and addition of 3 dB to the SIL. Suppose one trombone produces a sound level of 40dB. How loud are four trombones? Doubling the number of trombones to two adds 3dB, doubling again to four adds 3dB more so the new sound level is 46dB.
- Suppose the sound intensity is 100 W/m². What is the sound level? $I = 10 \log (100/10^{-12}) = 10 \log (10^{14}) = 10*14 = 140$ dB.
- Suppose the sound level is 110 dB. What is the sound intensity? 110 dB $= 10 \log (I/10^{-12})$. Divide both sides by 10 to get $11 = \log (I/10^{-12})$. Now take inverse log 11 (same as 10^{11}) to get $10^{11} = I/^{-12}$. Multiply both sides by $^{-12}$ to get 0.1 W/m² = I.
- A Sound Conversion web site that converts between sound level, sound pressure and sound intensity.

Both sound intensity (W/m²) and sound intensity level (SIL) are numbers that can be measured precisely in the laboratory (objective measurements). The human ear, however, is an imperfect measuring instrument. We hear better at a mid-range of frequencies than we do at very low or very high frequencies. The phon scale is a *subjective* measurement of loudness. This scale is arrived at by asking real humans to compare the loudness of different notes and an average is taken for many people (subjective). The units of the phon are the same as SIL units; the Decibels (dB).

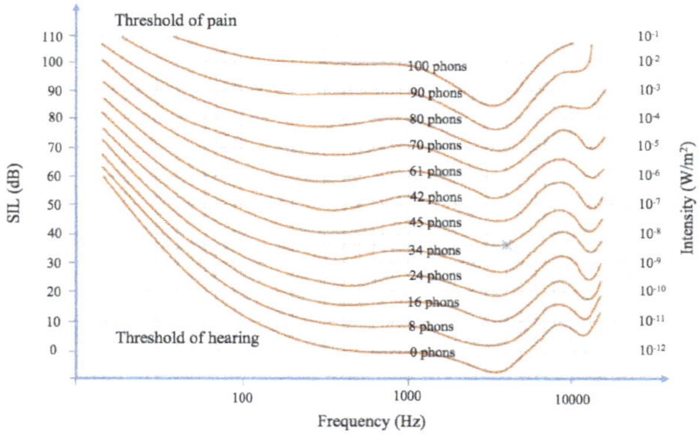

The diagram above (modified from an MIT OpenCourseWare graph) relates sound intensity level (SIL, measured in dB with laboratory instruments), pressure (measured in W/m² with laboratory instruments) and phons (human perception). A SIL of 110 dB is considered painful while a SIL of 0 is at the threshold of hearing. If our ears were the same as laboratory instruments the lines would go straight across. The phon scale and the SIL scale do give approximately the same number in dB but only for frequencies around 1000 Hz. In other words our subjective perception of loudness and the laboratory measurement agree but only for sounds with a frequency of 1000 Hz.

Notice there is a dip in all the curves between 1000 Hz and 5000 Hz indicating we are more sensitive to these frequencies and this is true for all loudness readings. For example suppose we perceive a sound at 4000 Hz to be 45 dB (phons) (labeled by a blue X in the diagram). The chart shows that at this loudness and frequency the dB reading in the laboratory is actually around 36 dB (dotted line to the SIL axis). So we perceive a sound of 36 dB (measured in the lab) as being much louder (45 dB) if it occurs at 4000 Hz. This is not surprising once you realize these are important frequencies for human speech; our hearing mechanism is built to hear human voices better than sound with much higher or much lower frequencies. This greater sensitivity around 3500 Hz is due to the tube resonance of the auditory canal.

It is also the case that intensity has an effect on perceived frequency; the same laboratory frequency will appear to be a slightly different frequency if the intensity is different. High frequencies are perceived to be a slightly higher pitch than normal if they are very loud. Low frequencies are perceived to be slightly lower than expected if they are very loud. Medium loudness doesn't change the perceived pitch very much.

The above curves are parallel to the frequency response curves of microphones and speakers. No microphone has the same sensitivity to all frequencies and no speaker reproduces all frequencies equally well.

Difference between Pitch and Loudness

Pitch and loudness seem to be same sometimes but they are not exactly the same. They differ on the basis of their tone quality. The pitch of a sound is our ear's response to the frequency of sound. Whereas loudness depends on the energy of the wave. In general, the pitch is the reason behind the difference in voice quality of different individuals.

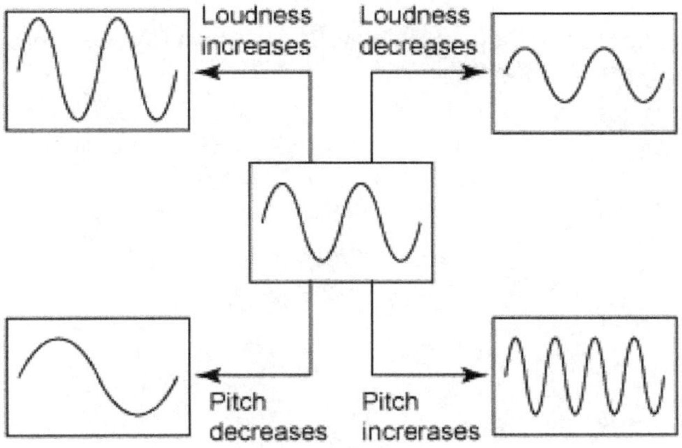

Difference between Pitch and Loudness

The pitch of a sound depends on the frequency while loudness of a sound depends on the amplitude of sound waves.

References

- Murdin, Paul (25 December 2008). Full Meridian of Glory: Perilous Adventures in the Competition to Measure the Earth. Springer Science & Business Media. pp. 35–36. ISBN 9780387755342
- McAdams, Stephen, and Albert Bregman (1979). "Hearing Musical Streams". Computer Music Journal 3, no. 4 (December): 26–43, 60
- Jones, S.; Longe, O.; Pato, M. V. (1998). "Auditory evoked potentials to abrupt pitch and timbre change of complex tones: electrophysiological evidence of streaming?". Electroencephalography and Clinical Neurophysiology. 108 (2): 131–142. doi:10.1016/s0168-5597(97)00077-4
- Corwin, J. (2009), The auditory system (PDF), archived(PDF) from the original on 2013-06-28, retrieved 2013-04-06
- Carroll C. Pratt, "The Spatial Character of High and Low Tones", Journal of Experimental Psychology 13 (1930): 278–85.Wessel, David (1979). "Low Dimensional Control of Musical Timbre". Computer Music Journal 3:45–52
- Massaro, D. W. (1972). "Preperceptual images, processing time, and perceptual units in auditory perception". Psychological Review. 79 (2): 124–145. doi:10.1037/h0032264
- Everest, F. (2001). The Master Handbook of Acoustics. New York: McGraw-Hill. pp. 262–263. ISBN 0-07-136097-2
- Cohen, D.; Dubnov, S. (1997), Gestalt phenomena in musical texture (PDF), archived (PDF) from the original on 2015-11-21, retrieved 2015-11-19
- Volland, Hans (1995). Handbook of Atmospheric Electrodynamics. Boca Raton: CRC Press. p. 22. ISBN 0-8493-8647-0
- Zwislocki, J. J. (1969). "Temporal summation of loudness: an analysis". The Journal of the Acoustical Society of America. 46 (2B): 431–441. Bibcode:1969ASAJ...46..431Z. doi:10.1121/1.1911708
- Dushaw, Brian D.; Worcester, P. F.; Cornuelle, B. D.; Howe, B. M. (1993). "On Equations for the Speed of Sound in Seawater". Journal of the Acoustical Society of America. 93 (1): 255–275. Bibcode:1993ASAJ...93..255D. doi:10.1121/1.405660
- Nemiroff, R.; Bonnell, J., eds. (19 August 2007). "A Sonic Boom". Astronomy Picture of the Day. NASA. Retrieved 24 October 2010
- Grey, John M. (1977). "Multidimensional Perceptual Scaling of Musical Timbres". The Journal of the Acoustical Society of America61(5):1270–77. doi:10.1121/1.381428
- Cozens, Peter (2006). The Darkest Days of the War: the Battles of Iuka and Corinth. Chapel Hill: The University of North Carolina Press. ISBN 0-8078-5783-1

6

Light

Light is an electromagnetic wave propagating in space with the speed of 3×10^8 m/s. The understanding of the different phenomena of light is approached from the wave-like or particle-like nature of light. The concepts of reflection and refraction of light are fundamental to the study of light, which have been discussed in detail in this book.

Light, electromagnetic radiation that can be detected by the human eye. Electromagnetic radiation occurs over an extremely wide range of wavelengths, from gamma rays with wavelengths less than about 1×10^{-11} metre to radio waves measured in metres. Within that broad spectrum the wavelengths visible to humans occupy a very narrow band, from about 700 nanometres (nm; billionths of a metre) for red light down to about 400 nm for violet light. The spectral regions adjacent to the visible band are often referred to as light also, infrared at the one end and ultraviolet at the other. The speed of light in a vacuum is a fundamental physical constant, the currently accepted value of which is exactly 299,792,458 metres per second, or about 186,282 miles per second.

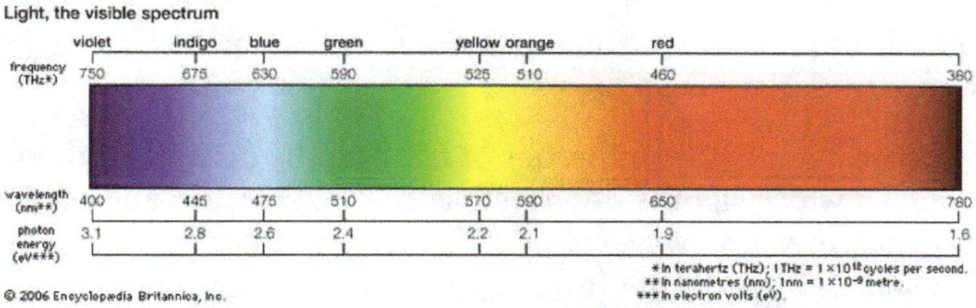

When white light is spread apart by a prism or a diffraction grating, the colours of the visible spectrum appear. The colours vary according to their wavelengths. Violet has the highest frequencies and shortest wavelengths, and red has the lowest frequencies and the longest wavelengths

No single answer to the question “What is light?” satisfies the many contexts in which light is experienced, explored, and exploited. The physicist is interested in the physical properties of light, the artist in an aesthetic appreciation of the visual world. Through the sense of sight, light is a primary tool for perceiving the world and communicating within it. Light from the Sun warms the Earth, drives global weather patterns, and initiates the life-sustaining process of photosynthesis. On the grandest scale, light’s interactions with matter have helped shape the structure of the universe. Indeed, light provides a window on the universe, from cosmological to atomic scales. Almost all of the information about the rest of the universe reaches Earth in the form of electromagnetic radiation. By interpreting that radiation, astronomers can glimpse the earliest epochs of the universe, measure the general expansion of the universe, and determine the chemical composition of stars and the interstellar medium. Just as the invention of the telescopedramatically broadened exploration of the universe, so too the invention of the microscope opened the intricate world of the cell. The analysis of the frequencies of light emitted and absorbed by atoms was a principal impetus for the development of quantum mechanics.

Atomic and molecular spectroscopies continue to be primary tools for probing the structure of matter, providing ultrasensitive tests of atomic and molecular models and contributing to studies of fundamental photochemical reactions.

The Sun shining from behind clouds

Light transmits spatial and temporal information. This property forms the basis of the fields of optics and optical communications and a myriad of related technologies, both mature and emerging. Technological applications based on the manipulations of light include lasers, holography, and fibre-optic telecommunications systems.

In most everyday circumstances, the properties of light can be derived from the theory of classical electromagnetism, in which light is described as coupled electric and magnetic fields propagatingthrough space as a traveling wave. However, this wave theory, developed in the mid-19th century, is not sufficient to explain the properties of light at very low intensities. At that level a quantumtheory is needed to explain the characteristics of light and to explain the interactions of light with atoms and molecules. In its simplest form, quantum theory describes light as consisting of discrete packets of energy, called photons. However, neither a classical wave model nor a classical particle model correctly describes light; light has a dual nature that is revealed only in quantum mechanics. This surprising wave-particle duality is shared by all of the primary constituents of nature (e.g., electrons have both particle-like and wavelike aspects). Since the mid-20th century, a more comprehensive theory of light, known as quantum electrodynamics (QED), has been regarded by physicists as complete. QED combines the ideas of classical electromagnetism, quantum mechanics, and the special theory of relativity.

This article focuses on the physical characteristics of light and the theoretical models that describe the nature of light. Its major themes include introductions to the fundamentals of geometrical optics, classical electromagnetic waves and the interference effects associated with those waves, and the foundational ideas of the quantum theory of light. The contemplation of the speed of light as measured in different reference frames was pivotal to the development of Albert Einstein's theory of special relativity in 1905.

Other Characteristics

The amplitude of a light wave is related to its intensity.

- Intensity is the absolute measure of a light wave's power density.
- Brightness is the relative intensity as perceived by the average human eye.

The frequency of a light wave is related to its color.

- Color is such a complex topic that it has its own section in this book.
- Monochromatic light is described by only one frequency.
    - Laser light is effectively monochromatic.
    - There are six simple, named colors in English (and many other languages) each associated with a band of monochromatic light. In order of increasing frequency they are red, orange, yellow, green, blue, and violet.
    - Light is sometimes also known as visible light to contrast it from "ultraviolet light" and "infrared light".
    - Other forms of electromagnetic radiation that are not visible to humans are sometimes also known informally as "light".
- Polychromatic light is described by many different frequencies.
    - Nearly every light source is polychromatic.
    - White light is polychromatic.

A graph of relative intensity vs. frequency is called a spectrum (plural: spectra). Although frequently associated with light, the term can be applied to any wave phenomena.

- A continuous spectrum is one in which every frequency is present within some range.
    - Blackbody radiators emit a continuous spectrum.
- A discrete spectrum is one in which only a well defined set of isolated frequencies are present. (A discrete spectrum is a finite collection of monochromatic light waves.)
    - The excited electrons in a gas emit a discrete spectrum.

The wavelength of a light wave is inversely proportional to its frequency.

- Light is often described by it's wavelength in a vacuum.
- Light ranges in wavelength from 400 nm on the violet end to 700 nm on the red end of the visible spectrum.

Electromagnetic Spectrum and Visible Light

Generally, EM radiation, or EMR (the designation "radiation" excludes static electric and magnetic and near fields), is classified by wavelength into radio, microwave, infrared, the visible region that we perceive as light, ultraviolet, X-rays and gamma rays.

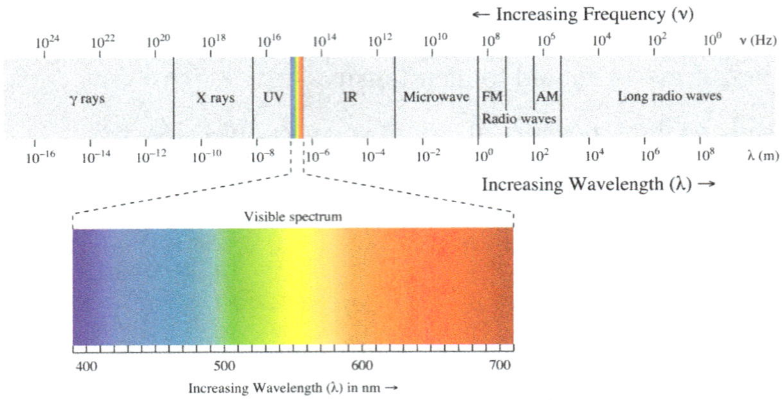

Electromagnetic spectrum with light highlighted

The behavior of EMR depends on its wavelength. Higher frequencies have shorter wavelengths, and lower frequencies have longer wavelengths. When EMR interacts with single atoms and molecules, its behavior depends on the amount of energy per quantum it carries.

EMR in the visible light region consists of quanta (called photons) that are at the lower end of the energies that are capable of causing electronic excitation within molecules, which leads to changes in the bonding or chemistry of the molecule. At the lower end of the visible light spectrum, EMR becomes invisible to humans (infrared) because its photons no longer have enough individual energy to cause a lasting molecular change (a change in conformation) in the visual molecule retinal in the human retina, which change triggers the sensation of vision.

There exist animals that are sensitive to various types of infrared, but not by means of quantum-absorption. Infrared sensing in snakes depends on a kind of natural thermal imaging, in which tiny packets of cellular water are raised in temperature by the infrared radiation. EMR in this range causes molecular vibration and heating effects, which is how these animals detect it.

Above the range of visible light, ultraviolet light becomes invisible to humans, mostly because it is absorbed by the cornea below 360 nanometers and the internal lens below 400. Furthermore, the rods and cones located in the retina of the human eye cannot detect the very short (below 360 nm) ultraviolet wavelengths and are in fact damaged by ultraviolet. Many animals with eyes that do not require lenses (such as insects and shrimp) are able to detect ultraviolet, by quantum photon-absorption mechanisms, in much the same chemical way that humans detect visible light.

Various sources define visible light as narrowly as 420 to 680 to as broadly as 380 to 800 nm. Under ideal laboratory conditions, people can see infrared up to at least 1050 nm; children and young adults may perceive ultraviolet wavelengths down to about 310 to 313 nm.

Plant growth is also affected by the color spectrum of light, a process known as photomorphogenesis.

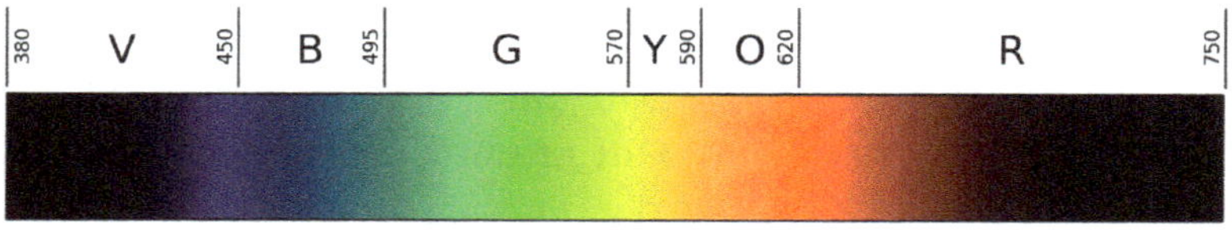

Speed of Light

The speed of light in a *vacuum* is defined to be exactly 299,792,458 *m/s* (approx. 186,282 miles per second). The fixed value of the speed of light in SI units results from the fact that the metre is now defined in terms of the speed of light. All forms of electromagnetic radiation move at exactly this same speed in vacuum.

Different physicists have attempted to measure the speed of light throughout history. Galileo attempted to measure the speed of light. An early experiment to measure the speed of light was conducted by Ole Rømer, a Danish physicist. Using a telescope, Rømer observed the motions of Jupiter and one of its moons, Io. Noting discrepancies in the apparent period of Io's orbit, he calculated that light takes about 22 minutes to traverse the diameter of Earth's orbit. However, its size was not known at that time. If Rømer had known the diameter of the Earth's orbit, he would have calculated a speed of 227,000,000 m/s.

Another, more accurate, measurement of the speed of light was performed in Europe by Hippolyte Fizeau in 1849. Fizeau directed a beam of light at a mirror several kilometers away. A rotating cog wheel was placed in the path of the light beam as it traveled from the source, to the mirror and then returned to its origin. Fizeau found that at a certain rate of rotation, the beam would pass through one gap in the wheel on the way out and the next gap on the way back. Knowing the distance to the mirror, the number of teeth on the wheel, and the rate of rotation, Fizeau was able to calculate the speed of light as 313,000,000 m/s.

Léon Foucault carried out an experiment which used rotating mirrors to obtain a value of 298,000,000 m/s in 1862. Albert A. Michelson conducted experiments on the speed of light from 1877 until his death in 1931. He refined Foucault's methods in 1926 using improved rotating mirrors to measure the time it took light to make a round trip from Mount Wilson to Mount San Antonio in California. The precise measurements yielded a speed of 299,796,000 m/s.

The effective velocity of light in various transparent substances containing ordinary matter, is less than in vacuum. For example, the speed of light in water is about 3/4 of that in vacuum.

Two independent teams of physicists were said to bring light to a "complete standstill" by passing it through a Bose–Einstein condensate of the element rubidium, one team at Harvard University and the Rowland Institute for Science in Cambridge, Massachusetts, and the other at the Harvard–Smithsonian Center for Astrophysics, also in Cambridge. However, the popular description of light being "stopped" in these experiments refers only to light being stored in the excited states of atoms, then re-emitted at an arbitrary later time, as stimulated by a second laser pulse. During the time it had "stopped" it had ceased to be light.

Propagation of Light

The word 'propagation' is sometimes used in scientific contexts to mean 'travel' or 'movement'. It is used in optics, which is the study of the physics of light and other forms of electromagnetic energy, in the context of describing energy moving ('propagating') in 'waves'.

Light is generally considered to travel through air in straight lines, only changing direction when it is reflected, scattered, or passes from one type of substance (called a medium) to another. For example, light changes direction slightly when it moves from air into water, or from air into glass, or vice-versa. This change in the direction of travel of light is due to refraction.

Light travels (propagates) from many different sources, including from large and powerful sources of illumination such as the sun or the main lights in a room. It is also reflected and scattered from many and various objects that are not sources of light - meaning that the light did not start at (originate from) that object but arrived at it from somewhere else, then left it again, perhaps in a different direction.

Eventually light reaches the surfaces of physical objects, e.g. it could come from a lamp in a light fitting in the ceiling of a room (the source of the light), propagate through the air in the room (a 'medium' through which the light passes) until it reaches a surface such as a table or other item of furniture.

When light reaches an object it can do one or some combination of absorption, reflection, scattering or refraction.

Optics

The study of light and the interaction of light and matter is termed optics. The observation and study of optical phenomena such as rainbows and the aurora borealis offer many clues as to the nature of light.

Refraction

An example of refraction of light. The straw appears bent, because of refraction of light as it enters liquid from air

A cloud illuminated by sunlight

Refraction is the bending of light rays when passing through a surface between one transparent material and another. It is described by Snell's Law:

$$n_1 \sin\theta_1 = n_2 \sin\theta_2$$

where θ_1 is the angle between the ray and the surface normal in the first medium, θ_2 is the angle between the ray and the surface normal in the second medium, and n_1 and n_2 are the indices of refraction, $n = 1$ in a vacuum and $n > 1$ in a transparent substance.

When a beam of light crosses the boundary between a vacuum and another medium, or between two different media, the wavelength of the light changes, but the frequency remains constant. If the beam of light is not orthogonal (or rather normal) to the boundary, the change in wavelength results in a change in the direction of the beam. This change of direction is known as refraction.

The refractive quality of lenses is frequently used to manipulate light in order to change the apparent size of images. Magnifying glasses, spectacles, contact lenses, microscopes and refracting telescopes are all examples of this manipulation.

Light Sources

There are many sources of light. A body at a given temperature emits a characteristic spectrum of black-body radiation. A simple thermal source is sunlight, the radiation emitted by the chromosphere of the Sun at around 6,000 kelvins (5,730 degrees Celsius; 10,340 degrees Fahrenheit) peaks in the visible region of the electromagnetic spectrum when plotted in wavelength units and roughly 44% of sunlight energy that reaches the ground is visible. Another example is incandescent light bulbs, which emit only around 10% of their energy as visible light and the remainder as infrared. A common thermal light source in history is the glowing solid particles in flames, but these also emit most of their radiation in the infrared, and only a fraction in the visible spectrum.

The peak of the blackbody spectrum is in the deep infrared, at about 10 micrometre wavelength, for relatively cool objects like human beings. As the temperature increases, the peak shifts to shorter wavelengths, producing first a red glow, then a white one, and finally a blue-white colour as the peak moves out of the visible part of the spectrum and into the ultraviolet. These colours can be seen when metal is heated to "red hot" or "white hot". Blue-white thermal emission is not often seen, except in stars (the commonly seen pure-blue colour in a gas flame or a welder's torch is in fact due to molecular emission, notably by CH radicals (emitting a wavelength band around 425 nm, and is not seen in stars or pure thermal radiation).

Atoms emit and absorb light at characteristic energies. This produces "emission lines" in the spectrum of each atom. Emission can be spontaneous, as in light-emitting diodes, gas discharge lamps (such as neon lamps and neon signs, mercury-vapor lamps, etc.), and flames (light from the hot gas itself—so, for example, sodium in a gas flame emits characteristic yellow light). Emission can also be stimulated, as in a laser or a microwave maser.

A city illuminated by colorful artificial lighting

Deceleration of a free charged particle, such as an electron, can produce visible radiation: cyclotron radiation, synchrotron radiation, and bremsstrahlung radiation are all examples of this. Particles moving through a medium faster than the speed of light in that medium can produce visible Cherenkov radiation. Certain chemicals produce visible radiation by chemoluminescence. In living things, this process is called bioluminescence. For example, fireflies produce light by this means, and boats moving through water can disturb plankton which produce a glowing wake.

Certain substances produce light when they are illuminated by more energetic radiation, a process known as fluorescence. Some substances emit light slowly after excitation by more energetic radiation. This is known as phosphorescence. Phosphorescent materials can also be excited by bombarding them with subatomic particles. Cathodoluminescence is one example. This mechanism is used in cathode ray tube television sets and computer monitors.

Certain other mechanisms can produce light:

- Bioluminescence
- Cherenkov radiation
- Electroluminescence
- Scintillation
- Sonoluminescence
- Triboluminescence

When the concept of light is intended to include very-high-energy photons (gamma rays), additional generation mechanisms include:

- Particle–antiparticle annihilation
- Radioactive decay

Units and Measures

Light is measured with two main alternative sets of units: radiometry consists of measurements of light power at all wavelengths, while photometry measures light with wavelength weighted with respect to a standardised model of human brightness perception. Photometry is useful, for example, to quantify Illumination (lighting) intended for human use. The SI units for both systems are summarised in the following tables.

Table- SI Radiometry Units

Quantity		Unit		Dimension	Notes
Name	Symbol	Name	Symbol	Symbol	
Radiant energy	Q_e	joule	J	$\mathbf{M{\cdot}L^{2}{\cdot}T^{-2}}$	Energy of electromagnetic radiation.
Radiant energy density	w_e	joule per cubic metre	J/m³	$\mathbf{M{\cdot}L^{-1}{\cdot}T^{-2}}$	Radiant energy per unit volume.

Radiant flux	Φ_e	watt	W = J/s	**M·L²·T⁻³**	Radiant energy emitted, reflected, transmitted or received, per unit time. This is sometimes also called "radiant power".
Spectral flux	$\Phi_{e,\nu}$ *or* $\Phi_{e,\lambda}$	watt per hertz *or* watt per metre	W/Hz *or* W/m	**$M{\cdot}L^2{\cdot}T^{-2}$** *or* **$M{\cdot}L{\cdot}T^{-3}$**	Radiant flux per unit frequency or wavelength. The latter is commonly measured in $W{\cdot}nm^{-1}$.
Radiant intensity	$I_{e,\Omega}$	watt per steradian	W/sr	**$M{\cdot}L^2{\cdot}T^{-3}$**	Radiant flux emitted, reflected, transmitted or received, per unit solid angle. This is a *directional* quantity.
Spectral intensity	$I_{e,\Omega,\nu}$ *or* $I_{e,\Omega,\lambda}$	watt per steradian per hertz *or* watt per steradian per metre	$W{\cdot}sr^{-1}{\cdot}Hz^{-1}$ *or* $W{\cdot}sr^{-1}{\cdot}m^{-1}$	**$M{\cdot}L^2{\cdot}T^{-2}$** *or* **$M{\cdot}L{\cdot}T^{-3}$**	Radiant intensity per unit frequency or wavelength. The latter is commonly measured in $W{\cdot}sr^{-1}{\cdot}nm^{-1}$. This is a *directional* quantity.
Radiance	$L_{e,\Omega}$	watt per steradian per square metre	$W{\cdot}sr^{-1}{\cdot}m^{-2}$	**$M{\cdot}T^{-3}$**	Radiant flux emitted, reflected, transmitted or received by a *surface*, per unit solid angle per unit projected area. This is a *directional* quantity. This is sometimes also confusingly called "intensity".
Spectral radiance	$L_{e,\Omega,\nu}$ *or* $L_{e,\Omega,\lambda}$	watt per steradian per square metre per hertz *or* watt per steradian per square metre, per metre	$W{\cdot}sr^{-1}{\cdot}m^{-2}{\cdot}Hz^{-1}$ *or* $W{\cdot}sr^{-1}{\cdot}m^{-3}$	**$M{\cdot}T^{-2}$** *or* **$M{\cdot}L^{-1}{\cdot}T^{-3}$**	Radiance of a *surface* per unit frequency or wavelength. The latter is commonly measured in $W{\cdot}sr^{-1}{\cdot}m^{-2}{\cdot}nm^{-1}$. This is a *directional* quantity. This is sometimes also confusingly called "spectral intensity".
Irradiance Flux density	E_e	watt per square metre	W/m^2	**$M{\cdot}T^{-3}$**	Radiant flux *received* by a *surface* per unit area. This is sometimes also confusingly called "intensity".
Spectral irradiance Spectral flux density	$E_{e,\nu}$ *or* $E_{e,\lambda}$	watt per square metre per hertz *or* watt per square metre, per metre	$W{\cdot}m^{-2}{\cdot}Hz^{-1}$ *or* W/m^3	**$M{\cdot}T^{-2}$** *or* **$M{\cdot}L^{-1}{\cdot}T^{-3}$**	Irradiance of a *surface* per unit frequency or wavelength. This is sometimes also confusingly called "spectral intensity". Non-SI units of spectral flux density include jansky (1 Jy = 10^{-26} $W{\cdot}m^{-2}{\cdot}Hz^{-1}$) and solar flux unit (1 sfu = 10^{-22} $W{\cdot}m^{-2}{\cdot}Hz^{-1}$ = 10^4 Jy).
Radiosity	J_e	watt per square metre	W/m^2	**$M{\cdot}T^{-3}$**	Radiant flux *leaving* (emitted, reflected and transmitted by) a *surface* per unit area. This is sometimes also confusingly called "intensity".
Spectral radiosity	$J_{e,\nu}$ *or* $J_{e,\lambda}$	watt per square metre per hertz *or* watt per square metre, per metre	$W{\cdot}m^{-2}{\cdot}Hz^{-1}$ *or* W/m^3	**$M{\cdot}T^{-2}$** *or* **$M{\cdot}L^{-1}{\cdot}T^{-3}$**	Radiosity of a *surface* per unit frequency or wavelength. The latter is commonly measured in $W{\cdot}m^{-2}{\cdot}nm^{-1}$. This is sometimes also confusingly called "spectral intensity".
Radiant exitance	M_e	watt per square metre	W/m^2	**$M{\cdot}T^{-3}$**	Radiant flux *emitted* by a *surface* per unit area. This is the emitted component of radiosity. "Radiant emittance" is an old term for this quantity. This is sometimes also confusingly called "intensity".

Spectral exitance	$M_{e,\nu}$ *or* $M_{e,\lambda}$	watt per square metre per hertz *or* watt per square metre, per metre	$W \cdot m^{-2} \cdot Hz^{-1}$ *or* W/m^3	$\mathbf{M \cdot T^{-2}}$ *or* $\mathbf{M \cdot L^{-1} \cdot T^{-3}}$	Radiant exitance of a *surface* per unit frequency or wavelength. The latter is commonly measured in $W \cdot m^{-2} \cdot nm^{-1}$. "Spectral emittance" is an old term for this quantity. This is sometimes also confusingly called "spectral intensity".
Radiant exposure	H_e	joule per square metre	J/m^2	$\mathbf{M \cdot T^{-2}}$	Radiant energy received by a *surface* per unit area, or equivalently irradiance of a *surface* integrated over time of irradiation. This is sometimes also called "radiant fluence".
Spectral exposure	$H_{e,\nu}$ *or* $H_{e,\lambda}$	joule per square metre per hertz *or* joule per square metre, per metre	$J \cdot m^{-2} \cdot Hz^{-1}$ *or* J/m^3	$\mathbf{M \cdot T^{-1}}$ *or* $\mathbf{M \cdot L^{-1} \cdot T^{-2}}$	Radiant exposure of a *surface* per unit frequency or wavelength. The latter is commonly measured in $J \cdot m^{-2} \cdot nm^{-1}$. This is sometimes also called "spectral fluence".
Hemispherical emissivity	ε			**1**	Radiant exitance of a *surface*, divided by that of a *black body* at the same temperature as that surface.
Spectral hemispherical emissivity	ε_ν *or* ε_λ			**1**	Spectral exitance of a *surface*, divided by that of a *black body* at the same temperature as that surface.
Directional emissivity	ε_Ω			**1**	Radiance *emitted* by a *surface*, divided by that emitted by a *black body* at the same temperature as that surface.
Spectral directional emissivity	$\varepsilon_{\Omega,\nu}$ *or* $\varepsilon_{\Omega,\lambda}$			**1**	Spectral radiance *emitted* by a *surface*, divided by that of a *black body* at the same temperature as that surface.
Hemispherical absorptance	A			**1**	Radiant flux *absorbed* by a *surface*, divided by that received by that surface. This should not be confused with "absorbance".
Spectral hemispherical absorptance	A_ν *or* A_λ			**1**	Spectral flux *absorbed* by a *surface*, divided by that received by that surface. This should not be confused with "spectral absorbance".
Directional absorptance	A_Ω			**1**	Radiance *absorbed* by a *surface*, divided by the radiance incident onto that surface. This should not be confused with "absorbance".
Spectral directional absorptance	$A_{\Omega,\nu}$ *or* $A_{\Omega,\lambda}$			**1**	Spectral radiance *absorbed* by a *surface*, divided by the spectral radiance incident onto that surface. This should not be confused with "spectral absorbance".
Hemispherical reflectance	R			**1**	Radiant flux *reflected* by a *surface*, divided by that received by that surface.
Spectral hemispherical reflectance	R_ν *or* R_λ			**1**	Spectral flux *reflected* by a *surface*, divided by that received by that surface.
Directional reflectance	R_Ω			**1**	Radiance *reflected* by a *surface*, divided by that received by that surface.

Spectral directional reflectance	$R_{\Omega,\nu}$ or $R_{\Omega,\lambda}$			**1**	Spectral radiance *reflected* by a *surface*, divided by that received by that surface.
Hemispherical transmittance	T			**1**	Radiant flux *transmitted* by a *surface*, divided by that received by that surface.
Spectral hemispherical transmittance	T_ν or T_λ			**1**	Spectral flux *transmitted* by a *surface*, divided by that received by that surface.
Directional transmittance	T_Ω			**1**	Radiance *transmitted* by a *surface*, divided by that received by that surface.
Spectral directional transmittance	$T_{\Omega,\nu}$ or $T_{\Omega,\lambda}$			**1**	Spectral radiance *transmitted* by a *surface*, divided by that received by that surface.
Hemispherical attenuation coefficient	μ	reciprocal metre	m^{-1}	$\mathbf{L}^{-1}$	Radiant flux *absorbed* and *scattered* by a *volume* per unit length, divided by that received by that volume.
Spectral hemispherical attenuation coefficient	μ_ν or μ_λ	reciprocal metre	m^{-1}	$\mathbf{L}^{-1}$	Spectral radiant flux *absorbed* and *scattered* by a *volume* per unit length, divided by that received by that volume.
Directional attenuation coefficient	μ_Ω	reciprocal metre	m^{-1}	$\mathbf{L}^{-1}$	Radiance *absorbed* and *scattered* by a *volume* per unit length, divided by that received by that volume.
Spectral directional attenuation coefficient	$\mu_{\Omega,\nu}$ or $\mu_{\Omega,\lambda}$	reciprocal metre	m^{-1}	$\mathbf{L}^{-1}$	Spectral radiance *absorbed* and *scattered* by a *volume* per unit length, divided by that received by that volume.

Table- SI photometry Quantities

Quantity		Unit		Dimension	Notes
Name	**Symbol**	**Name**	**Symbol**	**Symbol**	
Luminous energy	Q_ν	lumen second	lm·s	**T·J**	The lumen second is sometimes called the *talbot*.
Luminous flux / luminous power	Φ_ν	lumen (= cd·sr)	lm	**J**	Luminous energy per unit time
Luminous intensity	I_ν	candela (= lm/sr)	cd	**J**	Luminous flux per unit solid angle
Luminance	L_ν	candela per square metre	cd/m^2	$\mathbf{L}^{-2}\cdot\mathbf{J}$	Luminous flux per unit solid angle per unit *projected* source area. The candela per square metre is sometimes called the *nit*.
Illuminance	E_ν	lux (= lm/m^2)	lx	$\mathbf{L}^{-2}\cdot\mathbf{J}$	Luminous flux *incident* on a surface
Luminous exitance / luminous emittance	M_ν	lux	lx	$\mathbf{L}^{-2}\cdot\mathbf{J}$	Luminous flux *emitted* from a surface
Luminous exposure	H_ν	lux second	lx·s	$\mathbf{L}^{-2}\cdot\mathbf{T}\cdot\mathbf{J}$	Time-integrated illuminance

Luminous energy density	ω_v	lumen second per cubic metre	lm·s·m^{-3}	**L^{-3}·T·J**	
Luminous efficacy	η	lumen per watt	lm/W	**M^{-1}·L^{-2}·T^{3}·J**	Ratio of luminous flux to radiant flux or power consumption, depending on context
Luminous efficiency / luminous coefficient	V			**1**	Luminous efficacy normalized by the maximum possible efficacy

The photometry units are different from most systems of physical units in that they take into account how the human eye responds to light. The cone cells in the human eye are of three types which respond differently across the visible spectrum, and the cumulative response peaks at a wavelength of around 555 nm. Therefore, two sources of light which produce the same intensity (W/m^2) of visible light do not necessarily appear equally bright. The photometry units are designed to take this into account, and therefore are a better representation of how "bright" a light appears to be than raw intensity. They relate to raw power by a quantity called luminous efficacy, and are used for purposes like determining how to best achieve sufficient illumination for various tasks in indoor and outdoor settings. The illumination measured by a photocell sensor does not necessarily correspond to what is perceived by the human eye, and without filters which may be costly, photocells and charge-coupled devices (CCD) tend to respond to some infrared, ultraviolet or both.

Light Pressure

Light exerts physical pressure on objects in its path, a phenomenon which can be deduced by Maxwell's equations, but can be more easily explained by the particle nature of light: photons strike and transfer their momentum. Light pressure is equal to the power of the light beam divided by *c*, the speed of light. Due to the magnitude of *c*, the effect of light pressure is negligible for everyday objects. For example, a one-milliwatt laser pointer exerts a force of about 3.3 piconewtons on the object being illuminated; thus, one could lift a U.S. penny with laser pointers, but doing so would require about 30 billion 1-mW laser pointers. However, in nanometre-scale applications such as nanoelectromechanical systems (|NEMS), the effect of light pressure is more significant, and exploiting light pressure to drive NEMS mechanisms and to flip nanometre-scale physical switches in integrated circuits is an active area of research. At larger scales, light pressure can cause asteroids to spin faster, acting on their irregular shapes as on the vanes of a windmill. The possibility of making solar sails that would accelerate spaceships in space is also under investigation.

Although the motion of the Crookes radiometer was originally attributed to light pressure, this interpretation is incorrect; the characteristic Crookes rotation is the result of a partial vacuum. This should not be confused with the Nichols radiometer, in which the (slight) motion caused by torque (though not enough for full rotation against friction) *is* directly caused by light pressure. As a consequence of light pressure, Einstein in 1909 predicted the existence of "radiation friction" which would oppose the movement of matter. He wrote, "radiation will exert pressure on both sides of the plate. The forces of pressure exerted on the two sides are equal if the plate is at rest. However, if it is in motion, more radiation will be reflected on the surface that is ahead during the motion (front surface) than on the back surface. The backwardacting force of pressure exerted on the front surface is thus larger than the force of pressure acting on the back. Hence, as the resultant of the two forces, there remains a force that counteracts the motion of the plate and that increases with the velocity of the plate. We will call this resultant 'radiation friction' in brief."

Historical Theories about Light, in Chronological Order

Classical Greece and Hellenism

Empedocles postulated that everything was composed of four elements; fire, air, earth and water. He believed that Aphrodite made the human eye out of the four elements and that she lit the fire in the eye which shone out from the eye making sight possible. If this were true, then one could see during the night just as well as during the day, so Empedocles postulated an interaction between rays from the eyes and rays from a source such as the sun.

In about 300 BC, Euclid wrote *Optica*, in which he studied the properties of light. Euclid postulated that light travelled in straight lines and he described the laws of reflection and studied them mathematically. He questioned that sight is the result of a beam from the eye, for he asks how one sees the stars immediately, if one closes one's eyes, then opens them at night. If the beam from the eye travels infinitely fast this is not a problem.

In 55 BC, Lucretius, a Roman who carried on the ideas of earlier Greek atomists, wrote that "The light & heat of the sun; these are composed of minute atoms which, when they are shoved off, lose no time in shooting right across the interspace of air in the direction imparted by the shove." (from *On the nature of the Universe*). Despite being similar to later particle theories, Lucretius's views were not generally accepted. Ptolemy wrote about the refraction of light in his book *Optics*.

Classical India

In ancient India, the Hindu schools of Samkhya and Vaisheshika, from around the early centuries AD developed theories on light. According to the Samkhya school, light is one of the five fundamental "subtle" elements (*tanmatra*) out of which emerge the gross elements. The atomicity of these elements is not specifically mentioned and it appears that they were actually taken to be continuous. On the other hand, the Vaisheshika school gives an atomic theory of the physical world on the non-atomic ground of ether, space and time. The basic atoms are those of earth (*prthivi*), water (*pani*), fire (*agni*), and air (*vayu*) Light rays are taken to be a stream of high velocity of *tejas* (fire) atoms. The particles of light can exhibit different characteristics depending on the speed and the arrangements of the *tejas* atoms. The *Vishnu Purana* refers to sunlight as "the seven rays of the sun".

The Indian Buddhists, such as Dignāga in the 5th century and Dharmakirti in the 7th century, developed a type of atomism that is a philosophy about reality being composed of atomic entities that are momentary flashes of light or energy. They viewed light as being an atomic entity equivalent to energy.

Descartes

René Descartes (1596–1650) held that light was a mechanical property of the luminous body, rejecting the "forms" of Ibn al-Haytham and Witelo as well as the "species" of Bacon, Grosseteste, and Kepler. In 1637 he published a theory of the refraction of light that assumed, incorrectly, that light travelled faster in a denser medium than in a less dense medium. Descartes arrived at this conclusion by analogy with the behaviour of sound waves. Although Descartes was incorrect about the relative speeds, he was correct in assuming that light behaved like a wave and in concluding that refraction could be explained by the speed of light in different media.

Descartes is not the first to use the mechanical analogies but because he clearly asserts that light is only a mechanical property of the luminous body and the transmitting medium, Descartes' theory of light is regarded as the start of modern physical optics.

Particle Theory

Pierre Gassendi

Pierre Gassendi (1592–1655), an atomist, proposed a particle theory of light which was published posthumously in the 1660s. Isaac Newton studied Gassendi's work at an early age, and preferred his view to Descartes' theory of the *plenum*. He stated in his *Hypothesis of Light* of 1675 that light was composed of corpuscles (particles of matter) which were emitted in all directions from a source. One of Newton's arguments against the wave nature of light was that waves were known to bend around obstacles, while light travelled only in straight lines. He did, however, explain the phenomenon of the diffraction of light (which had been observed by Francesco Grimaldi) by allowing that a light particle could create a localised wave in the aether.

Newton's theory could be used to predict the reflection of light, but could only explain refraction by incorrectly assuming that light accelerated upon entering a denser medium because the gravitational pull was greater. Newton published the final version of his theory in his *Opticks* of 1704. His reputation helped the particle theory of light to hold sway during the 18th century. The particle theory of light led Laplace to argue that a body could be so massive that light could not escape from it. In other words, it would become what is now called a black hole. Laplace withdrew his suggestion later, after a wave theory of light became firmly established as the model for light (as has been explained, neither a particle or wave theory is fully correct). A translation of Newton's essay on light appears in *The large scale structure of space-time,* by Stephen Hawking and George F. R. Ellis.

The fact that light could be polarized was for the first time qualitatively explained by Newton using the particle theory. Étienne-Louis Malus in 1810 created a mathematical particle theory of polarization. Jean-Baptiste Biot in 1812 showed that this theory explained all known phenomena of light polarization. At that time the polarization was considered as the proof of the particle theory.

Wave Theory

Christiaan Huygens

To explain the origin of colors, Robert Hooke (1635-1703) developed a "pulse theory" and compared the spreading of light to that of waves in water in his 1665 work *Micrographia* ("Observation IX"). In 1672 Hooke suggested that light's vibrations could be perpendicular to the direction of propagation. Christiaan Huygens (1629-1695) worked out a mathematical wave theory of light in 1678, and published it in his *Treatise on light* in 1690. He proposed that light was emitted in all directions as a series of waves in a medium called the *Luminiferous ether*. As waves are not affected by gravity, it was assumed that they slowed down upon entering a denser medium.

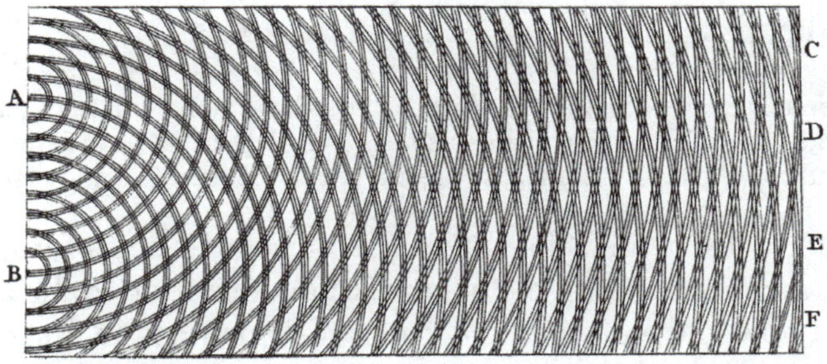

Thomas Young's sketch of a double-slit experiment showing diffraction.
Young's experiments supported the theory that light consists of waves

The wave theory predicted that light waves could interfere with each other like sound waves (as noted around 1800 by Thomas Young). Young showed by means of a diffraction experiment that light behaved as waves. He also proposed that different colours were caused by different wavelengths of light, and explained colour vision in terms of three-coloured receptors in the eye. Another supporter of the wave theory was Leonhard Euler. He argued in *Nova theoria lucis et colorum* (1746) that diffraction could more easily be explained by a wave theory. In 1816 André-Marie Ampère gave Augustin-Jean Fresnel an idea that the polarization of light can be explained by the wave theory if light were a transverse wave.

Later, Fresnel independently worked out his own wave theory of light, and presented it to the Académie des Sciences in 1817. Siméon Denis Poisson added to Fresnel's mathematical work to produce a convincing argument in favour of the wave theory, helping to overturn Newton's corpuscular theory. By the year 1821, Fresnel was able to show via mathematical methods that polarisation could be explained by the wave theory of light and only if light was entirely transverse, with no longitudinal vibration whatsoever.

The weakness of the wave theory was that light waves, like sound waves, would need a medium for transmission. The existence of the hypothetical substance *luminiferous aether* proposed by Huygens in 1678 was cast into strong doubt in the late nineteenth century by the Michelson–Morley experiment.

Newton's corpuscular theory implied that light would travel faster in a denser medium, while the wave theory of Huygens and others implied the opposite. At that time, the speed of light could not be measured accurately enough to decide which theory was correct. The first to make a sufficiently accurate measurement was Léon Foucault, in 1850. His result supported the wave theory, and the classical particle theory was finally abandoned, only to partly re-emerge in the 20th century.

Electromagnetic Theory

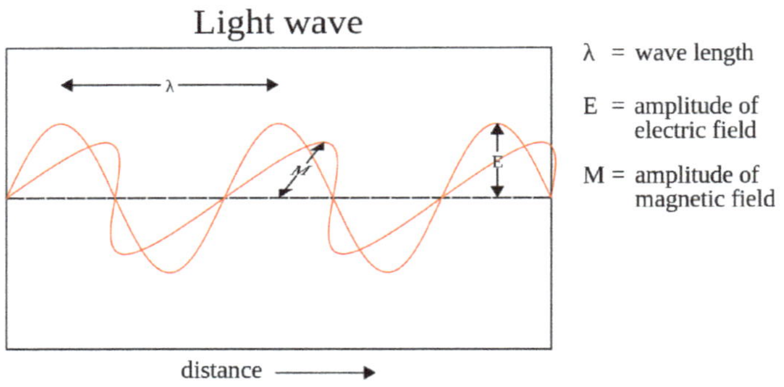

A 3–dimensional rendering of linearly polarised light wave frozen in time and showing the two oscillating components of light; an electric field and a magnetic field perpendicular to each other and to the direction of motion (a transverse wave)

In 1845, Michael Faraday discovered that the plane of polarisation of linearly polarised light is rotated when the light rays travel along the magnetic field direction in the presence of a transparent dielectric, an effect now known as Faraday rotation. This was the first evidence that light was related to electromagnetism. In 1846 he speculated that light might be some form of disturbance propagating along magnetic field lines. Faraday proposed in 1847 that light was a high-frequency electromagnetic vibration, which could propagate even in the absence of a medium such as the ether.

Faraday's work inspired James Clerk Maxwell to study electromagnetic radiation and light. Maxwell discovered that self-propagating electromagnetic waves would travel through space at a constant speed, which happened to be equal to the previously measured speed of light. From this,

Maxwell concluded that light was a form of electromagnetic radiation: he first stated this result in 1862 in *On Physical Lines of Force*. In 1873, he published *A Treatise on Electricity and Magnetism*, which contained a full mathematical description of the behaviour of electric and magnetic fields, still known as Maxwell's equations. Soon after, Heinrich Hertz confirmed Maxwell's theory experimentally by generating and detecting radio waves in the laboratory, and demonstrating that these waves behaved exactly like visible light, exhibiting properties such as reflection, refraction, diffraction, and interference. Maxwell's theory and Hertz's experiments led directly to the development of modern radio, radar, television, electromagnetic imaging, and wireless communications.

In the quantum theory, photons are seen as wave packets of the waves described in the classical theory of Maxwell. The quantum theory was needed to explain effects even with visual light that Maxwell's classical theory could not (such as spectral lines).

Quantum Theory

In 1900 Max Planck, attempting to explain black body radiation suggested that although light, such as baseballs or ocean waves). was a wave, these waves could gain or lose energy only in finite amounts related to their frequency. Planck called these "lumps" of light energy "quanta" (from a Latin word for "how much"). In 1905, Albert Einstein used the idea of light quanta to explain the photoelectric effect, and suggested that these light quanta had a "real" existence. In 1923 Arthur Holly Compton showed that the wavelength shift seen when low intensity X-rays scattered from electrons (so called Compton scattering) could be explained by a particle-theory of X-rays, but not a wave theory. In 1926 Gilbert N. Lewis named these light quanta particles photons.

Eventually the modern theory of quantum mechanics came to picture light as (in some sense) *both* a particle and a wave, and (in another sense), as a phenomenon which is *neither* a particle nor a wave (which actually are macroscopic phenomenaInstead, modern physics sees light as something that can be described sometimes with mathematics appropriate to one type of macroscopic metaphor (particles), and sometimes another macroscopic metaphor (water waves), but is actually something that cannot be fully imagined. As in the case for radio waves and the X-rays involved in Compton scattering, physicists have noted that electromagnetic radiation tends to behave more like a classical wave at lower frequencies, but more like a classical particle at higher frequencies, but never completely loses all qualities of one or the other. Visible light, which occupies a middle ground in frequency, can easily be shown in experiments to be describable using either a wave or particle model, or sometimes both.

In February 2018, scientists reported, for the first time, the discovery of a new form of light, which may involve polaritons, that could be useful in the development of quantum computers.

Photon

The photon is the fundamental particle of visible light.

In some ways, visible light behaves like a wave phenomenon, but in other respects it acts like a stream of high-speed, submicroscopic particles. Isaac Newton was one of the first scientists to theorize that light consists of particles. Modern physicists have demonstrated that the energy in any electromagnetic field is made up of discrete packets. The term photon (meaning "visible-light particle") was coined for these energy packets. Particle-like behavior is not restricted to the visible-light portion of the electromagnetic radiation spectrum, however. Radio waves, infrared rays, visible light, ultraviolet rays, X rays, and gamma rays all consist of photons, each of which contains a particular amount of energy that depends on the wavelength.

Photons travel through empty space at a speed of approximately 186,282miles (299,792 kilometers) per second. This is true no matter what the electromagnetic wavelength. In media other than a vacuum, the speed is reduced. For example, visible light travels more slowly through glass than through outer space. Radio waves travel more slowly through the polyethylene in a transmission line than they do through the atmosphere. The ratio of the speed of the photons in a particular medium to their speed in a vacuum is called the velocity factor . This factor is always between 0 and 1 (or 0 and 100 percent), and it depends to some extent on the wavelength.

The shorter the wavelength of an electromagnetic disturbance, the more energy each photon contains. In fact, this relationship is so precise that a mathematical formula applies. If e represents the energy (the unit of measurement is the joule) contained in each photon and s represents the electromagnetic wavelength (in meters), then

$$e = hc / s$$

where h is Planck's constant (approximately equal to6.626 times 10^{-34} joule-second) and c is the speed of electromagnetic-field propagation in the medium in question (approximately 2.998 times 10^{8} meters per second in a vacuum). A simpler formula applies to frequency. If f represents the frequency of an electromagnetic field (in hertz), then

$$e = hf$$

The energy contained in a single photon does not depend on the intensity of the radiation. At any specific wavelength -- say, the wavelength of light emitted by a helium-neon laser -- every photon contains exactly the same amount of energy, whether the source appears as dim as a candle or as bright as the sun. The brilliance or intensity is a function of the number of photons striking a given surface area per unit time.

Speed of Light

Speed of light, speed at which light waves propagate through different materials. In particular, the value for the speed of light in a vacuum is now defined as exactly 299,792,458 metres per second.

The speed of light is considered a fundamental constant of nature. Its significance is far broader than its role in describing a property of electromagnetic waves. It serves as the single limiting velocity in the universe, being an upper bound to the propagation speed of signals and to the speeds of all material particles. In the famous relativity equation, $E = mc^2$, the speed of light (c) serves as a constant of proportionality, linking the formerly disparate concepts of mass (m) and energy (E).

Numerical Value, Notation, and Units

The speed of light in vacuum is usually denoted by a lowercase *c*, for "constant" or the Latin *celeritas* (meaning "swiftness, celerity"). In 1856, Wilhelm Eduard Weber and Rudolf Kohlrausch had used *c* for a different constant later shown to equal √2 times the speed of light in vacuum. Historically, the symbol *V* was used as an alternative symbol for the speed of light, introduced by James Clerk Maxwell in 1865. In 1894, Paul Drude redefined *c* with its modern meaning. Einstein used *V* in his original German-language papers on special relativity in 1905, but in 1907 he switched to *c*, which by then had become the standard symbol for the speed of light.

Sometimes *c* is used for the speed of waves in *any* material medium, and c_0 for the speed of light in vacuum. This subscripted notation, which is endorsed in official SI literature, has the same form as other related constants: namely, μ_0 for the vacuum permeability or magnetic constant, ε_0 for the vacuum permittivity or electric constant, and Z_0 for the impedance of free space. This article uses *c* exclusively for the speed of light in vacuum.

Since 1983, the metre has been defined in the International System of Units (SI) as the distance light travels in vacuum in $1/299\,792\,458$ of a second. This definition fixes the speed of light in vacuum at exactly 299,792,458 m/s. As a dimensional physical constant, the numerical value of *c* is different for different unit systems. In branches of physics in which *c* appears often, such as in relativity, it is common to use systems of natural units of measurement or the geometrized unit system where $c = 1$. Using these units, *c* does not appear explicitly because multiplication or division by 1 does not affect the result.

Fundamental Role in Physics

The speed at which light waves propagate in vacuum is independent both of the motion of the wave source and of the inertial frame of reference of the observer. This invariance of the speed of light was postulated by Einstein in 1905, after being motivated by Maxwell's theory of electromagnetism and the lack of evidence for the luminiferous aether; it has since been consistently confirmed by many experiments. It is only possible to verify experimentally that the two-way speed of light (for example, from a source to a mirror and back again) is frame-independent, because it is impossible to measure the one-way speed of light (for example, from a source to a distant detector) without some convention as to how clocks at the source and at the detector should be synchronized. However, by adopting Einstein synchronization for the clocks, the one-way speed of light becomes equal to the two-way speed of light by definition. The special theory of relativity explores the consequences of this invariance of *c* with the assumption that the laws of physics are the same in all inertial frames of reference. One consequence is that *c* is the speed at which all massless particles and waves, including light, must travel in vacuum.

Special relativity has many counterintuitive and experimentally verified implications. These include the equivalence of mass and energy ($E = mc^2$), length contraction (moving objects shorten), and time dilation (moving clocks run more slowly). The factor γ by which lengths contract and times dilate is known as the Lorentz factor and is given by $\gamma = (1 - v^2/c^2)^{-1/2}$, where v is the speed of the object. The difference of γ from 1 is negligible for speeds much slower than *c*, such as most everyday speeds—in which case special relativity is closely approximated by Galilean relativity—but it increases at relativistic speeds and diverges to infinity as v approaches *c*. For example, a time dilation factor of $\gamma = 2$ occurs at a relative velocity of 86.6% of the speed of light ($v = .866c$). Similarly, a time dilation factor of $\gamma = 10$ occurs at $v = 99.5\%\ c$.

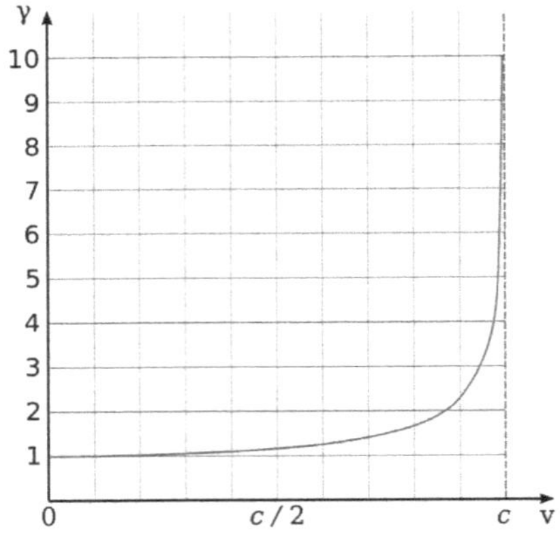

The Lorentz factor γ as a function of velocity. It starts at 1 and approaches infinity as v approaches c

The results of special relativity can be summarized by treating space and time as a unified structure known as spacetime (with c relating the units of space and time), and requiring that physical theories satisfy a special symmetry called Lorentz invariance, whose mathematical formulation contains the parameter c. Lorentz invariance is an almost universal assumption for modern physical theories, such as quantum electrodynamics, quantum chromodynamics, the Standard Model of particle physics, and general relativity. As such, the parameter c is ubiquitous in modern physics, appearing in many contexts that are unrelated to light. For example, general relativity predicts that c is also the speed of gravity and of gravitational waves. In non-inertial frames of reference (gravitationally curved spacetime or accelerated reference frames), the *local* speed of light is constant and equal to c, but the speed of light along a trajectory of finite length can differ from c, depending on how distances and times are defined.

It is generally assumed that fundamental constants such as c have the same value throughout spacetime, meaning that they do not depend on location and do not vary with time. However, it has been suggested in various theories that the speed of light may have changed over time. No conclusive evidence for such changes has been found, but they remain the subject of ongoing research.

It also is generally assumed that the speed of light is isotropic, meaning that it has the same value regardless of the direction in which it is measured. Observations of the emissions from nuclear energy levels as a function of the orientation of the emitting nuclei in a magnetic field, and of rotating optical resonators have put stringent limits on the possible two-way anisotropy.

Upper Limit on Speeds

According to special relativity, the energy of an object with rest mass m and speed v is given by γmc^2, where γ is the Lorentz factor defined above. When v is zero, γ is equal to one, giving rise to the famous $E = mc^2$ formula for mass–energy equivalence. The γ factor approaches infinity as v approaches c, and it would take an infinite amount of energy to accelerate an object with mass to the speed of light. The speed of light is the upper limit for the speeds of objects with positive rest mass, and individual photons cannot travel faster than the speed of light. This is experimentally established in many tests of relativistic energy and momentum.

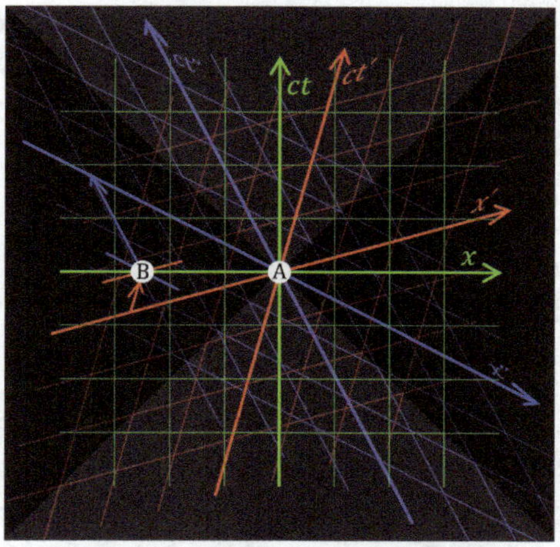

Event A precedes B in the red frame, is simultaneous with B in the green frame, and follows B in the blue frame

More generally, it is normally impossible for information or energy to travel faster than *c*. One argument for this follows from the counter-intuitive implication of special relativity known as the relativity of simultaneity. If the spatial distance between two events A and B is greater than the time interval between them multiplied by *c* then there are frames of reference in which A precedes B, others in which B precedes A, and others in which they are simultaneous. As a result, if something were travelling faster than *c* relative to an inertial frame of reference, it would be travelling backwards in time relative to another frame, and causality would be violated. In such a frame of reference, an "effect" could be observed before its "cause". Such a violation of causality has never been recorded, and would lead to paradoxes such as the tachyonic antitelephone.

Faster-than-light Observations and Experiments

There are situations in which it may seem that matter, energy, or information travels at speeds greater than *c*, but they do not. For example, as is discussed in the propagation of light in a medium section below, many wave velocities can exceed *c*. For example, the phase velocity of X-rays through most glasses can routinely exceed *c*, but phase velocity does not determine the velocity at which waves convey information.

If a laser beam is swept quickly across a distant object, the spot of light can move faster than *c*, although the initial movement of the spot is delayed because of the time it takes light to get to the distant object at the speed *c*. However, the only physical entities that are moving are the laser and its emitted light, which travels at the speed *c* from the laser to the various positions of the spot. Similarly, a shadow projected onto a distant object can be made to move faster than *c*, after a delay in time. In neither case does any matter, energy, or information travel faster than light.

The rate of change in the distance between two objects in a frame of reference with respect to which both are moving (their closing speed) may have a value in excess of *c*. However, this does not represent the speed of any single object as measured in a single inertial frame.

Certain quantum effects appear to be transmitted instantaneously and therefore faster than *c*, as in the EPR paradox. An example involves the quantum states of two particles that can be entangled. Until either of the particles is observed, they exist in a superposition of two quantum states. If the particles are separated and one particle's quantum state is observed, the other particle's quantum state is determined instantaneously (i.e., faster than light could travel from one particle to the other). However, it is impossible to control which quantum state the first particle will take on when it is observed, so information cannot be transmitted in this manner.

Another quantum effect that predicts the occurrence of faster-than-light speeds is called the Hartman effect: under certain conditions the time needed for a virtual particle to tunnel through a barrier is constant, regardless of the thickness of the barrier. This could result in a virtual particle crossing a large gap faster-than-light. However, no information can be sent using this effect.

So-called superluminal motion is seen in certain astronomical objects, such as the relativistic jets of radio galaxies and quasars. However, these jets are not moving at speeds in excess of the speed of light: the apparent superluminal motion is a projection effect caused by objects moving near the speed of light and approaching Earth at a small angle to the line of sight: since the light which was emitted when the jet was farther away took longer to reach the Earth, the time between two successive observations corresponds to a longer time between the instants at which the light rays were emitted.

In models of the expanding universe, the farther galaxies are from each other, the faster they drift apart. This receding is not due to motion *through* space, but rather to the expansion of space itself. For example, galaxies far away from Earth appear to be moving away from the Earth with a speed proportional to their distances. Beyond a boundary called the Hubble sphere, the rate at which their distance from Earth increases becomes greater than the speed of light.

Propagation of Light

In classical physics, light is described as a type of electromagnetic wave. The classical behaviour of the electromagnetic field is described by Maxwell's equations, which predict that the speed c with which electromagnetic waves (such as light) propagate through the vacuum is related to the distributed capacitance and inductance of the vacuum otherwise known as the electric constant ε_0 and the magnetic constant μ_0 by the equation

$$c = \frac{1}{\sqrt{\varepsilon_0 \mu_0}}.$$

In modern quantum physics, the electromagnetic field is described by the theory of quantum electrodynamics (QED). In this theory, light is described by the fundamental excitations (or quanta) of the electromagnetic field, called photons. In QED, photons are massless particles and thus, according to special relativity, they travel at the speed of light in vacuum.

Extensions of QED in which the photon has a mass have been considered. In such a theory, its speed would depend on its frequency, and the invariant speed c of special relativity would then be the upper limit of the speed of light in vacuum. No variation of the speed of light with frequency has been observed in rigorous testing, putting stringent limits on the mass of the photon. The limit obtained depends on the model used: if the massive photon is described by Proca theory, the experimental upper bound for its mass is about 10^{-57} grams; if photon mass is generated by a Higgs mechanism, the experimental upper limit is less sharp, $m \leq 10^{-14}$ eV/c^2 (roughly 2×10^{-47} g).

Another reason for the speed of light to vary with its frequency would be the failure of special relativity to apply to arbitrarily small scales, as predicted by some proposed theories of quantum gravity. In 2009, the observation of the spectrum of gamma-ray burst GRB 090510did not find any difference in the speeds of photons of different energies, confirming that Lorentz invariance is verified at least down to the scale of the Planck length ($l_P = \sqrt{\hbar G/c^3} \approx 1.6163 \times 10^{-35}$ m) divided by 1.2.

In a Medium

In a medium, light usually does not propagate at a speed equal to *c*; further, different types of light wave will travel at different speeds. The speed at which the individual crests and troughs of a plane wave (a wave filling the whole space, with only one frequency) propagate is called the phase velocity v_p. An actual physical signal with a finite extent (a pulse of light) travels at a different speed. The largest part of the pulse travels at the group velocity v_g, and its earliest part travels at the front velocity v_f.

The phase velocity is important in determining how a light wave travels through a material or from one material to another. It is often represented in terms of a *refractive index*. The refractive index of a material is defined as the ratio of *c* to the phase velocity v_p in the material: larger indices of refraction indicate lower speeds. The refractive index of a material may depend on the light's frequency, intensity, polarization, or direction of propagation; in many cases, though, it can be treated as a material-dependent constant. The refractive index of air is approximately 1.0003. Denser media, such as water, glass, and diamond, have refractive indexes of around 1.3, 1.5 and 2.4, respectively, for visible light. In exotic materials like Bose–Einstein condensates near absolute zero, the effective speed of light may be only a few metres per second. However, this represents absorption and re-radiation delay between atoms, as do all slower-than-*c* speeds in material substances. As an extreme example of light "slowing" in matter, two independent teams of physicists claimed to bring light to a "complete standstill" by passing it through a Bose–Einstein condensate of the element rubidium, one team at Harvard University and the Rowland Institute for Science in Cambridge, Mass., and the other at the Harvard–Smithsonian Center for Astrophysics, also in Cambridge. However, the popular description of light being "stopped" in these experiments refers only to light being stored in the excited states of atoms, then re-emitted at an arbitrarily later time, as stimulated by a second laser pulse. During the time it had "stopped," it had ceased to be light. This type of behaviour is generally microscopically true of all transparent media which "slow" the speed of light.

In transparent materials, the refractive index generally is greater than 1, meaning that the phase velocity is less than *c*. In other materials, it is possible for the refractive index to become smaller than 1 for some frequencies; in some exotic materials it is even possible for the index of refraction to become negative. The requirement that causality is not violated implies that the real and imaginary parts of the dielectric constant of any material, corresponding respectively to the index of refraction and to the attenuation coefficient, are linked by the Kramers–Kronig relations. In practical terms, this means that in a material with refractive index less than 1, the absorption of the wave is so quick that no signal can be sent faster than *c*.

A pulse with different group and phase velocities (which occurs if the phase velocity is not the same for all the frequencies of the pulse) smears out over time, a process known as dispersion. Certain materials have an exceptionally low (or even zero) group velocity for light waves, a phenomenon called slow light, which has been confirmed in various experiments. The opposite, group velocities exceeding *c*, has also been shown in experiment. It should even be possible for the group velocity to become infinite or negative, with pulses travelling instantaneously or backwards in time.

None of these options, however, allow information to be transmitted faster than *c*. It is impossible to transmit information with a light pulse any faster than the speed of the earliest part of the pulse (the front velocity). It can be shown that this is (under certain assumptions) always equal to *c*.

It is possible for a particle to travel through a medium faster than the phase velocity of light in that medium (but still slower than *c*). When a charged particle does that in a dielectricmaterial, the electromagnetic equivalent of a shock wave, known as Cherenkov radiation, is emitted.

Practical Effects of Finiteness

The speed of light is of relevance to communications: the one-way and round-trip delay timeare greater than zero. This applies from small to astronomical scales. On the other hand, some techniques depend on the finite speed of light, for example in distance measurements.

Small Scales

In supercomputers, the speed of light imposes a limit on how quickly data can be sent between processors. If a processor operates at 1 gigahertz, a signal can only travel a maximum of about 30 centimetres (1 ft) in a single cycle. Processors must therefore be placed close to each other to minimize communication latencies; this can cause difficulty with cooling. If clock frequencies continue to increase, the speed of light will eventually become a limiting factor for the internal design of single chips.

Large Distances on Earth

G the Earth is about 40,075 km and that *c* is about 300,000 km/s, the theoretical shortest time for a piece of information to travel half the globe along the surface is about 67 milliseconds. When light is travelling around the globe in an optical fibre, the actual transit time is longer, in part because the speed of light is slower by about 35% in an optical fibre, depending on its refractive index *n*. Furthermore, straight lines rarely occur in global communications situations, and delays are created when the signal passes through an electronic switch or signal regenerator.

Spaceflights and Astronomy

Similarly, communications between the Earth and spacecraft are not instantaneous. There is a brief delay from the source to the receiver, which becomes more noticeable as distances increase. This delay was significant for communications between ground control and Apollo 8 when it became the first manned spacecraft to orbit the Moon: for every question, the ground control station had to wait at least three seconds for the answer to arrive. The communications delay between Earth and Mars can vary between five and twenty minutes depending upon the relative positions of the two planets. As a consequence of this, if a robot on the surface of Mars were to encounter a problem, its human controllers would not be aware of it until at least five minutes later, and possibly up to twenty minutes later; it would then take a further five to twenty minutes for instructions to travel from Earth to Mars.

NASA must wait several hours for information from a probe orbiting Jupiter, and if it needs to correct a navigation error, the fix will not arrive at the spacecraft for an equal amount of time, creating a risk of the correction not arriving in time.

Receiving light and other signals from distant astronomical sources can even take much longer. For example, it has taken 13 billion (13×10^9) years for light to travel to Earth from the faraway galaxies viewed in the Hubble Ultra Deep Field images. Those photographs, taken today, capture images of the galaxies as they appeared 13 billion years ago, when the universe was less than a billion years old. The fact that more distant objects appear to be younger, due to the finite speed of light, allows astronomers to infer the evolution of stars, of galaxies, and of the universe itself.

Astronomical distances are sometimes expressed in light-years, especially in popular sciencepublications and media. A light-year is the distance light travels in one year, around 9461 billion kilometres, 5879 billion miles, or 0.3066 parsecs. In round figures, a light year is nearly 10 trillion kilometres or nearly 6 trillion miles. Proxima Centauri, the closest star to Earth after the Sun, is around 4.2 light-years away.

Distance Measurement

Radar systems measure the distance to a target by the time it takes a radio-wave pulse to return to the radar antenna after being reflected by the target: the distance to the target is half the round-trip transit time multiplied by the speed of light. A Global Positioning System (GPS) receiver measures its distance to GPS satellites based on how long it takes for a radio signal to arrive from each satellite, and from these distances calculates the receiver's position. Because light travels about 300,000 kilometres (186000 mi) in one second, these measurements of small fractions of a second must be very precise. The Lunar Laser Ranging Experiment, radar astronomy and the Deep Space Network determine distances to the Moon, planets and spacecraft, respectively, by measuring round-trip transit times.

High-frequency Trading

The speed of light has become important in high-frequency trading, where traders seek to gain minute advantages by delivering their trades to exchanges fractions of a second ahead of other traders. For example, traders have been switching to microwave communications between trading hubs, because of the advantage which microwaves travelling at near to the speed of light in air, have over fibre optic signals which travel 30–40% slower at the speed of light through glass.

Measurement

There are different ways to determine the value of *c*. One way is to measure the actual speed at which light waves propagate, which can be done in various astronomical and earth-based setups. However, it is also possible to determine *c* from other physical laws where it appears, for example, by determining the values of the electromagnetic constants ε_0 and μ_0 and using their relation to *c*. Historically, the most accurate results have been obtained by separately determining the frequency and wavelength of a light beam, with their product equalling *c*.

In 1983 the metre was defined as "the length of the path travelled by light in vacuum during a time interval of 1⁄299 792 458 of a second", fixing the value of the speed of light at 299,792,458 m/s by definition. Consequently, accurate measurements of the speed of light yield an accurate realization of the metre rather than an accurate value of *c*.

Astronomical Measurements

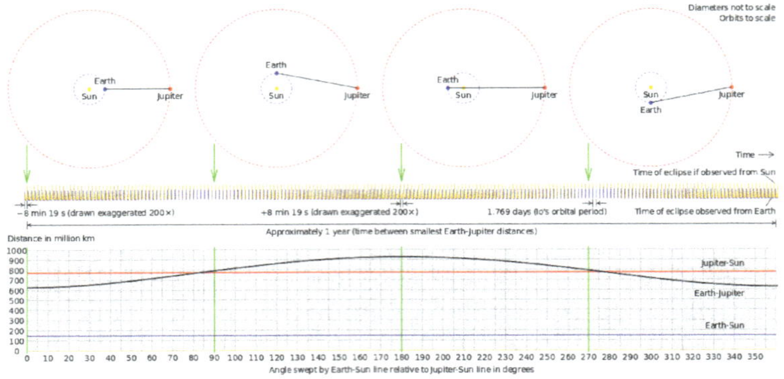

Measurement of the speed of light using the eclipse of Io by Jupiter

Outer space is a convenient setting for measuring the speed of light because of its large scale and nearly perfect vacuum. Typically, one measures the time needed for light to traverse some reference distance in the solar system, such as the radius of the Earth's orbit. Historically, such measurements could be made fairly accurately, compared to how accurately the length of the reference distance is known in Earth-based units. It is customary to express the results in astronomical units (AU) per day.

Ole Christensen Rømer used an astronomical measurement to make the first quantitative estimate of the speed of light. When measured from Earth, the periods of moons orbiting a distant planet are shorter when the Earth is approaching the planet than when the Earth is receding from it. The distance travelled by light from the planet (or its moon) to Earth is shorter when the Earth is at the point in its orbit that is closest to its planet than when the Earth is at the farthest point in its orbit, the difference in distance being the diameter of the Earth's orbit around the Sun. The observed change in the moon's orbital period is caused by the difference in the time it takes light to traverse the shorter or longer distance. Rømer observed this effect for Jupiter's innermost moon Io and deduced that light takes 22 minutes to cross the diameter of the Earth's orbit.

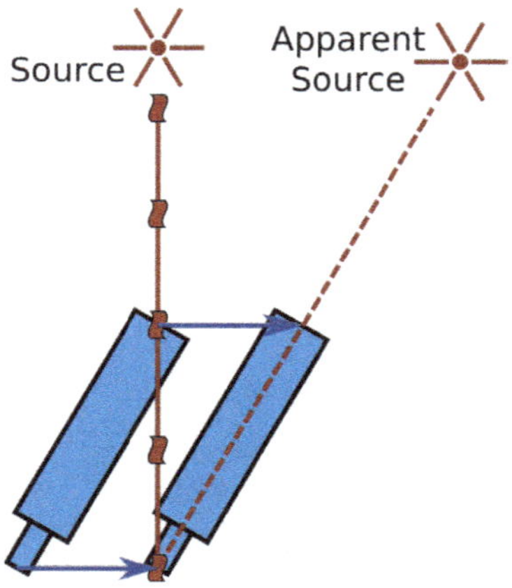

Aberration of light: light from a distant source appears to be from a different location for a moving telescope due to the finite speed of light

Another method is to use the aberration of light, discovered and explained by James Bradley in the 18th century. This effect results from the vector addition of the velocity of light arriving from a distant source (such as a star) and the velocity of its observer. A moving observer thus sees the light coming from a slightly different direction and consequently sees the source at a position shifted from its original position. Since the direction of the Earth's velocity changes continuously as the Earth orbits the Sun, this effect causes the apparent position of stars to move around. From the angular difference in the position of stars (maximally 20.5 arcseconds) it is possible to express the speed of light in terms of the Earth's velocity around the Sun, which with the known length of a year can be converted to the time needed to travel from the Sun to the Earth. In 1729, Bradley used this method to derive that light travelled 10,210 times faster than the Earth in its orbit (the modern figure is 10,066 times faster) or, equivalently, that it would take light 8 minutes 12 seconds to travel from the Sun to the Earth.

Astronomical Unit

An astronomical unit (AU) is approximately the average distance between the Earth and Sun. It was redefined in 2012 as exactly 149,597,870,700 m. Previously the AU was not based on the International System of Units but in terms of the gravitational force exerted by the Sun in the framework of classical mechanics. The current definition uses the recommended value in metres for the previous definition of the astronomical unit, which was determined by measurement. This redefinition is analogous to that of the metre, and likewise has the effect of fixing the speed of light to an exact value in astronomical units per second (via the exact speed of light in metres per second).

Previously, the inverse of *c* expressed in seconds per astronomical unit was measured by comparing the time for radio signals to reach different spacecraft in the Solar System, with their position calculated from the gravitational effects of the Sun and various planets. By combining many such measurements, a best fit value for the light time per unit distance could be obtained. For example, in 2009, the best estimate, as approved by the International Astronomical Union (IAU), was:

light time for unit distance: $t_{au} = 499.004\ 783\ 836(10)$ s

$c = 0.00200398880410(4)$ AU/s $= 173.144\ 632\ 674(3)$ AU/day.

The relative uncertainty in these measurements is 0.02 parts per billion (2×10^{-11}), equivalent to the uncertainty in Earth-based measurements of length by interferometry.Since the metre is defined to be the length travelled by light in a certain time interval, the measurement of the light time in terms of the previous definition of the astronomical unit can also be interpreted as measuring the length of an AU (old definition) in metres.

Time of Flight Techniques

A method of measuring the speed of light is to measure the time needed for light to travel to a mirror at a known distance and back. This is the working principle behind the Fizeau–Foucault apparatus developed by Hippolyte Fizeau and Léon Foucault.

The setup as used by Fizeau consists of a beam of light directed at a mirror 8 kilometres (5 mi) away. On the way from the source to the mirror, the beam passes through a rotating cogwheel. At a certain rate of rotation, the beam passes through one gap on the way out and another on the way back, but at slightly higher or lower rates, the beam strikes a tooth and does not pass through the wheel. Knowing the distance between the wheel and the mirror, the number of teeth on the wheel, and the rate of rotation, the speed of light can be calculated.

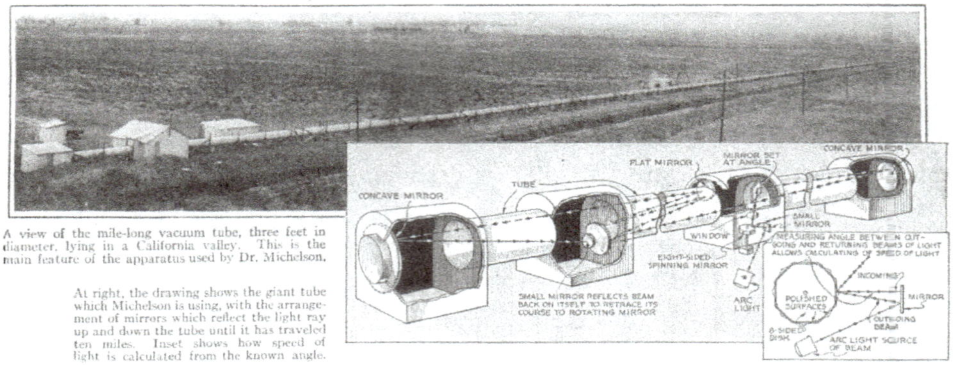

One of the last and most accurate time of flight measurements, Michelson, Pease and Pearson's 1930–35 experiment used a rotating mirror and a one-mile (1.6 km) long vacuum chamber which the light beam traversed 10 times. It achieved accuracy of ±11 km/s

The method of Foucault replaces the cogwheel by a rotating mirror. Because the mirror keeps rotating while the light travels to the distant mirror and back, the light is reflected from the rotating mirror at a different angle on its way out than it is on its way back. From this difference in angle, the known speed of rotation and the distance to the distant mirror the speed of light may be calculated.

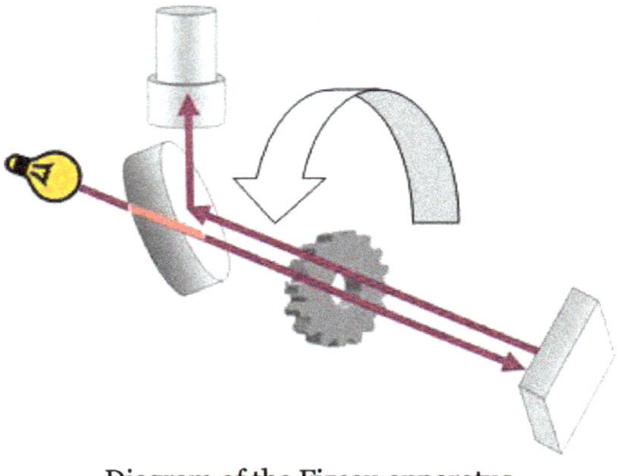

Diagram of the Fizeau apparatus

Nowadays, using oscilloscopes with time resolutions of less than one nanosecond, the speed of light can be directly measured by timing the delay of a light pulse from a laser or an LED reflected from a mirror. This method is less precise (with errors of the order of 1%) than other modern techniques, but it is sometimes used as a laboratory experiment in college physics classes.

Electromagnetic Constants

An option for deriving c that does not directly depend on a measurement of the propagation of electromagnetic waves is to use the relation between c and the vacuum permittivity ε_0 and vacuum permeability μ_0 established by Maxwell's theory: $c^2 = 1/(\varepsilon_0\mu_0)$. The vacuum permittivity may be determined by measuring the capacitance and dimensions of a capacitor, whereas the value of the vacuum permeability is fixed at exactly $4\pi \times 10^{-7}$ H·m^{-1} through the definition of the ampere. Rosa and Dorsey used this method in 1907 to find a value of 299,710±22 km/s.

Cavity Resonance

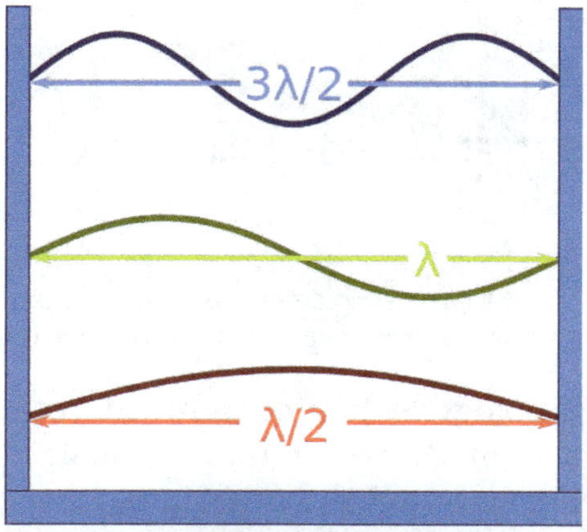

Electromagnetic standing waves in a cavity

Another way to measure the speed of light is to independently measure the frequency f and wavelength λ of an electromagnetic wave in vacuum. The value of c can then be found by using the relation $c = f\lambda$. One option is to measure the resonance frequency of a cavity resonator. If the dimensions of the resonance cavity are also known, these can be used to determine the wavelength of the wave. In 1946, Louis Essen and A.C. Gordon-Smith established the frequency for a variety of normal modes of microwaves of a microwave cavityof precisely known dimensions. The dimensions were established to an accuracy of about ±0.8 μm using gauges calibrated by interferometry. As the wavelength of the modes was known from the geometry of the cavity and from electromagnetic theory, knowledge of the associated frequencies enabled a calculation of the speed of light.

The Essen–Gordon-Smith result, 299,792±9 km/s, was substantially more precise than those found by optical techniques. By 1950, repeated measurements by Essen established a result of 299,792.5±3.0 km/s.

A household demonstration of this technique is possible, using a microwave oven and food such as marshmallows or margarine: if the turntable is removed so that the food does not move, it will cook the fastest at the antinodes (the points at which the wave amplitude is the greatest), where it will begin to melt. The distance between two such spots is half the wavelength of the microwaves; by measuring this distance and multiplying the wavelength by the microwave frequency (usually displayed on the back of the oven, typically 2450 MHz), the value of c can be calculated, "often with less than 5% error".

Interferometry

Interferometry is another method to find the wavelength of electromagnetic radiation for determining the speed of light. A coherentbeam of light (e.g. from a laser), with a known frequency (f), is split to follow two paths and then recombined. By adjusting the path length while observing the interference pattern and carefully measuring the change in path length, the wavelength of the light (λ) can be determined. The speed of light is then calculated using the equation $c = \lambda f$.

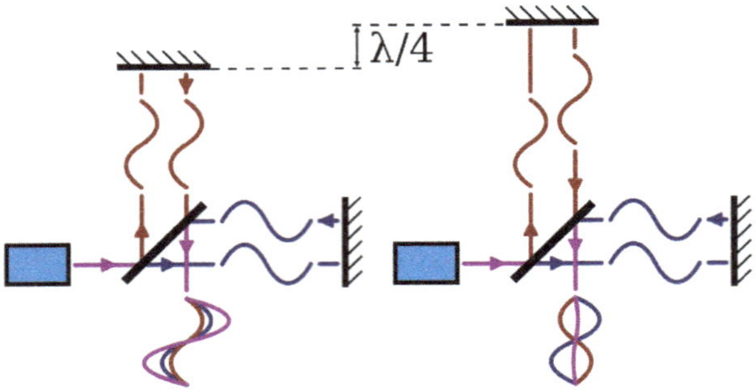

An interferometric determination of length.
Left: constructive interference; Right: destructive interference

Before the advent of laser technology, coherent radio sources were used for interferometry measurements of the speed of light. However interferometric determination of wavelength becomes less precise with wavelength and the experiments were thus limited in precision by the long wavelength (~0.4 cm (0.16 in)) of the radiowaves. The precision can be improved by using light with a shorter wavelength, but then it becomes difficult to directly measure the frequency of the light. One way around this problem is to start with a low frequency signal of which the frequency can be precisely measured, and from this signal progressively synthesize higher frequency signals whose frequency can then be linked to the original signal. A laser can then be locked to the frequency, and its wavelength can be determined using interferometry. This technique was due to a group at the National Bureau of Standards (NBS) (which later became NIST). They used it in 1972 to measure the speed of light in vacuum with a fractional uncertainty of 3.5×10^{-9}.

Light-year

The distance light travels in the course of a year is called a light-year. A light-year is a measure of both time and distance. It is not as hard to understand as it seems. Think of it this way: Light travels from the moon to our eyes in about 1 second, which means the moon is about 1 light-second away. Sunlight takes about 8 minutes to reach our eyes, so the sun is about 8 light-minutes away. Light from the nearest star system, Alpha Centauri, is requires roughly 4.3 years to get here, so that star system is said to be 4.3 light-years away.

“To obtain an idea of the size of a light-year, take the circumference of the Earth (24,900 miles), lay it out in a straight line, multiply the length of the line by 7.5 (the corresponding distance is one light-second), then place 31.6 million similar lines end to end,” NASA’s Glenn Research center writes on its website. “The resulting distance is almost 6 trillion (6,000,000,000,000) miles.”

Stars and other objects beyond our solar system lie anywhere from a few light-years to a few billion light-years away. Thus, when astronomers study objects that lie a light-year away or more, they are seeing it as existed at the time that light left it, not as it would appear if they stood near its surface today. In this sense, everything we see in the distant universe is, literally, history.

This principle allows astronomers to see how the universe as it looked after the Big Bang, which took place about 13.8 billion years ago. Examining objects that are, say, 10 billion light-years away, we see them as they looked 10 billion years ago, relatively soon after the beginning of the universe, rather than how they appear today.

Is the Speed of Light Really Constant?

Light travels in waves, and, like sound, can be slowed depending on what it is traveling through. Nothing can outpace light in a vacuum. However, if a region contains any matter, even dust, light can bend when it comes in contact with the particles, which results in a decrease in speed.

Light traveling through Earth's atmosphere moves almost as fast as light in a vacuum, while light passing through a diamond is slowed to less than half that speed. Still, it travels through the gem at over 277 million mph (almost 124,000 km/s).

Is the Speed of Light Slowing Down?

Modern physics rests on the foundational notion that the speed of light is a constant, which in a vacuum is 186,000 miles per second (299,792 km/s). Einstein established this within his theory of general relativity, first developed in 1906 when he was just 26 years-old. But what if it doesn't? A few albeit controversial incidents in recent years challenge the idea that light always travels at a constant speed. And in fact, we've known for a long time that there are several phenomena that travel faster than light, without violating the theory of relativity.

For instance, whereas traveling faster than sound creates a sonic boom, traveling faster than light creates a "luminal boom." Russian scientist Pavel Alekseyevich Cherenkov discovered this in 1934, which won him the Nobel Prize in Physics in 1958. Cherenkov radiation can be observed in the core of a nuclear reactor. When the core is submerged in water to cool it, electrons move through the water faster than the speed of light, causing a luminal boom.

On another front, while no particle with mass can travel faster than light, the fabric of space can and does. According to Inflation Theory, immediately after the Big Bang, the universe doubled in size and then doubled again, in less than a trillionth of a trillionth of a second, much faster than the speed of light. More recently, astronomers have discovered that some galaxies, the distant ones anyway, move away from us faster than light speed, supposedly, pushed along by dark energy. The best estimate for the rate of acceleration for the universe is 68 kilometers per second per mega-parsec.

Quantum entanglement is another example of a faster-than-light interaction that doesn't violate Einstein's theory. When two particles are entangled, one can travel to its partner instantaneously, even if its mate is on the other side of the universe. Einstein called this, "Spooky action at a distance." The last example is a theoretical one (at least for now). If we were somehow able to warp or fold space-time, such as with a wormhole, it would allow a spacecraft to pass instantaneously from one side of space to another.

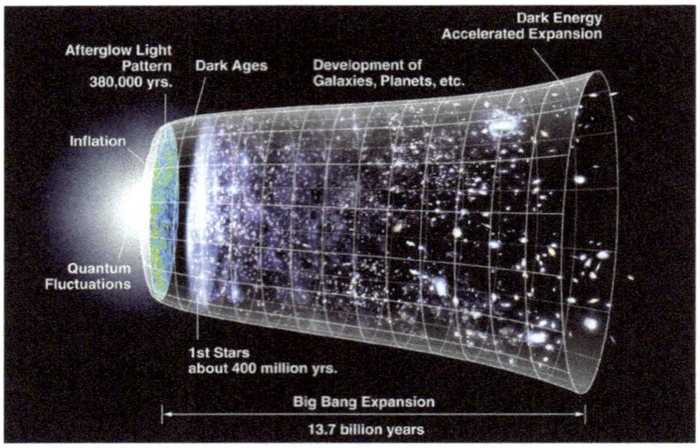

Einstein says that light acts pretty much the same throughout the universe. There's a problem though. Today, scientists marvel at just how homogenous the universe is. One way we can tell, is by investigating the cosmic microwave background (CMB). This is essentially the light left over from the Big Bang, located in every corner of the universe.

No matter where you examine it, it's always the same temperature, -454 F° (-270 C°). If that's the case and light travels at a constant speed, how could it have made it from one edge of the universe to the other? To date, scientists have no idea, other than to say, some peculiar conditions must have existed in that early "inflation field."

The idea of light slowing down over time was first proposed by Professor João Magueijo, from Imperial College London and his colleague, Dr. Niayesh Afshordi, of the Perimeter Institute in Canada. Their paper was submitted to Astrophysics in late 1998 and published shortly thereafter. Unfortunately, the proper instrumentation necessary to investigate the CMB to search for clues supporting it, wasn't available at the time.

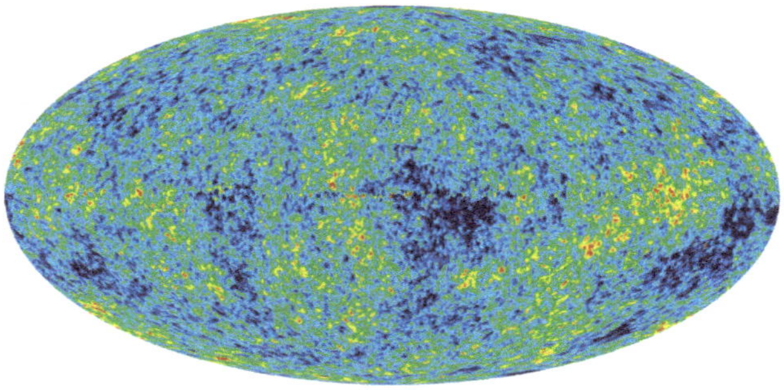

The cosmic microwave background

Magueijo and Afshordi eliminated the inflation field altogether. Instead, they argue that the intense heat that existed when the universe was young, ten thousand trillion trillion C°, allowed particles—

including photons (light particles), to move at an infinite speed. Light therefore traveled to every point in the universe, causing a uniformity in the CMB that we can observe today. "We can say what the fluctuations in the early universe would have looked like," Afshordi told The Guardian, "and these are the fluctuations that grow to form planets, stars, and galaxies." An experiment the following year lent credence to Magueijo and Afshordi's theory.

Reflection of Light

When a ray of light falls on a polished and shining surface of an object, then it is sent back in the same medium. The process of sending back the light rays after falling on the shining surface of an object is called reflection of light.

Reflection of Light from a Plane Mirror

When a ray of light (AO) falls on a plane mirror at the point O then it gets reflected by the mirror in the direction OB. Now we shall discuss some important terms related with the reflection of light by a mirror.

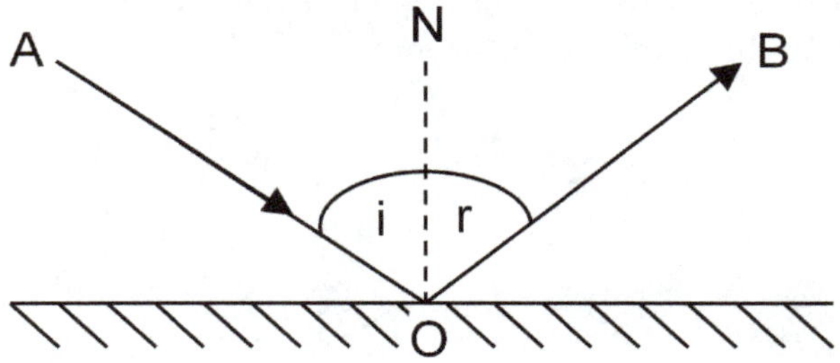

Incident Ray

The ray of light which falls on the surface of mirror is called incident ray. In figure AO is the incident ray of light.

Point of Incidence

The point at the surface of mirror at which the incident ray falls is called the point of incidence. In figure the point O on the surface of mirror is the point of incident.

Reflected Ray

When the incident ray falls on the surface of a mirror, the mirror sends it back in another direction (in same medium). The ray of light which is sent back by the mirror is called reflected ray.

Normal

Normal may be defined as a line which is perpendicular to the mirror at the point of incidence. In figure line ON is normal to the mirror at point O. the normal is always represented by a dotted line.

Angle of Incidence

The angle of incidence may be defined as the angle which is made by the incident ray with normal at the point of incident. It is denoted by the letter i. In figure the angle AON between the incident ray AO and the normal ON is the angle of incidence.

Angle of Reflection

The angle of reflection may be defined as the angle which is made by the reflected ray with normal at the point of reflection. It is denoted by the letter r. In figure the angle NOB between the reflected ray OB and the normal ON is the angle of incidence.

Reflection of light is either *specular* (mirror-like) or *diffuse* (retaining the energy, but losing the image) depending on the nature of the interface. In specular reflection the phase of the reflected waves depends on the choice of the origin of coordinates, but the relative phase between s and p (TE and TM) polarizations is fixed by the properties of the media and of the interface between them.

A mirror provides the most common model for specular light reflection, and typically consists of a glass sheet with a metallic coating where the significant reflection occurs. Reflection is enhanced in metals by suppression of wave propagation beyond their skin depths. Reflection also occurs at the surface of transparent media, such as water or glass.

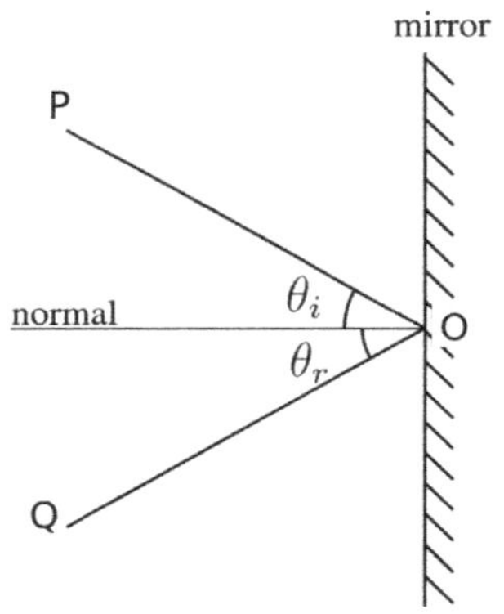

Diagram of specular reflection

In the diagram, a light ray PO strikes a vertical mirror at point O, and the reflected ray is OQ. By projecting an imaginary line through point O perpendicular to the mirror, known as the *normal*, we can measure the *angle of incidence*, θ_i and the *angle of reflection*, θ_r. The *law of reflection* states that $\theta_i = \theta_r$, *or in other words, the angle of incidence equals the angle of reflection.*

In fact, reflection of light may occur whenever light travels from a medium of a given refractive index into a medium with a different refractive index. In the most general case, a certain fraction of the light is reflected from the interface, and the remainder is refracted. Solving Maxwell's equations for a light ray striking a boundary allows the derivation of the Fresnel equations, which can be used to predict how much of the light is reflected, and how much is refracted in a given situation. This is analogous to the way impedance mismatch in an electric circuit causes reflection of signals. Total internal reflection of light from a denser medium occurs if the angle of incidence is greater than the critical angle.

Total internal reflection is used as a means of focusing waves that cannot effectively be reflected by common means. X-ray telescopes are constructed by creating a converging "tunnel" for the waves. As the waves interact at low angle with the surface of this tunnel they are reflected toward the focus point (or toward another interaction with the tunnel surface, eventually being directed to the detector at the focus). A conventional reflector would be useless as the X-rays would simply pass through the intended reflector.

When light reflects off a material denser (with higher refractive index) than the external medium, it undergoes a phase inversion. In contrast, a less dense, lower refractive index material will reflect light in phase. This is an important principle in the field of thin-film optics.

Specular reflection forms images. Reflection from a flat surface forms a mirror image, which appears to be reversed from left to right because we compare the image we see to what we would see if we were rotated into the position of the image. Specular reflection at a curved surface forms an image which may be magnified or demagnified; curved mirrors have optical power. Such mirrors may have surfaces that are spherical or parabolic.

Laws of Reflection

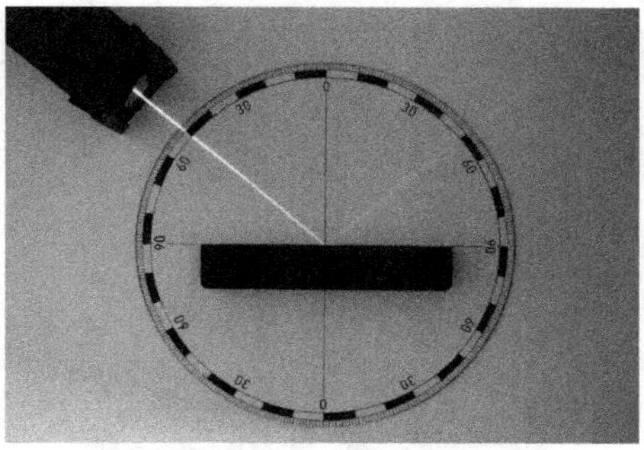

An example of the law of reflection

If the reflecting surface is very smooth, the reflection of light that occurs is called specular or regular reflection. The laws of reflection are as follows:

1. The incident ray, the reflected ray and the normal to the reflection surface at the point of the incidence lie in the same plane.
2. The angle which the incident ray makes with the normal is equal to the angle which the reflected ray makes to the same normal.
3. The reflected ray and the incident ray are on the opposite sides of the normal.

These three laws can all be derived from the Fresnel equations.

Mechanism

In classical electrodynamics, light is considered as an electromagnetic wave, which is described by Maxwell's equations. Light waves incident on a material induce small oscillations of polarisation in the individual atoms (or oscillation of electrons, in metals), causing each particle to radiate a small secondary wave in all directions, like a dipole antenna. All these waves add up to give specular reflection and refraction, according to the Huygens–Fresnel principle.

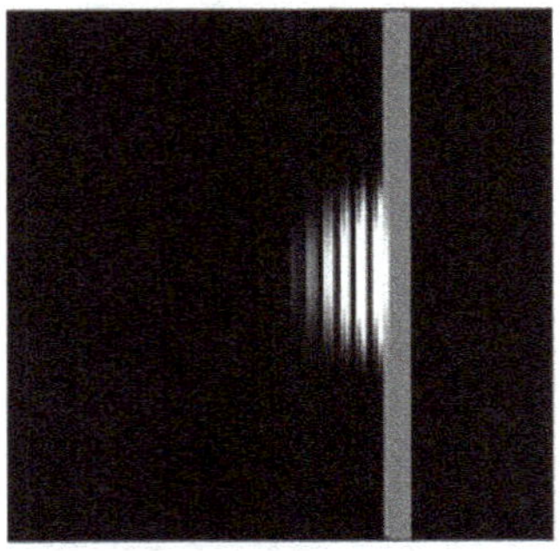

2D simulation: reflection of a quantum particle. White blur represents the probability distribution of finding a particle in a given place if measured

In the case of dielectrics such as glass, the electric field of the light acts on the electrons in the material, and the moving electrons generate fields and become new radiators. The refracted light in the glass is the combination of the forward radiation of the electrons and the incident light. The reflected light is the combination of the backward radiation of all of the electrons.

In metals, electrons with no binding energy are called free electrons. When these electrons oscillate with the incident light, the phase difference between their radiation field and the incident field is π (180°), so the forward radiation cancels the incident light, and backward radiation is just the reflected light.

Diffuse Reflection

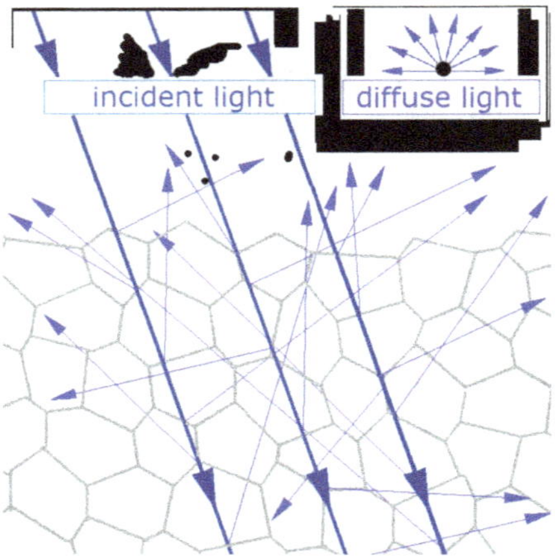

General scattering mechanism which gives diffuse reflection by a solid surface

When light strikes the surface of a (non-metallic) material it bounces off in all directions due to multiple reflections by the microscopic irregularities *inside* the material (e.g. the grain boundaries of a polycrystalline material, or the cell or fiberboundaries of an organic material) and by its surface, if it is rough. Thus, an 'image' is not formed. This is called *diffuse reflection*. The exact form of the reflection depends on the structure of the material. One common model for diffuse reflection is Lambertian reflectance, in which the light is reflected with equal luminance (in photometry) or radiance (in radiometry) in all directions, as defined by Lambert's cosine law.

The light sent to our eyes by most of the objects we see is due to diffuse reflection from their surface, so that this is our primary mechanism of physical observation.

Retroreflection

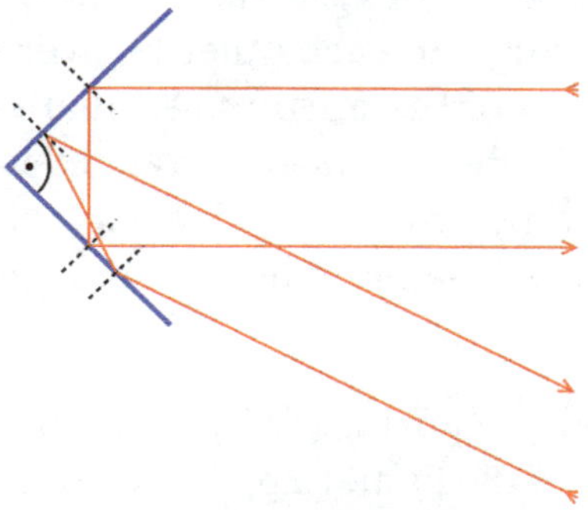

Working principle of a corner reflector

Some surfaces exhibit *retroreflection*. The structure of these surfaces is such that light is returned in the direction from which it came.

When flying over clouds illuminated by sunlight the region seen around the aircraft's shadow will appear brighter, and a similar effect may be seen from dew on grass. This partial retro-reflection is created by the refractive properties of the curved droplet's surface and reflective properties at the backside of the droplet.

Some animals' retinas act as retroreflectors, as this effectively improves the animals' night vision. Since the lenses of their eyes modify reciprocally the paths of the incoming and outgoing light the effect is that the eyes act as a strong retroreflector, sometimes seen at night when walking in wildlands with a flashlight.

A simple retroreflector can be made by placing three ordinary mirrors mutually perpendicular to one another (a corner reflector). The image produced is the inverse of one produced by a single mirror. A surface can be made partially retroreflective by depositing a layer of tiny refractive spheres on it or by creating small pyramid like structures. In both cases internal reflection causes the light to be reflected back to where it originated. This is used to make traffic signs and automobile license plates reflect light mostly back in the direction from which it came. In this application perfect retroreflection is not desired, since the light would then be directed back into the headlights of an oncoming car rather than to the driver's eyes.

Multiple Reflections

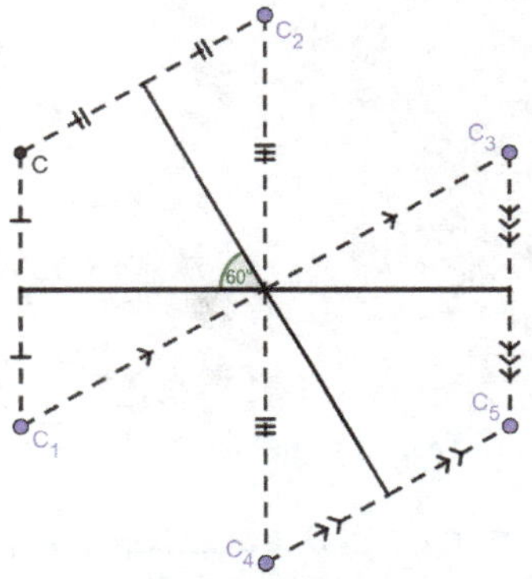

Multiple reflections in two plane mirrors at a 60° angle

When light reflects off a mirror, one image appears. Two mirrors placed exactly face to face give the appearance of an infinite number of images along a straight line. The multiple images seen between two mirrors that sit at an angle to each other lie over a circle. The center of that circle is located at the imaginary intersection of the mirrors. A square of four mirrors placed face to face give the appearance of an infinite number of images arranged in a plane. The multiple images seen between four mirrors assembling a pyramid, in which each pair of mirrors sits an angle to each other, lie over a sphere. If the base of the pyramid is rectangle shaped, the images spread over a section of a torus.

Note that these are theoretical ideals, requiring perfect alignment of perfectly smooth, perfectly flat perfect reflectors that absorb none of the light. In practice, these situations can only be approached but not achieved because the effects of any surface imperfections in the reflectors propagate and magnify, absorption gradually extinguishes the image, and any observing equipment (biological or technological) will interfere.

Complex Conjugate Reflection

In this process (which is also known as phase conjugation), light bounces exactly back in the direction from which it came due to a nonlinear optical process. Not only the direction of the light is reversed, but the actual wavefronts are reversed as well. A conjugate reflector can be used to remove aberrations from a beam by reflecting it and then passing the reflection through the aberrating optics a second time.

Other Types of Reflection

Neutron Reflection

Materials that reflect neutrons, for example beryllium, are used in nuclear reactors and nuclear weapons. In the physical and biological sciences, the reflection of neutrons off of atoms within a material is commonly used to determine the material's internal structure.

Sound Reflection

Sound diffusion panel for high frequencies

When a longitudinal sound wave strikes a flat surface, sound is reflected in a coherent manner provided that the dimension of the reflective surface is large compared to the wavelength of the sound. Note that audible sound has a very wide frequency range (from 20 to about 17000 Hz), and thus a very wide range of wavelengths (from about 20 mm to 17 m). As a result, the overall nature of the reflection varies according to the texture and structure of the surface. For example, porous materials will absorb some energy, and rough materials (where rough is relative to the wavelength) tend to reflect in many directions—to scatter the energy, rather than to reflect it coherently. This leads into the field of architectural acoustics, because the nature of these reflections is critical to the auditory feel of a space. In the theory of exterior noise mitigation, reflective surface size mildly detracts from the concept of a noise barrier by reflecting some of the sound into the opposite direction. Sound reflection can affect the acoustic space.

Seismic Reflection

Seismic waves produced by earthquakes or other sources (such as explosions) may be reflected by layers within the Earth. Study of the deep reflections of waves generated by earthquakes has allowed seismologists to determine the layered structure of the Earth. Shallower reflections are used in reflection seismology to study the Earth's crust generally, and in particular to prospect for petroleum and natural gas deposits.

Specular Reflection

There are two types of reflection: specular and diffuse. Specular reflection is a type of surface reflectance often described as a mirror-like reflection of light from the surface. In specular reflection, the incident light is reflected into a single outgoing direction. The name specular is derived from the Latin word speculum, meaning mirror.

Specular reflection obeys the Laws of Reflection and the light is reflected from a smooth surface at a definite angle. Specular reflection occurs for all polished and smooth surfaces, like mirrors. The angle of reflection in specular reflection is equal to the angle of incidence. Rough or unpolished surfaces are not good specular reflectors as they do not reflect light at the same angle as the angle of incidence.

Mirrors that are used for specular reflection are usually made of metals such as aluminum or silver. Specular reflection is measured using incidence reflectometers by scanning a variable wavelength source.

Background

When light hits a surface, there are three possible outcomes. Light may be absorbed by the material, light may be transmitted through the surface, or light may be reflected. Materials often show some mix of these behaviors, with the proportion of light that goes to each depending on the properties of the material, the wavelength of the light, and the angle of incidence. For most interfaces between materials, the fraction of the light that is reflected increases with increasingangle of incidence θ_i.

Reflected light can be divided into two sub-types, specular reflection and diffuse reflection. Specular reflection reflects all light which arrives from a given direction at the same angle, whereas diffuse reflection reflects that light in a broad range of directions. An example of the distinction between specular and diffuse reflection would be glossy and matte paints. Matte paints have almost exclusively diffuse reflection, while glossy paints have both specular and diffuse reflection. A surface built from a non-absorbing powder, such as plaster, can be a nearly perfect diffuser, whereas polished metallic objects can specularly reflect light very efficiently. The reflecting material of mirrors is usually aluminum or silver.

Law of Reflection

The law of reflection describes the angle of reflected light: the angle of incident light is the same as the angle of the reflected light.

The law of reflection arises from diffraction of a plane wave with small wavelength on a flat boundary: when the boundary size is much larger than the wavelength, then electrons of the boundary are seen oscillating exactly in phase only from one direction – the specular direction. If a mirror becomes very small compared to the wavelength, the law of reflection no longer holds, and the behavior of light is more complicated.

Vector Formulation

The law of reflection can also be equivalently expressed using linear algebra. The direction of a reflected ray is determined by the vector of incidence and the surface normal vector. Given an incident direction $\hat{\mathbf{d}}_i$ from the surface to the light source and the surface normal direction $\hat{\mathbf{d}}_n$, the specularly reflected direction $\hat{\mathbf{d}}_s$ (all unit vectors) is:

$$\hat{\mathbf{d}}_s = 2\left(\hat{\mathbf{d}}_n \cdot \hat{\mathbf{d}}_i\right)\hat{\mathbf{d}}_n - \hat{\mathbf{d}}_i,$$

where $\hat{\mathbf{d}}_n \cdot \hat{\mathbf{d}}_i$ is a scalar obtained with the dot product. Different authors may define the incident and reflection directions with different signs. Assuming these Euclidean vectors are represented in column form, the equation can be equivalently expressed as a matrix-vector multiplication:

$$\hat{\mathbf{d}}_s = \mathbf{R}\,\hat{\mathbf{d}}_i,$$

where $\mathbf{R}$ is the so-called Householder transformation matrix, defined as:

$$\mathbf{R} = \mathbf{I} - 2\hat{\mathbf{d}}_n\hat{\mathbf{d}}_n^T;$$

T denotes transposition and $\mathbf{I}$ is the identity matrix.

Reflectivity

Reflectivity is the ratio of the power of the reflected wave to that of the incident wave. It is a function of the wavelength of radiation, and is related to the refractive index of the material as expressed by Fresnel's equations. In regions of the electromagnetic spectrum in which absorption by the material is significant, it is related to the electronic absorption spectrum through the imaginary component of the complex refractive index. The electronic absorption spectrum of an opaque material, which is difficult or impossible to measure directly, may therefore be indirectly determined from the reflection spectrum by a Kramers-Kronig transform. The polarization of the reflected light depends on the symmetry of the arrangement of the incident probing light with respect to the absorbing transitions dipole moments in the material.

Measurement of specular reflection is performed with normal or varying incidence reflection spectrophotometers (*reflectometer*) using a scanning variable-wavelength light source. Lower quality measurements using a glossmeter quantify the glossy appearance of a surface in gloss units.

Consequences

Internal Reflection

When light is propagating in a material and strikes an interface with a material of lower index of refraction, some of the light is reflected. If the angle of incidence is greater than the critical angle, total internal reflection occurs: all of the light is reflected. The critical angle can be shown to be given by

$$\theta_{crit} = \arcsin\left(\frac{n_2}{n_1}\right).$$

Polarization

When light strikes an interface between two materials, the reflected light is generally partially polarized. However, if the light strikes the interface at Brewster's angle, the reflected light is *completely* linearly polarized parallel to the interface. Brewster's angle is given by

$$\theta_B = \arctan\left(\frac{n_2}{n_1}\right).$$

Reflected Images

The image in a flat mirror has these features:

- It is the same distance behind the mirror as the object is in front.
- It is the same size as the object.
- It is the right way up (erect).
- It is reversed.
- It is *virtual*, meaning that the image appears to be behind the mirror, and cannot be projected onto a screen.

The reversal of images by a plane mirror is perceived differently depending on the circumstances. In many cases, the image in a mirror appears to be reversed from left to right. If a flat mirror is mounted on the ceiling it can appear to reverse *up* and *down* if a person stands under it and looks up at it. Similarly a car turning *left* will still appear to be turning *left*in the rear view mirror for the driver of a car in front of it. The reversal of directions, or lack thereof, depends on how the directions are defined. More specifically a mirror changes the handedness of the coordinate system, one axis of the coordinate system appears to be reversed, and the chirality of the image may change. For example, the image of a right shoe will look like a left shoe.

Examples

Esplanade of the Trocadero in Paris after rain. The layer of water exhibits specular reflection, reflecting an image of the Eiffel Tower and other objects

A classic example of specular reflection is a mirror, which is specifically designed for specular reflection.

In addition to visible light, specular reflection can be observed in the ionospheric reflection of radiowavesand the reflection of radio- or microwave radarsignals by flying objects. The measurement technique of x-ray reflectivity exploits specular reflectivity to study thin films and interfaces with sub-nanometer resolution, using either modern laboratory sources or synchrotron x-rays.

Non-electromagnetic waves can also exhibit specular reflection, as in acoustic mirrors which reflect sound, and atomic mirrors, which reflect neutral atoms. For the efficient reflection of atoms from a solid-state mirror, very cold atoms and/or grazing incidence are used in order to provide significant quantum reflection; ridged mirrors are used to enhance the specular reflection of atoms. Neutron reflectometry utilizes specular reflection to study material surfaces and thin film interfaces in an analogous fashion to x-ray reflectivity.

Diffuse Reflection

Diffuse reflection is uniform reflection of light with no directional dependence for the viewer, e.g., a matte surface such as cardboard. Diffuse reflection originates from a combination of internal scattering of light, i.e. the light is absorbed and then re-emitted, and external scattering from the rough surface of the object. An illumination model must handle both direct diffuse reflection, i.e., light coming directly from a source to a surface and then reflected to the viewer, and indirect diffuse reflection (or diffuse interreflections), that is light coming from a source, being reflected to a surface, then reflected to another surface, etc, and finally to the viewer.

Some local illumination models, such as those that are usually used in many scan-line rendering and ray tracing systems, handle the diffuse interreflection contribution by simply assuming it has uniform intensity in all directions. This constant term is called the “ambient” contribution and it is considered a non-directional background light. Radiosity explicitly computes the diffuse interreflection term.

Diffuse Reflection of Ambient Light

If we assume a uniform intensity of ambient light, Ia, then the intensity of the diffuse reflection at any point on the surface: I = ka * Ia, where ka = the ambient diffuse coefficient of reflection with: 0.0 <= ka <= 1.0. For colored surfaces there are three ka values, one for red, green, and blue. We are approximating the ambient diffuse part of the reflection intensity function $R(\phi,\theta,\lambda)$ by the constant ka. For a dull, non-reflecting surface ka is close to 0.0 and for a highly reflective surface, such as a mirror, ka is close to 1.0.

Diffuse Reflection of Point Light Sources

Consider a point source and assume the source is far from the surface so that incident light rays are parallel. Note: if the light source is near the surface then the light rays are not parallel as below:

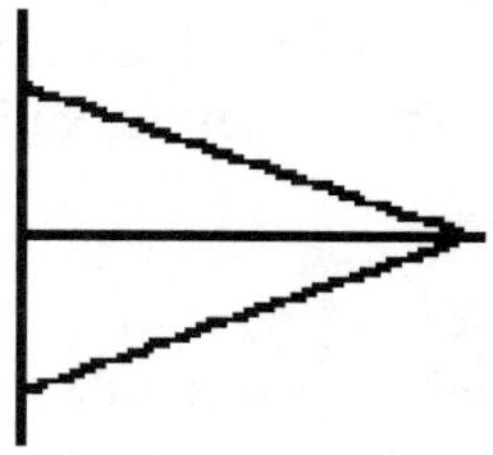

For a point light far from the surface, we use Lambert's cosine law which is as follows: the intensity of reflected light is f(angle of illumination). This can be qualitatively seen if we look at the two cases below:

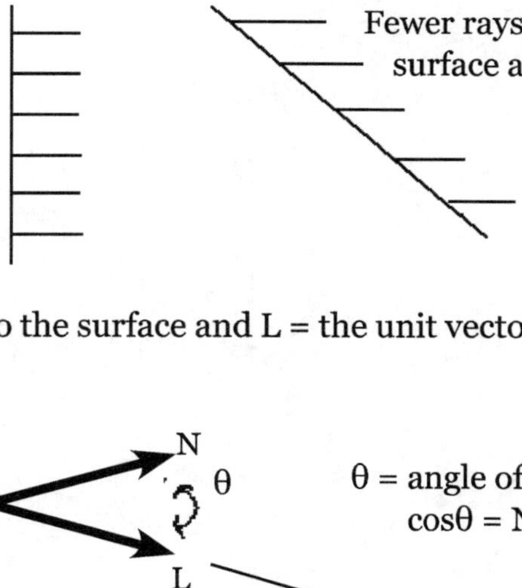

Define N = the unit normal to the surface and L = the unit vector which is the direction to the light source

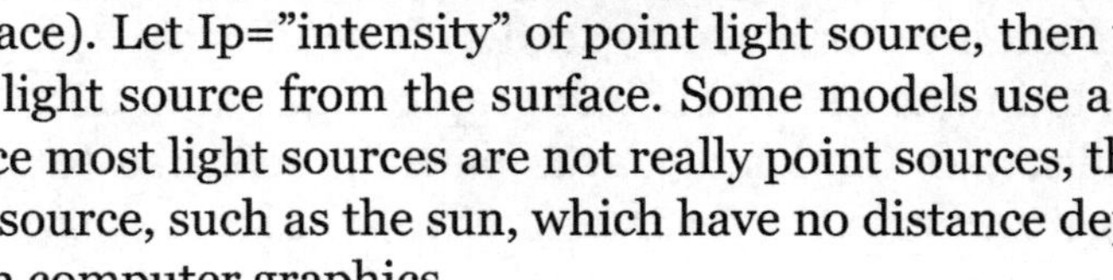

Then Lambert's law states that the reflected light is proportional to cos θ (with θ <= 90 else the light is behind the surface). Let Ip="intensity" of point light source, then the brightness is $f(1/d^2)$ where d="distance" of light source from the surface. Some models use a d distance dependence rather than d^2, and since most light sources are not really point sources, this works. Some models define a "distant" light source, such as the sun, which have no distance dependence. Here is more information on lights in computer graphics.

Then I = [(kd * Ip) / (D)] * (N � L) with D = the distance function (< = 1 for a distant light source, d or d^2 for other models and kd = the diffuse coefficient of reflection with: 0.0 <= kd <= 1.0. For colored surfaces there are three kd values, one for red, green, and blue. Some models also add a constant term d_0 to prevent d the denominator from going to zero. Frequently ka is set to be the same as kd. Note that the vectors N and L are always normalized.

Then including ambient light we get a simple two term illumination model:

I = ka * Ia + [(kd * Ip) / (d)] * (N � L)

If we consider color then we should have an explicit wavelength dependence. But we usually make the approximation of sampling the color space at only three colors, i.e., "red", "green", and "blue", without really defining what wavelengths these are. Then we will have 3 equations like the above: one for Red, Green, and Blue. For example, a blue surface might have $kd^R = kd^G = 0$, $Kd^B = 0.8$. Note that the intensity, I, should always have a value between 0.0 and 1.0. It may be necessary to clamp the computed values so that they remain in the valid range.

Mechanism

Diffuse reflection from solids is generally not due to surface roughness. A flat surface is indeed required to give specular reflection, but it does not prevent diffuse reflection. A piece of highly polished white marble remains white; no amount of polishing will turn it into a mirror. Polishing produces some specular reflection, but the remaining light continues to be diffusely reflected.

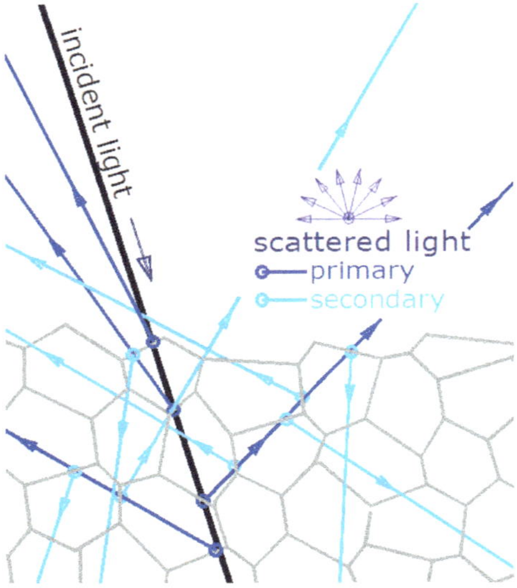

Figure– General mechanism of diffuse reflection by a solid surface (refraction phenomena not represented)

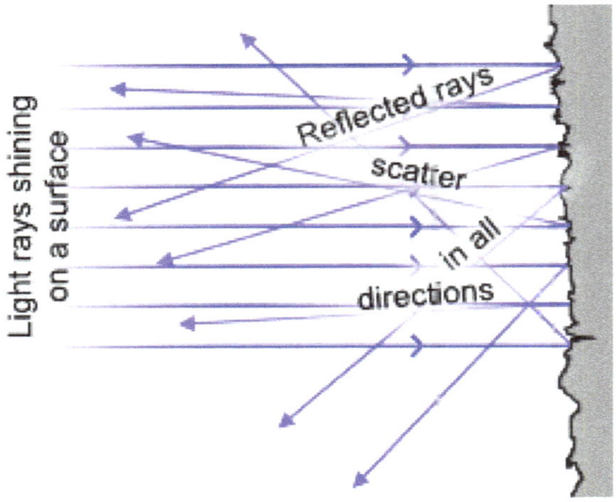

Figure – Diffuse reflection from an irregular surface

The most general mechanism by which a surface gives diffuse reflection does not involve *exactly* the surface: most of the light is contributed by scattering centers beneath the surface, as illustrated in figure. If one were to imagine that the figure represents snow, and that the polygons are its (transparent) ice crystallites, an impinging ray is partially reflected (a few percent) by the first particle, enters in it, is again reflected by the interface with the second particle, enters in it, impinges on the third, and so on, generating a series of "primary" scattered rays in random directions, which, in turn, through the same mechanism, generate a large number of "secondary" scattered rays, which generate "tertiary" rays, and so forth. All these rays walk through the snow crystallytes, which do not absorb light, until they arrive at the surface and exit in random directions. The result is that the light that was sent out is returned in all directions, so that snow is white despite being made of transparent material (ice crystals).

For simplicity, "reflections" are spoken of here, but more generally the interface between the small particles that constitute many materials is irregular on a scale comparable with light wavelength, so diffuse light is generated at each interface, rather than a single reflected ray, but the story can be told the same way.

This mechanism is very general, because almost all common materials are made of "small things" held together. Mineral materials are generally polycrystalline: one can describe them as made of a 3D mosaic of small, irregularly shaped defective crystals. Organic materials are usually composed of fibers or cells, with their membranes and their complex internal structure. And each interface, inhomogeneity or imperfection can deviate, reflect or scatter light, reproducing the above mechanism.

Few materials do not cause diffuse reflection: among these are metals, which do not allow light to enter; gases, liquids, glass, and transparent plastics (which have a liquid-like amorphous microscopic structure); single crystals, such as some gems or a salt crystal; and some very special materials, such as the tissues which make the cornea and the lens of an eye. These materials can reflect diffusely, however, if their surface is microscopically rough, like in a frost glass (Figure), or, of course, if their homogeneous structure deteriorates, as in cataracts of the eye lens.

A surface may also exhibit both specular and diffuse reflection, as is the case, for example, of glossy paints as used in home painting, which give also a fraction of specular reflection, while matte paints give almost exclusively diffuse reflection.

Most materials can give some specular reflection, provided that their surface can be polished to eliminate irregularities comparable with the light wavelength (a fraction of a micrometer). Depending on the material and surface roughness, reflection may be mostly specular, mostly diffuse, or anywhere in between. A few materials, like liquids and glasses, lack the internal subdivisions which produce the subsurface scattering mechanism described above, and so give *only* specular reflection. Among common materials, only polished metals can reflect light specularly with high efficiency, as in aluminum or silver usually used in mirrors. All other common materials, even when perfectly polished, usually give not more than a few percent specular reflection, except in particular cases, such as grazing angle reflection by a lake, or the *total reflection* of a glass prism, or when structured in certain complex configurations such as the silvery skin of many fish species or the reflective surface of a dielectric mirror. Diffuse reflection can be highly efficient, as in white materials, due to the summing up of the many subsurface reflections.

Colored Objects

Up to this point white objects have been discussed, which do not absorb light. But the above scheme continues to be valid in the case that the material is absorbent. In this case, diffused rays will lose some wavelengths during their walk in the material, and will emerge colored.

Diffusion affects the color of objects in a substantial manner because it determines the average path of light in the material, and hence to which extent the various wavelengths are absorbed. Red ink looks black when it stays in its bottle. Its vivid color is only perceived when it is placed on a scattering material (e.g. paper). This is so because light's path through the paper fibers (and through the ink) is only a fraction of millimeter long. However, light from the bottle has crossed several centimeters of ink and has been heavily absorbed, even in its red wavelengths.

And, when a colored object has both diffuse and specular reflection, usually only the diffuse component is colored. A cherry reflects diffusely red light, absorbs all other colors and has a specular reflection which is essentially white (if the incident light is white light). This is quite general, because, except for metals, the reflectivity of most materials depends on their refractive index, which varies little with the wavelength (though it is this variation that causes the chromatic dispersion in a prism), so that all colors are reflected nearly with the same intensity. Reflections from different origin, instead, may be colored: metallic reflections, such as in gold or copper, or interferential reflections: iridescences, peacock feathers, butterfly wings, beetle elytra, or the antireflection coating of a lens.

Importance for Vision

Looking at one's surrounding environment, the vast majority of visible objects are seen primarily by diffuse reflection from their surface. This holds with few exceptions, such as glass, reflective liquids, polished or smooth metals, glossy objects, and objects that themselves emit light: the Sun, lamps, and computer screens (which, however, emit *diffuse* light). Outdoors it is the same, with perhaps the exception of a transparent water stream or of the iridescent colors of a beetle. Additionally, Rayleigh scattering is responsible for the blue color of the sky, and Mie scattering for the white color of the water droplets of clouds.

Light scattered from the surfaces of objects is by far the primary light which humans visually observe.

Interreflection

Diffuse interreflection is a process whereby light reflected from an object strikes other objects in the surrounding area, illuminating them. Diffuse interreflection specifically describes light reflected from objects which are not shiny or specular. In real life terms what this means is that light is reflected off non-shiny surfaces such as the ground, walls, or fabric, to reach areas not directly in view of a light source. If the diffuse surface is colored, the reflected light is also colored, resulting in similar coloration of surrounding objects.

In 3D computer graphics, diffuse interreflection is an important component of global illumination. There are a number of ways to model diffuse interreflection when rendering a scene. Radiosity and photon mapping are two commonly used methods.

Applications of Specular and Diffuse Reflection

There are several interesting applications of this distinction between specular and diffuse reflection. One application pertains to the relative difficulty of night driving on a wet asphalt roadway compared to a dry asphalt roadway. Most drivers are aware of the fact that driving at night on a wet roadway results in an annoying glare from oncoming headlights. The glare is the result of the specular reflection of the beam of light from an oncoming car. Normally a roadway would cause diffuse reflection due to its rough surface. But if the surface is wet, water can fill in the crevices and smooth out the surface. Rays of light from the beam of an oncoming car hit this smooth surface, undergo specular reflection and remain concentrated in a beam. The driver perceives an annoying glare caused by this concentrated beam of reflected light.

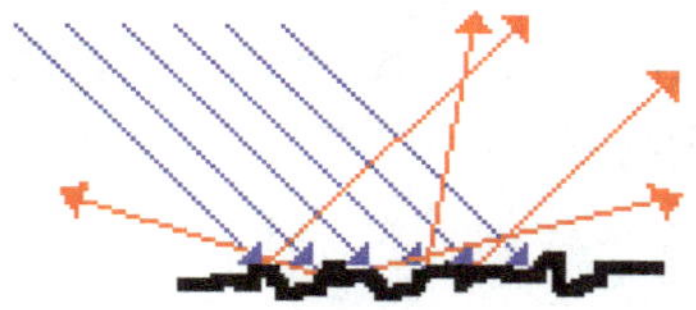

A dry asphalt roadway diffuses incident light

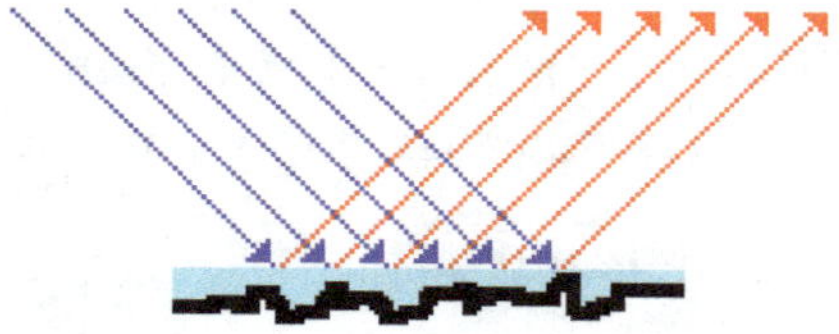

When wet, water fills in the crevices, resulting in specular reflaction and a glare

Photograph of Mount Moran in the Grand Teton National Park in Wyoming

A second application of the distinction between diffuse and specular reflection pertains to the field of photography. Many people have witnessed in person or have seen a photograph of a beautiful nature scene captured by a photographer who set up the shot with a calm body of water in the foreground. The water (if calm) provides for the specular reflection of light from the subject of the photograph. Light from the subject can reach the camera lens directly or it can take a longer path in which it reflects off the water before traveling to the lens. Since the light reflecting off the water undergoes specular reflection, the incident rays remain concentrated (instead of diffusing). The light is thus able to travel together to the lens of the camera and produce an image (an exact replica) of the subject which is strong enough to perceive in the photograph.

Refraction

Refraction is the bending of light (it also happens with sound, water and other waves) as it passes from one transparent substance into another.

This bending by refraction makes it possible for us to have lenses, magnifying glasses, prisms and rainbows. Even our eyes depend upon this bending of light. Without refraction, we wouldn't be able to focus light onto our retina.

Change of Speed Causes Change of Direction

Light refracts whenever it travels at an angle into a substance with a different refractive index (optical density).

This change of direction is caused by a change in speed. For example, when light travels from air into water, it slows down, causing it to continue to travel at a different angle or direction.

How much does light bend?

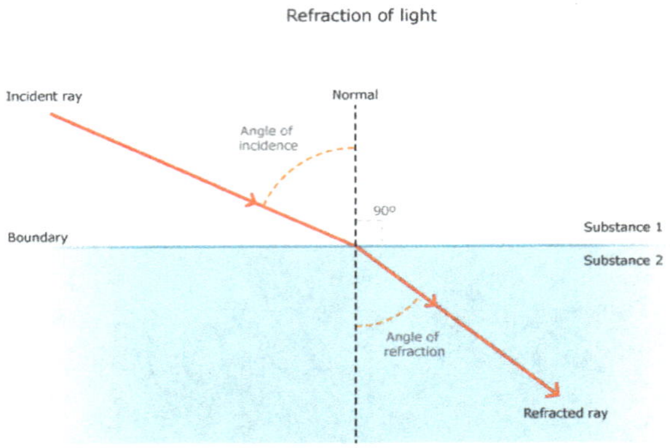

Refraction of light in water

When light travels from air into water, it slows down, causing it to change direction slightly. This change of direction is called refraction. When light enters a more dense substance (higher refractive index), it 'bends' more towards the normal line.

The amount of bending depends on two things:

- Change in speed – if a substance causes the light to speed up or slow down more, it will refract (bend) more.
- Angle of the incident ray – if the light is entering the substance at a greater angle, the amount of refraction will also be more noticeable. On the other hand, if the light is entering the new substance from straight on (at 90° to the surface), the light will still slow down, but it won't change direction at all.

Refractive Index of some Transparent Substances

Substance	Refractive index	Speed of light in substance (x 1,000,000 m/s)	Angle of refraction if incident ray enters substance at 20°
Air	1.00	300	20
Water	1.33	226	14.9
Glass	1.5	200	13.2
Diamond	2.4	125	8.2

All angles are measured from an imaginary line drawn at 90° to the surface of the two substances This line is drawn as a dotted line and is called the normal.

If light enters any substance with a *higher* refractive index (such as from air into glass) it slows down. The light bends *towards* the normal line.

If light travels enters into a substance with a *lower* refractive index (such as from water into air) it speeds up. The light bends *away* from the normal line.

A higher refractive index shows that light will slow down and change direction more as it enters the substance.

Explanation

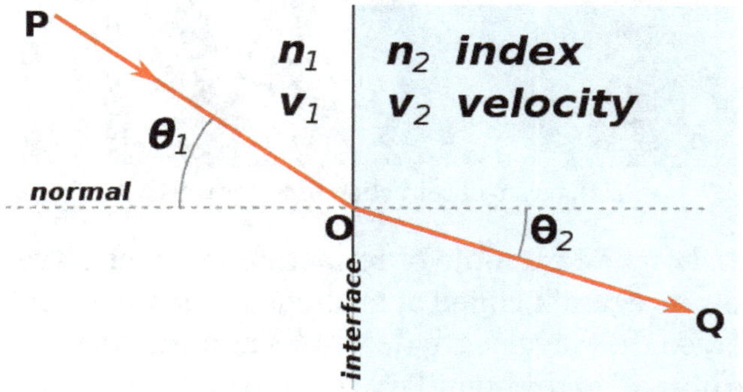

Refraction of light at the interface between two media of different refractive indices, with $n_2 > n_1$. Since the phase velocity is lower in the second medium ($v_2 < v_1$), the angle of refraction θ_2 is less than the angle of incidence θ_1

In optics, refraction is a phenomenon that often occurs when waves travel from a medium with a given refractive index to a medium with another at an oblique angle. At the boundary between the media, the wave's phase velocity is altered, usually causing a change in direction. Its wavelength increases or decreases, but its frequency remains constant. For example, a light ray will refract as it enters and leaves glass, as there is a change in refractive index. A ray traveling along the normal (perpendicular to the boundary) will suffer change in speed, but not direction. Refraction still occurs in this case (by Snell's Law as angle of incidence will be 0°). Understanding of this concept led to the invention of lensesand the refracting telescope.

Refraction can be seen when looking into a bowl of water. Air has a refractive index of about 1.0003, and water has a refractive index of about 1.3333. If a person looks at a straight object, such as a pencil or straw, which is placed at a slant, partially in the water, the object appears to bend at the water's surface. This is due to the bending of light rays as they move from the water to the air. Once the rays reach the eye, the eye traces them back as straight lines (lines of sight). The lines of sight (shown as dashed lines) intersect at a higher position than where the actual rays originated. This causes the pencil to appear higher and the water to appear shallower than it really is. The depth that the water appears to be when viewed from above is known as the *apparent depth*. This is an important consideration for spearfishing from the surface because it will make the target fish appear to be in a different place, and the fisher must aim lower to catch the fish. Conversely, an object above the water has a higher *apparent height* when viewed from below the water. The opposite correction must be made by an archer fish. For small angles of incidence (measured from the normal, when sin θ is approximately the same as tan θ), the ratio of apparent to real depth is the ratio of the refractive indexes of air to that of water. But, as the angle of incidence approaches 90°, the apparent depth approaches zero, albeit reflection increases, which limits observation at high angles of incidence. Conversely, the apparent height approaches infinity as the angle of incidence (from below) increases, but even earlier, as the angle of total internal reflection is approached, albeit the image also fades from view as this limit is approached.

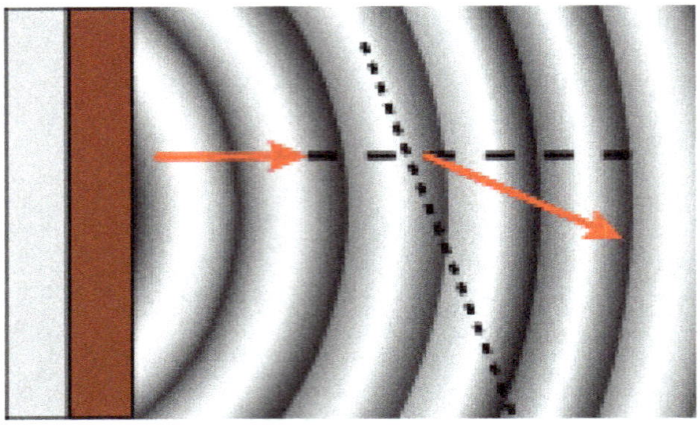

Diagram of refraction of water waves

The diagram on the right shows an example of refraction in water waves. Ripples travel from the left and pass over a shallower region inclined at an angle to the wavefront. The waves travel slower in the more shallow water, so the wavelength decreases and the wave bends at the boundary. The dotted line represents the normal to the boundary. The dashed line represents the original direction of the waves. This phenomenon explains why waves on a shoreline tend to strike the shore close to a perpendicular angle. As the waves travel from deep water into shallower water near the shore, they are refracted from their original direction of travel to an angle more normal to the shoreline. Refraction is also responsible for rainbows and for the splitting of white light into a rainbow-spectrum as it passes through a glass prism. Glass has a higher refractive index than air. When a beam of white light passes from air into a material having an index of refraction that varies with frequency, a phenomenon known as dispersion occurs, in which different coloured components of the white light are refracted at different angles, i.e., they bend by different amounts at the interface, so that they become separated. The different colors correspond to different frequencies.

While refraction allows for phenomena such as rainbows, it may also produce peculiar optical phenomena, such as mirages and Fata Morgana. These are caused by the change of the refractive index of air with temperature.

The refractive index of materials can also be nonlinear, as occurs with the Kerr effect when high intensity light leads to a refractive index proportional to the intensity of the incident light.

Recently, some metamaterials have been created that have a negative refractive index. With meta-materials, we can also obtain total refraction phenomena when the wave impedances of the two media are matched. There is then no reflected wave.

Also, since refraction can make objects appear closer than they are, it is responsible for allowing water to magnify objects. First, as light is entering a drop of water, it slows down. If the water's surface is not flat, then the light will be bent into a new path. This round shape will bend the light outwards and as it spreads out, the image you see gets larger.

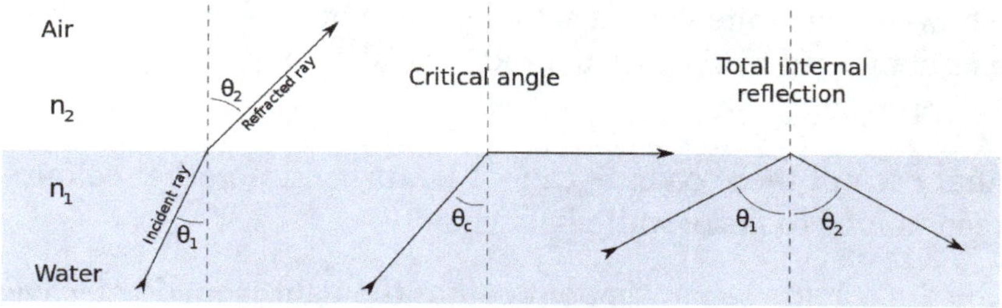

Refraction of light at the interface between two media

An analogy that is often put forward to explain the refraction of light is as follows: "Imagine a marching band as it marches at an oblique angle from a pavement (a fast medium) into mud (a slower medium). The marchers on the side that runs into the mud first will slow down first. This causes the whole band to pivot slightly toward the normal (make a smaller angle from the normal)."

Why refraction occurs when light travels from a medium with a given refractive index to a medium with another, can be explained by the path integral formulation of quantum mechanics (the complete method was developed in 1948 by Richard Feynman). Feynman humorously explained it himself in the recording.

The effects of refraction between materials can be minimised through index matching, the close matching of their respective indices of refraction.

Refraction can Create a Spectrum

Isaac Newton performed a famous experiment using a triangular block of glass called a prism. He used sunlight shining in through his window to create a spectrum of colours on the opposite side of his room.

Prism

When white light shines through a prism, each colour refracts at a slightly different angle. Violet light refracts slightly more than red light. A prism can be used to show the seven colours of the spectrum that make up white light.

This experiment showed that white light is actually made of all the colours of the rainbow. These seven colours are remembered by the acronym ROY G BIV – red, orange, yellow, green, blue, indigo and violet.

Newton showed that each of these colours cannot be turned into other colours. He also showed that they can be recombined to make white light again.

The explanation for the colours separating out is that the light is made of waves. Red light has a longer wavelength than violet light. The refractive index for red light in glass is slightly different than for violet light. Violet light slows down even more than red light, so it is refracted at a slightly greater angle.

The refractive index of red light in glass is 1.513. The refractive index of violet light is 1.532. This slight difference is enough for the shorter wavelengths of light to be refracted more.

References

- Michelson,, A. A. (January 1927). "Measurements of the velocity of light between Mount Wilson and Mount San Antonio". Astrophysical Journal. 65: 1. Bibcode:1927ApJ....65....1M. doi:10.1086/143021. Retrieved 12 March 2014
- "Scientific Method, Statistical Method and the Speed of Light". Statistical Science. 15 (3): 254–278. 2000. doi:10.1214/ss/1009212817. MR 1847825
- Harvard News Office (2001-01-24). "Harvard Gazette: Researchers now able to stop, restart light". News.harvard.edu. Archived from the original on 28 October 2011. Retrieved 2011-11-08
- Ptolemy and A. Mark Smith (1996). Ptolemy's Theory of Visual Perception: An English Translation of the Optics with Introduction and Commentary. Diane Publishing. p. 23. ISBN 0-87169-862-5
- Tang, Hong (1 October 2009). "May The Force of Light Be With You". IEEE Spectrum. 46 (10): 46–51. doi:10.1109/MSPEC.2009.5268000
- Laufer, Gabriel (13 July 1996). Introduction to Optics and Lasers in Engineering. Cambridge University Press. p. 11. ISBN 978-0-521-45233-5. Retrieved 20 October 2013
- Theories of light, from Descartes to Newton A. I. Sabra CUP Archive,1981 pg 48 ISBN 0-521-28436-8, ISBN 978-0-521-28436-3
- Nichols, E.F; Hull, G.F. (1903). "The Pressure due to Radiation". The Astrophysical Journal. 17 (5): 315–351. Bibcode:1903ApJ....17..315N. doi:10.1086/141035
- Terrell, J (1959). "Invisibility of the Lorentz Contraction". Physical Review. 116 (4): 1041–5. Bibcode:1959PhRv..116.1041T. doi:10.1103/PhysRev.116.1041
- Bradt, Hale (2004). Astronomy Methods: A Physical Approach to Astronomical Observations. Cambridge University Press. p. 26. ISBN 978-0-521-53551-9. Retrieved 20 October 2013
- Fokko Jan Dijksterhuis, Lenses and Waves: Christiaan Huygens and the Mathematical Science of Optics in the 17th Century, Kluwer Academic Publishers, 2004, ISBN 1-4020-2697-8
- Penrose, R (1959). "The Apparent Shape of a Relativistically Moving Sphere". Proceedings of the Cambridge Philosophical Society. 55 (01): 137–9. Bibcode:1959PCPS...55..137P. doi:10.1017/S0305004100033776
- Ohannesian, Lena; Streeter, Anthony (9 November 2001). Handbook of Pharmaceutical Analysis. CRC Press. p. 187. ISBN 978-0-8247-4194-5. Retrieved 20 October 2013
- David Cassidy; Gerald Holton; James Rutherford (2002). Understanding Physics. Birkhäuser. ISBN 0-387-98756-8

Permissions

All chapters in this book are published with permission under the Creative Commons Attribution Share Alike License or equivalent. Every chapter published in this book has been scrutinized by our experts. Their significance has been extensively debated. The topics covered herein carry significant information for a comprehensive understanding. They may even be implemented as practical applications or may be referred to as a beginning point for further studies.

We would like to thank the editorial team for lending their expertise to make the book truly unique. They have played a crucial role in the development of this book. Without their invaluable contributions this book wouldn't have been possible. They have made vital efforts to compile up to date information on the varied aspects of this subject to make this book a valuable addition to the collection of many professionals and students.

This book was conceptualized with the vision of imparting up-to-date and integrated information in this field. To ensure the same, a matchless editorial board was set up. Every individual on the board went through rigorous rounds of assessment to prove their worth. After which they invested a large part of their time researching and compiling the most relevant data for our readers.

The editorial board has been involved in producing this book since its inception. They have spent rigorous hours researching and exploring the diverse topics which have resulted in the successful publishing of this book. They have passed on their knowledge of decades through this book. To expedite this challenging task, the publisher supported the team at every step. A small team of assistant editors was also appointed to further simplify the editing procedure and attain best results for the readers.

Apart from the editorial board, the designing team has also invested a significant amount of their time in understanding the subject and creating the most relevant covers. They scrutinized every image to scout for the most suitable representation of the subject and create an appropriate cover for the book.

The publishing team has been an ardent support to the editorial, designing and production team. Their endless efforts to recruit the best for this project, has resulted in the accomplishment of this book. They are a veteran in the field of academics and their pool of knowledge is as vast as their experience in printing. Their expertise and guidance has proved useful at every step. Their uncompromising quality standards have made this book an exceptional effort. Their encouragement from time to time has been an inspiration for everyone.

The publisher and the editorial board hope that this book will prove to be a valuable piece of knowledge for students, practitioners and scholars across the globe.

Index

CPSIA information can be obtained
at www.ICGtesting.com
Printed in the USA
BVHW011052110719
553192BV00003B/7/P